Mud Volcanoes, Geodynamics and Seismicity

NATO Science Series

A Series presenting the results of scientific meetings supported under the NATO Science Programme.

The Series is published by IOS Press, Amsterdam, and Springer (formerly Kluwer Academic Publishers) in conjunction with the NATO Public Diplomacy Division.

Sub-Series

I. Life and Behavioural Sciences	IOS Press
II. Mathematics, Physics and Chemistry	Springer (formerly Kluwer Academic Publishers)
III. Computer and Systems Science	IOS Press
IV. Earth and Environmental Sciences	Springer (formerly Kluwer Academic Publishers)

The NATO Science Series continues the series of books published formerly as the NATO ASI Series.

The NATO Science Programme offers support for collaboration in civil science between scientists of countries of the Euro-Atlantic Partnership Council. The types of scientific meeting generally supported are "Advanced Study Institutes" and "Advanced Research Workshops", and the NATO Science Series collects together the results of these meetings. The meetings are co-organized by scientists from NATO countries and scientists from NATO's Partner countries – countries of the CIS and Central and Eastern Europe.

Advanced Study Institutes are high-level tutorial courses offering in-depth study of latest advances in a field.
Advanced Research Workshops are expert meetings aimed at critical assessment of a field, and identification of directions for future action.

As a consequence of the restructuring of the NATO Science Programme in 1999, the NATO Science Series was re-organized to the four sub-series noted above. Please consult the following web sites for information on previous volumes published in the Series.

http://www.nato.int/science
http://www.springeronline.com
http://www.iospress.nl

Series IV: Earth and Environmental Series – Vol. 51

Mud Volcanoes, Geodynamics and Seismicity

edited by

Giovanni Martinelli
ARPA, Environmental Protection Agency of Emilia-Romagna Region,
Reggio Emilia, Italy

and

Behrouz Panahi
Geology Institute,
Azerbaijan National Academy of Sciences,
Baku, Azerbaijan

Springer

Proceedings of the NATO Advanced Research Workshop on
Mud Volcanism, Geodynamics and Seismicity
Baku, Azerbaijan
20–22 May 2003

A C.I.P. Catalogue record for this book is available from the Library of Congress.

ISBN 1-4020-3203-X (PB)
ISBN 1-4020-3202-1 (HB)
ISBN 1-4020-3204-8 (e-book)

Published by Springer,
P.O. Box 17, 3300 AA Dordrecht, The Netherlands.

Sold and distributed in North, Central and South America
by Springer,
101 Philip Drive, Norwell, MA 02061, U.S.A.

In all other countries, sold and distributed
by Springer,
P.O. Box 322, 3300 AH Dordrecht, The Netherlands.

Printed on acid-free paper

CONTENTS

CONTRIBUTING AUTHORS

Ben Avraham, Zvi	Department of Geophysics and Planetary Sciences, Tel Aviv University, Tel Aviv 69978, Israel, Fax: +2721 6503783 e-mail: Zvi@terra.tau.ac.il
Abegg Friedrich	GEOMAR, Wischlofstr. 1-3, 24148, Kiel, Germany, Phone + 494316002321 e-mail: fabegg@ifm-geomar.de
Albarello Dario	Department of Earth Sciences, University of Siena, Via Laterina, 8, Siena, 53100 , Italy Phone: +390577 233825/22 Fax: +390577 233820 e-mail: albarello@unisi.it
Aliyev Chingiz S.	Geology Institute, Azerbaijan National Academy of Sciences, H. Javid Ave., 29A, 370143 Baku, Azerbaijan.
Baciu Calin	Babes-Bolyai University, Department of GeologyM. Kogalniceanu str. 1, RO-3400 Cluj-Napoca, Romania, Phone: 0040.64.405300 ext. 5227 Fax: 0040.64.191906 e-mail: cbaciu@bioge.ubbcluj.ro
Bedir Mourad	Georesources Laboratory, National Institute of Scientific Research and Technics, Route touristique de Soliman B.P. 95 Hamman Lif Tunis, 2050 Tunisia, Fax: 00216 430934 e-mail: Mourad.bedir@inrst.rnrt.tn
Brown Kevin M.	SCRIPPS Institution of Oceanography, University of California San Diego 9500 Gilman Drive, La Jolla CA 92093-0244, USA e-mail: kmbrown@ucsd.edu
Brueckmann Warner	Sonderforschungsbereich 574, Christian-Albrechts-University of Kiel, Wischhofstr. 1-3, 24148 Kiel, Germany Phone: +494316002819 GEOMAR Research Centre, Wischhofstr. 1-3, 24148 Kiel, Germany e-mail: wbrueckmann@ifm-geomar.de

Bunyapanasarn Ted	Department of Geological Sciences, San Diego State University 5500 Campanile Dr, San Diego, CA 92182, USA, Phone: 619-594-3455 Fax: 619-594-4372
Chen Ju-Chin	Institute of Oceanography, National Taiwan University No.1, Sec.4, Roosevelt Road, Taipei 106, Taiwan, R.O.C. e-mail: jcchen@iodec1.oc.ntu.edu.tw
Chen Yue-Gau	Department of Geosciences, National Taiwan University No.1, Sec.4, Roosevelt Road, Taipei 106, Taiwan, R.O.C. e-mail: ygchen@ccms.ntu.edu.tw
Clennell Ben M.	CSIRO Petroleum, Australian Resources Research Centre, 26 Dick Perry Avenue, Kensington, WA 6151, Perth Australia Phone +61864368500 e-mail: Ben.Clennell@csiro.au
Dadomo Andrea	Dept. of Geological Sciences and Geotechnology, University of Milan, Piazza delle Scienze, 4, 20126 Milan, Italy
Delisle George	Bundesanstalt fur Geowissenschaften und Rohstoffe (BGR), Stilleweg 2, D-30655 Hannover Germany Phone: +49 511 643 2627 Fax: +49 511 643 2304 e-mail: G.Delisle@bgr.de
Etiope Giuseppe	Istituto Nazionale di Geofisica e Vulcanologia via Vigna Murata 605, 00143 Roma, Italy, Phone: +39 0651860394/341 Fax: +39 0651860338 e-mail: Etiope@ingv.it
Fekete Noemi	Sonderforschungsbereich 574, Christian-Albrechts-University of Kiel, Wischhofstr. 1-3, 24148 Kiel, Germany
Feyzullayev Akper	Geology Institute, Azerbaijan National Academy of Sciences 29-A, H. Javid Ave., Baku, Azerbaijan, 370143, Phone: +99412 330141 Fax: +99412 975285 e-mail: akperf@yahoo.com

Garagash Igor A.	Schmidt United Istitute of Physics of the Earth, Russian Academy of Sciences, 10 Bolshaya Gruzinskaya, 123995 Moscow, Russia
Gokhberg Mikhail	Schmidt United Institute of Physics of the Earth, Russian Academy of Sciences, 10 Bolshaya Gruzinskaya, 123995 Moscow, Russia, Phone: +7095 2545370 e-mail: gmb@uipe-ras.scgis.ru
Hepp Daniel A.	Sonderforschungsbereich 574, Christian-Albrechts-University of Kiel, Wischhofstr. 1-3, 24148 Kiel, Germany
Huehnerbach Veit	Southampton Oceanography Centre, European Way, Southampton, SO14 3ZH, UK
Isayeva Mayya I.	Geology Institute, Azerbaijan National Academy of Sciences, H. Javid 29-A Av., 370143 Baku Azerbaijan
Ismail-zadeh Tofik	Schmidt United Institute of Physics of the Earth, Russian Academy of Sciences, 10 Bolshaya Gruzinskaya, 123995 Moscow, Russia Fedynsky Centre GEON, Russian Ministry of Natural Resources, 4 Chisty per., 119034 Moscow, Russia, Phone: +095 2014468 Fax: +095 2014637 e-mail: moscow@geon.msk.su
Judd Alan	School of Marine Science & Technology, Ridley Building, University of Newcastle upon Tyne, Newcastle upon Tyne, NE1 7RU, U.K. e-mail: alan.judd@ncl.ac.uk
Kadirov Fakhretdin A.	Geology Institute, Azerbaijan National Academy of Sciences, H. Javid Prospekt, 29A, 370143 Baku, Azerbaijan.
Khavroshkin O.B.	Schmidt United Institute of Physics of the Earth, Russian Academy of Sciences, 10 Bolshaya Gruzinskaya, 123995 Moscow, Russia

Kociu Siasi	Department of Geology and Geophysics, University of Missouri-Rolla 125 Mc-Nutt Hall Rolla-Missouri 65409 USA, Phone + 573 341 4616 e-mail: kocius@umr.edu
Kolosnitsyn Nikolat I.	Schmidt United Institute of Physics of the Earth, Russian Academy of Sciences, 10 Bolshaya Gruzinskaya, 123995 Moscow, Russia
Kondorskaya Nadejda V.	Schmidt United Institute of Physics of the Earth, Russian Academy of Sciences, 10 Bolshaya Gruzinskaya, 123995 Moscow, Russia
Kopf Achim	Research Center Ocean Margins, University of Bremen, Klangenfurter Strasse, 28359 Bremen, Germany, e-mail: akopf@uni-bremen.de
Kreiter Stefan	Technische Universität Berlin, Mueller-Breslau-Str., 10623 Berlin, Germany
Kutterolf Steffen	Sonderforschungsbereich 574, Christian-Albrechts-University of Kiel, Wischhofstr. 1-3, 24148 Kiel, Germany
Levin Leonid	Fedynsky Centre GEON, Russian Ministry of Natural Resources, 4, Chisty per., 119034 Moscow, Russia, Phone: +095 4380837 Fax: +095 2014637 e-mail: moscow@geon.msk.su
Martinelli Giovanni	ARPA, Environmental Protection Agency of Emilia-Romagna Region, Via Amendola 2, 42100 Reggio Emilia, Italy, Phone: +39 059 223963 Fax: +39 0522 330 546 e-mail: giovanni.martinelli15@tin.it
Masson Douglas G.	Southampton Oceanography Centre, European Way, Southampton, SO14 3ZH, UK
Mellors Robert	Department of Geological Sciences, San Diego State University 5500 Campanile D., San Diego, CA 92182, USA, Phone: 619-594-3455 Fax: 619-594-4372 e-mail: rmellors@geology.sdsu.edu

Milkov Alexei	BP America, Exploration and Production Technology Group Room 818, Westlake 4, Building, 200 Westlake park Blvd. Houston, TX 77079 Phone: 281 3362806 Fax: 281 3667416 e-mail: alexei.milkov@bp.com
Moerz Tobias	GEOMAR Research Center for Marine Geoscience Wischhofstr. 1-3, Geb. 8A-214, D-24148 Kiel, Germany, Phone: +49-(0)431-600-2566 Fax: +49-(0)431-600-2941 e-mail: tmoerz@geomar.de
Panahi Behrouz	Geology Institute, Azerbaijan National Academy of Sciences 29-A, H. Javid Ave., Baku 370143, Azerbajian, Phone: +99412 330141 ext. 107 Fax: +99412 975285 e-mail: bpanahi@gia.ab.az
Ranero Cesar R.	Sonderforschungsbereich 574, Christian-Albrechts-University of Kiel, Wischhofstr. 1-3, 24148 Kiel, Germany GEOMAR, Wischhofstr. 1-3, 24148 Kiel, Germany Phone +494316002279 e-mail: cranero@geomar.de
Reshef Moshe	Department of Geophysics and Planetary Sciences, Tel Aviv University, Tel Aviv 69978, Israel
Rogozhin Eugeny	Schmidt United Institute of Physics of the Earth, Russian Academy of Sciences, 10 Bolshaya Gruzinskaya, 123995 Moscow, Russia Phone: +095 2545370 Fax: +095 2549088 e-mail: eurog@uipe-ras.scgis.ru
Sahling Heiko	Sonderforschungsbereich 574, Christian-Albrechts-University of Kiel, Wischhofstr. 1-3, 24148 Kiel, Germany
Smith George	Department of Geological Sciences, University of Cape Town, Rondebosch 7700, South Africa

Schmidt Mark	Sonderforschungsbereich 574, Christian-Albrechts-University of Kiel, Wischhofstr. 1-3, 24148 Kiel, Germany Institute of Geosciences, University of Kiel, Ludwig-Meyn-Str. 10, 24118 Kiel, Germany
Solodilov Leonid	Fedynsky Centre GEON, Russian Ministry of Natural Resources, 4 Chisty per., 119034 Moscow, Phone: +095 2015780 Fax: +095 2014637 e-mail: moscow@geon.msk.su
Song, Sheng Rong	Department of Geosciences, National Taiwan University No.1, Sec.4, Roosevelt Road, Taipei 106, Taiwan, R.O.C. e-mail: srsong@ccms.ntu.edu.tw
Spiess Volkhard	Research Center Ocean Margins, University of Bremen, Klagenfurter Str., 28359 Bremen, Germany
Starodub Nikolay	Department of Biochemistry of Sensory and Regulatory Systems of A. V. Palladin Institute of Biochemistry of National Academy of Sciences, 9 Leontovicha Str., 01030 Kiev, Ukraine, Phone: +380-44-229 47 43 Fax: +38044 2296365 e-mail: nstarodub@hotmail.com
Suess Erwin	Sonderforschungsbereich 574, Christian-Albrechts-University of Kiel, Wischhofstr. 1-3, 24148 Kiel, Germany GEOMAR Research Centre, Wischhofstr. 1-3, 24148 Kiel, Germany Phone: + 494316002232 e-mail: esuess@geomar.de
Tsyplakov V. V.	Schmidt United Institute of Physics the Earth, Russian Academy of Sciences, 10 Bolshaya Gruzinskaya, 123995 Moscow, Russia
Urdukhanov E. M.-K.	Schmidt United Institute of Physics of the Earth, Russian Academy of Sciences, 10 Bolshaya Gruzinskaya, 123995 Moscow, Russia

Vidmont Natalya	Schmidt United Institute of Physics of the Earth, Russian Academy of Sciences, 10 Bolshaya Gruzinskaya, 123995 Moscow, Russia, Phone: +7095 2545370 e-mail: eurog@uipe-ras.scgis.ru
Weinrebe Wilhelm	Sonderforschungsbereich 574, Christian-Albrechts-University of Kiel, Wischhofstr. 1-3, 24148 Kiel, Germany GEOMAR Research Centre, Wischhofstr. 1-3, 24148 Kiel, Germany Phone: + 494316002281 e-mail: wweinrebe@ifm-geomar.de
Yang, Tsanyao Frank	Department of Geosciences, National Taiwan University No.1, Sec.4, Roosevelt Road, Taipei 106, Taiwan, R.O.C. e-mail: tyyang@ntu.edu.tw
Yeh, Gau-Hua	Department of Geosciences, National Taiwan University No.1, Sec.4, Roosevelt Road, Taipei 106, Taiwan, R.O.C.

PREFACE

1. PURPOSE OF PRESENT BOOK

During the period May 19-26, 2003 the NATO Advanced Research Workshop (ARW) "Mud volcanism, Geodynamics and Seismicity" was held in Baku. Participants coming from USA, Germany, France, Italy, Portugal, Russian Federation, Ukraine, Romania, Georgia, UK, Israel, Azerbaijan, Tunisia have discussed about different geodynamic features of mud volcanism and participated to field trips oriented to a better knowledge of mud volcanic features. The Meeting focused on many features of mud volcanism occurrence and related geodynamic topics. The purpose of present book is to collect contributions discussed during the Meeting and to fill a marked editorial gap on mud volcanism. Mud volcanism was to date described by local monographies or by articles published by scientific journals. In particular no books were published on topics able to highlight the link among mud volcanism, geodynamics and seismicity.

Mud volcano of Nirano (Northern Italy). Engraving from Stoppani A. (1871), Corso di Geologia, Milan, Bernardoni G. and Brigola G. Publishers.

2. WHY MUD VOLCANOES ARE GEOLOGICALLY IMPORTANT ?

Mud volcanoes have attracted the attention of earth scientists for many years. Due to their importance in hydrocarbon research, a consistent progress in the knowledge of mud volcanism took place in the past twenty years. Mud extrusion is a well-known phenomenon occurring in geological environments where fluid-rich, fine grained sediments ascend within a lithologic succession due to their buoyancy. Mud extrusion has been recognized to be generally related to hydrocarbon deposits, seismotectonic activity and to orogenic belts. Mud volcanism occurs both onshore and offshore particularly in compressional tectonic environments. The size of mud volcanoes reaches up to some kilometres in diameter and up to about several 100 meters in height. Known submarine mud volcanoes occur in many locations across the world: Barbados Islands, Gulf of Mexico, Norwegian Sea, Offshore Greece, Offshore Crete, Offshore Cyprus, Black Sea, Offshore Nigeria, Caspian Sea. In other geologically similar areas submarine mud volcanoes have been inferred.

Mud volcanism affects inland areas as well. It has been observed in: Alaska, Azerbaijan, Barbados Ridge, Black Sea region, Borneo, Caspian Sea region, China, Ecuador, Georgia, Greece, Greenland, India, Iran, Italy, Java, Kyrghyzstan, Mexico, Mississippi Delta, Myanmar, Netherlands, New Guinea, New Zealand, Pakistan, Panama, Roumania, Russia, Sakhalin region, Spain, Sumatra, Taiwan, Tanganyika, Tanzania, Timor, Trinidad, Tunisia, Turkmenistan, Ukraine, United Kingdom, Venezuela etc. Other geologically similar areas host less explored inland mud volcanic processes.

The main features of mud volcanism are: a) a link with rapidly deposited and overpressured clayey sequences, b) the presence of gaseous (mainly methane) and liquid fluids (brackish water) which make possible extrusive activities, c) a relation with regional seismotectonic features.

Methane outgassing is likely to contribute significantly to the planet global methane budget although its relation with climate dynamics is still poorly quantified. Gas hydrates are often associated to mud volcanism, and intense research activity is devoted to evaluate their contribution to the total methane budget. Consistent experimental evidence, mainly coming from geochemical analyses and radioisotope studies, suggests that brackish waters linked to mud extrusion are representative of confined or semiconfined deep reservoirs.

In compressional zones, where mud volcanism is more common, dewatering processes due to fluid flow caused by mud extrusion exceed the compaction-driven pore fluid expulsion. This peculiar behaviour can significantly influence regional hydrogeochemical features. Geochemical fluctuations possibly linked to flow rate variations due to crustal deformation processes have also been recorded.

Mud volcanoes of Otman-Bozdag, Touragay and Kichikdag in Azerbaijan. Engraving from Stoppani A. (1871), Corso di Geologia, Milan, Bernardoni G. and Brigola G. Publishers.

3. WHY MUD VOLCANOES ARE IMPORTANT IN GEOPHYSICS?

In spite of the comparatively large amount of data collected by scientists working in different disciplines - geophysics, geochemistry, structural geology, marine geology, hydrocarbon geosciences, etc., a comprehensive and complete understanding of mud volcanic phenomena is still lacking. One of the most important gaps of knowledge is connected to the role played by active tectonics and in particular by crustal deformation processes and seismogeneis on the fluid dynamics of mud volcanism. Both theoretical considerations and experimental evidence suggest that firm relationships between these processes actually exist and that mud volcanoes act as natural strain meters. In general, a correct mechanical modelling of subsoil mechanical behaviour should take into account the coupling between gaseous / fluid and solid phases. Poroelastic constitutive equations represent the simplest theoretical framework for such a modelling. In this formulation, subsurface pore pressure of the gaseous/liquid phase and volumetric strain in the solid phase are considered mechanically interrelated. This mechanical coupling is enhanced in the undrained conditions typical of confined reservoirs. As a consequence of such coupling, strain variations in the solid structure of the crust reflect in variations of pore pressure. Realistic evaluations of the relevant phenomenological coefficients suggest that volumetric strain variations of the order of 10-6 can produce a pressure increase of the order 0.01 MPa (equivalent to a piezometric variation of 1 m of water).

A possible non-linear mechanical coupling between solid and gaseous/ fluid phases has been also recognised in the field of fracture dynamics. It is well known, in fact, that the failure of rocks is controlled by the effective stress which results from the combination of pore pressure and environmental stress. Consistent experimental results show that pore pressure plays a primary role in earthquake triggering. On the other hand, it is well known that coseismic and post seismic stress redistribution dramatically affects groundwater circulation. These considerations suggest that a careful monitoring of mud volcanoes, being representative of deep seated confined aquifers sensitive to strain field fluctuations, can be an invaluable tool for monitoring ongoing tectonic and seismogenic processes. The effectiveness of this is shown by experimental evidence, which in the past decades allowed to recognize a significant connection of cyclic mud volcanic activity with earth tides and a relation between paroxistic extrusive phenomena and regional seismic activity. The latter also suggest that a systematic monitoring of mud volcanism could supply new useful information for seismic hazard assessment. In fact, short/ medium term information on strain field fluctuations would allow to gain new insights on post seismic stress redistribution and pre-seismic preparatory processes.

Only a satisfactory comprehension of the complex physicochemical interactions of the fluid and gaseous phases with fracture dynamics and strain variations which give rise to mud volcanism will allow the full exploitation of mud volcanism as a natural strain meter. To this extent, the physicochemical dynamics of mud volcanism and its relationship with other geophysical phenomena must be carefully scrutinized from both the theoretical and experimental points of view.

4. WHAT YOU'LL FIND IN THE BOOK

Some articles are devoted to seismic and seismic hazard topics (see also contributions by Rogozhin, Levin et al., Kociu, Ismail-zadeh and Isayeva), other communications were devoted to mud volcanism geology (see also contributions by Bedir, Ben-Avraham et al., Milkov, Moerz et al.), to geodynamic implications of mud volcanism (see also contributions by Baciu and Etiope, Panahi), to the contribution of mud volcanic gases to greenhouse effect (see also Etiope, Judd) to monitoring techniques of mud volcanism (see also contributions by Delisle, Garagash et al., Khavroshkin et al., Martinelli and Dadomo, Mellors et al.), to geochemical features of mud volcanoes (Martinelli and Dadomo, Starodub, Yeh et al.) and to physical models of mud volcanoes (see also contributions by Albarello, Feyzullayev et al., Kopf et al.).

The Authors of present contributions were involved in scientific debates during the ARW held in Baku in 2003 and agreed on some findings on mud volcanism.

5. ASSESSED FINDINGS ON MUD VOLCANOES

The scientific debate among participants coming from various disciplines allowed to confirm that mud volcanoes are:

1) located in areas of continental margins
2) connected with rapidly deposited clay sequences
3) connected with hydrocarbon reservoirs
4) affected by a gaseous component whose origin is sometimes syngenetic to extruded sediments and sometimes deeper
5) able to originate a continuous flux of CH4 to atmosphere of about 3-4 Tg/year
6) subjected to possible fluctuations in their extrusive activity due to seismotectonic features of the areas of occurrence.

Mud volcano of Otman-Bozdag in Azerbaijan. Engraving from Stoppani A. (1871), Corso di Geologia, Milan, Bernardoni G. and Brigola G. Publishers.

6. RECOMMENDATIONS

Discussions among Authors evidenced that in spite of the relatively high amount of collected data an exhaustive physical model of mud volcanism is still lacking. A more reliable physical modelling will be probably reached by means of a broader utilization of monitoring techniques. During the ARW some monitoring techniques were described. They include seismic networks, hydrophones, pressure and temperature sensors, geochemical sensors, geoelectrical sensors, satellites. Discussions evidenced an increasing in the new discoveries of mud volcanoes from hydrocarbon geophysical prospections but a better knowledge of mud volcanism is also depending on complete mapping of known phenomena and the realization of an Atlas of mud volcanoes of the world have been solicited by participants. Satellite monitoring of gaseous emissions and of crustal deformations has also been defined as very desirable for monitoring of possible fluctuations of mud volcanic extrusive activity in connection with geodynamic events. Discussions offered the chance to participants to project cooperative researches in the field of geochemical, geophysical and seismologic monitoring of mud volcanoes in different places of the world with particular reference to Azerbaijan.

Antonio Stoppani (Lecco 1824 – Milan 1891) in his book "Corso di Geologia"(1871) described mud volcanoes of Azerbajian on information by Hermann Abich (Berlin 1806 – Gratz 1886). Abich, geologist and volcanologist, lived in Tbilisi visited Azerbaijan and wrote in 1863 "Ueber eine Caspischen Meere erschienene Insel, nebst Beitraegen zur Kenntraiss der Schlammvulkane der Caspischen Region", Mem. Acad. Imp. Sci. St. Petersbourg, VII series , VI, n.3.

Giovanni Martinelli, Reggio Emilia
Behrouz Panahi, Baku
May 2004

Chapter 1
MUD VOLCANIC GEOLOGY

NEW SEISMIC NEOGENE CLAY DIAPIRS AND HYDROCARBON IMPLICATIONS IN THE NORTH-EASTERN AFRICAN MARGIN OF TUNISIA

Mourad Bedir
Georesource Laboratory. Institut National de Recherche Scientifique et Technique, BP 95, 2050 Hammam-Lif, Tunisia.

Abstract: Subsurface and surface geological and geophysical studies performed on the Neogene series in the northern and oriental regions of Tunisia has allowed us to highlight the basin structuring and evolution. Neogene sequence deposits are distributed across graben, half graben, platforms, folds and syncline basins. Miocene, Pliocene and Quaternary series are made up of several thousand meters of mainly thick clay and sandstone packages. Clay and mud diapirs were recognized on a small scale in the Miocene outcrops of the Northern Atlassic and Sahel regions and in the Gulf of Hammamet, where they are well developed in the sub-surface and visible in the seismic scale. Clay-kinesis is induced by the lithostatic pressure of Plio-Quaternary thick deposits and by the transtensive and transpressive movements of North-South Grombalia-Enfidha-El Jem and East-West Hammamet-Maamoura, Kuriat and Boumerdès-Mahdia flower fault corridors. The occurrence of diapirs is fossilized on their flanks by synsedimentary Miocene, Pliocene and Quaternary high-angle subsiding graben and syncline basins accompanied by downlap prograding turbiditic sequences onlapped by aggrading and retrograding system tracts and pinch-outs. Structural and stratigraphic unconformities are located on the flank diapirs. The clay diapirism accompanied the right and left lateral formation of basins and subsidence migration along the strike-slip fault corridors. The tops of diapir structures are marked by toplap erosional surfaces of the Upper Miocene, Pliocene or Quaternary sequences. Basin modeling is presented to visualize the structurating, the mechanisms and the distribution of clay diapirs around the basin and the tectonic structures. These Neogene diapirs were recognized and described for the first time in Tunisia in 1997. The existence of mud diapirism in the Miocene oil field area of the gulf of Hammamet allows us to devise a new geodynamic model comprising important petroleum implications in terms of structuration, system tracts, migration fluids, traps and seals.

Key words: mud diapirism, seismic profiles, oil prospection, Tunisia, stratigraphy

G. Martinelli and B. Panahi (eds.), Mud Volcanoes, Geodynamics and Seismicity, 1–15

1. INTRODUCTION

Geological and geophysical studies carried out on the Neogene deposits of the Eastern margins of Tunisia (Bédir, 1985; Bédir and Bobier, 1987; Bédir, 1995; Gaaloul, 1995; Gaaloul et al., 1997) and the northern regions (Figs. 1 and 2) by sequence stratigraphy, seismic stratigraphy and tectonics have allowed the recognisition of the sequence and basin structuring in relation to the problematics of the formation and the evolution of Miocene (Bédir and Tlig, 1992b ; Bedir, 1993a; Bédir, 1993b; Bédir and Tlig, 1993c; Bédir et al., 1996; EL Manaa, et al., 1997), Pliocene and Quaternary basins (M'ridekh, 1994; M'ridekh et al., 1995; Bédir et al., 1997).

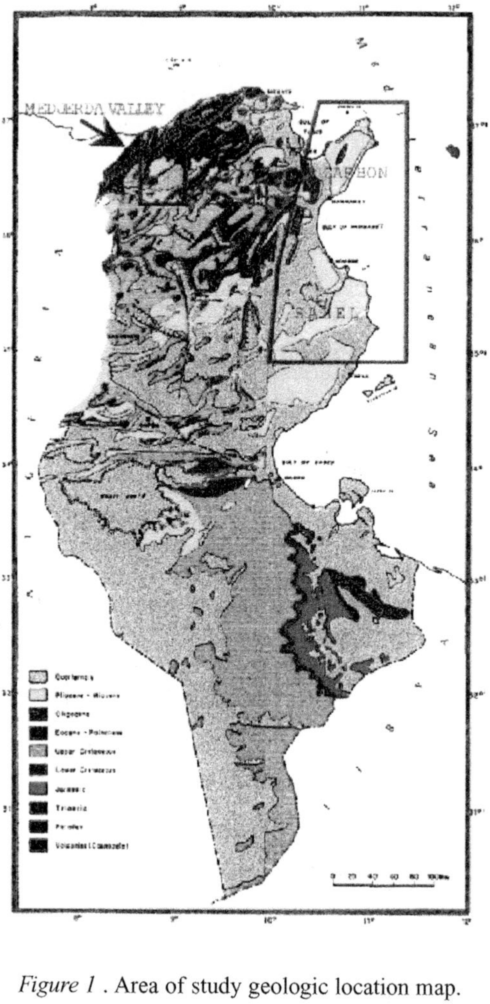

Figure 1 . Area of study geologic location map.

For the geophysical study, seismic reflexion profiles covering the Onshore and Offshore zones of the gulf of Hammamet and the Sahel from many seismic data since 1972 until 1990 (Bédir, 1995) were used for the Neogene horizon correlations, isochrone and isopach mapping, seismic stratigraphy and seismic tectonic analyses. The calibration of Miocene and Pliocene seismic horizons was established by thirty petroleum wells (Fig. 2).

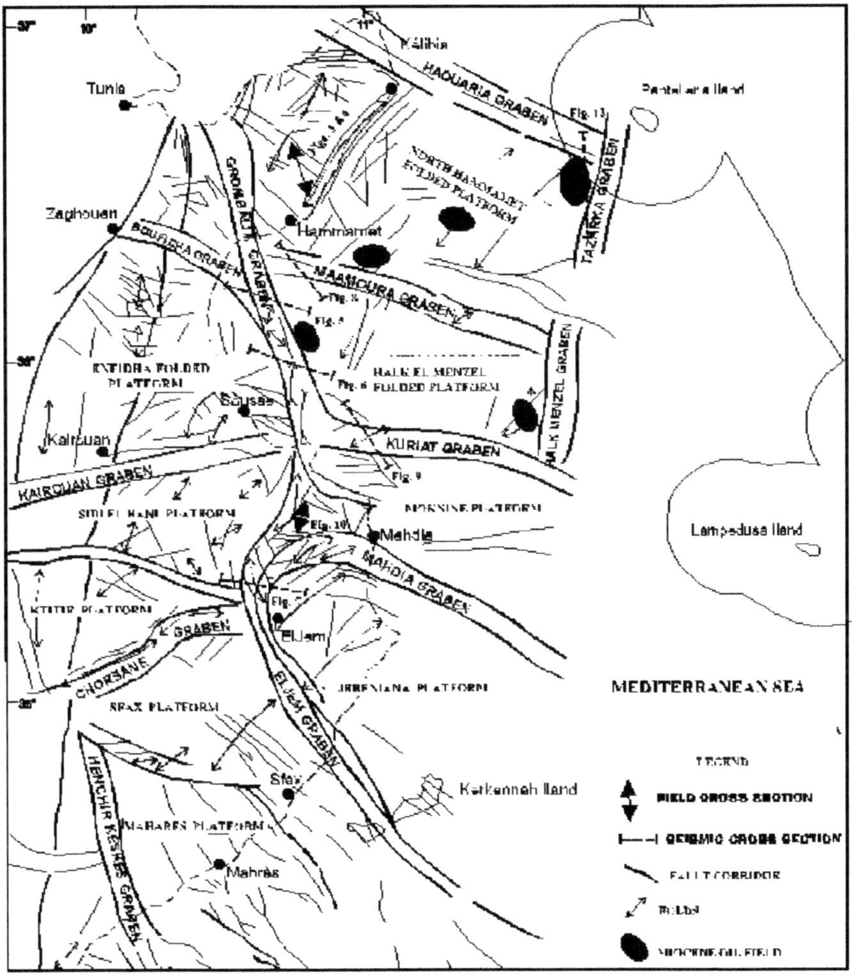

Figure 2. Subsurface Tectonic Framework map of the Neogene deposits of the Eastern Margin of Tunisia. Used Petroleum wells, presented seismic lines and outcrop observations.

In this work, we have focused the new model that considers the importance of the clay movements in the geodynamic context of Neogene deposits at their basin scale. The clay diapiric and intrusive structures in seismic and outcrop scales were identified and we have tried to explain their occurrence in relation to Neogene geodynamic context and mechanisms. This basin structuring related to the clay-kinesis must be taken into consideration for petroleum prospectivity in these regions well-known for their Miocene oil field reservoirs.

2. LITHOSTRATIGRAPHIC CHARACTERISTICS

Neogene deposits consist of several hundred meters of alternating packages of clays, marls, Lignites and sandstones from Miocene, Pliocene and Quaternary series (Figs. 3 and 4).

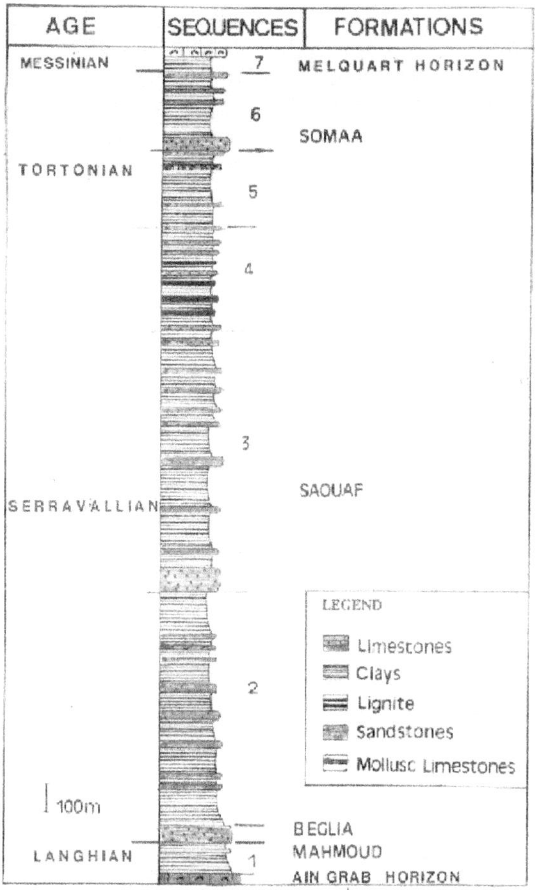

Figure 3. Synthetic stratigraphic and sequence deposit column of the Middle and Upper Miocene of the Cap Bon, the gulf of Hammamet and the Sahel regions.

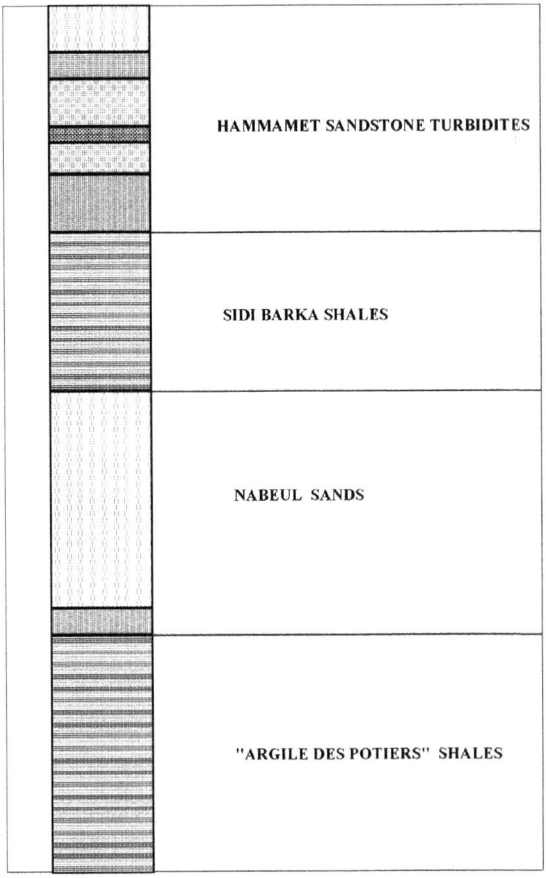

Figure 4. Synthetic stratigraphic column of the Pliocene deposits of the Cap Bon (Colleuil, 1976, Ben Salem, 1992, Damak-Derbal, 1993 and M'ridekh, 1994).

These deposits show important changes in lithological facies and thicknesses. They are very thick in the graben and syncline structures of the gulf of Hammamet, the Sahel (Bédir, 1995; Bédir et al., 1996; EL Manaa et al., 1997) and to the Northwest Medjerda and Kechabta basins (Biély et al., 1972; Burollet, 1956). Particularly, Miocene and Pliocene clays and sandstones are very thick in the Eastern margin of the Cap Bon and the Sahel zone comprising sandstone turbidite sequences outcropping in the Jebel Abderrahman (Bédir, 1993a; Bédir and Tlig, 1993c; Bédir et al., 1996) and in Pliocene deposits of the coastal Nabeul and Hammamet regions (Ben Salem, 1992; Damak-Derbel, 1993; Damak-Derbel and Zaghbib-Turki, 1995).

This fact is demonstrated by the Jiriba 1 well (JRB 1) that crossed a thousand meters of exclusively Pliocene and Quaternary deposits in the North-South Enfidha graben (Fig. 5).

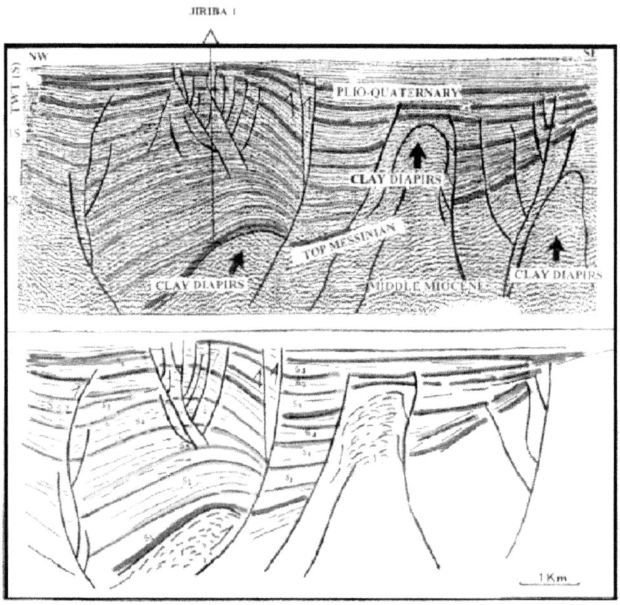

Figure 5. Seismic section of the North-South Enfidha Flower fault corridor showing the Jiriba 1 well location, Pliocene and Quaternary sequence deposits, intra-basin folds and synclines with Miocene and Pliocene Clay Diapirs.

The appearence of subsurface clay diapiric structures in the gulf of Hammamet and in the Sahel grabens and the syncline depocenters is intimately related to the thickness of Miocene and Pliocene clay packages. In the Medjerda north zone, the Neogene deposits are also very thick and consist of clay packages with intercalated sands and sandstones classified as the Medjerda Group (Biély et al., 1972). Hence, the lithologic nature and thickness of Neogene deposits due to the great tectonic subsidence rates are responsible for the occurrence of intrusive and diapiric clay structures.

3. TECTONICS AND STRUCTURING FRAMEWORK

The Eastern margin of Tunisia comprises the Cap Bon and the gulf of Hammamet zones to the North and the Sahel regions to the South (Figs. 1 and 2). This margin is structured by North-South and East-West subsurface strike-slip fault corridors that limit platform, graben, syncline and fold structures (Bédir, 1995). Neogene sequence deposits are distributed around these structures where the thickness of sedimentary series is very important in the syndepositional subsiding zones of grabens and synclines (Bédir, 1995; Bédir et al., 1996; Ben Salem, 1992).

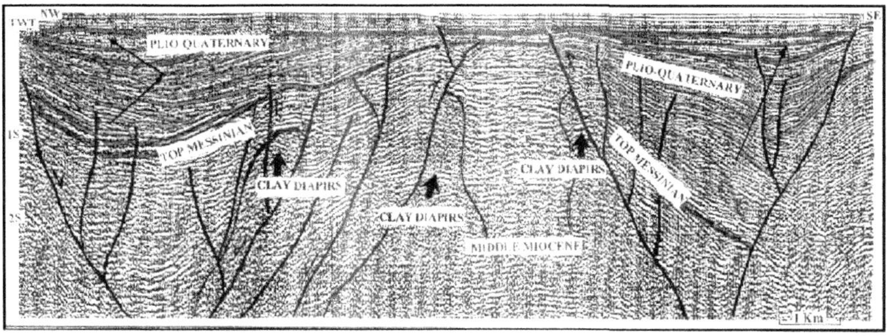

Figure 6 . Seismic section of the North-South Enfidha Flower fault corridor showing Pliocene and Quaternary folds and synclines with Miocene Clay Diapirs, Depocenter Basin Migrations and Downlap prograding sequence deposits.

The main tectonic structures constituting the background and the guide of the Neogene sequence deposit distribution are represented by the North-South Grombalia-Enfidha-El Jem and Halk El Menzel flower fault corridors (Figs. 5,6 and 7) and the East-West flower fault corridors of El Haouaria, Hammamet-Maamoura (Fig. 8), Kuriat (Fig. 9) and Boumerdès-Mahdia. These fault corridors particularly limit the tilted Miocene platform blocks of Halk El Menzel, Moknine and Jebéniana. Inside these corridors, Neogene distensive and compressive structures of grabens, en echelon folds and synclines (Figs. 5, 6, 7 and 9) take place.

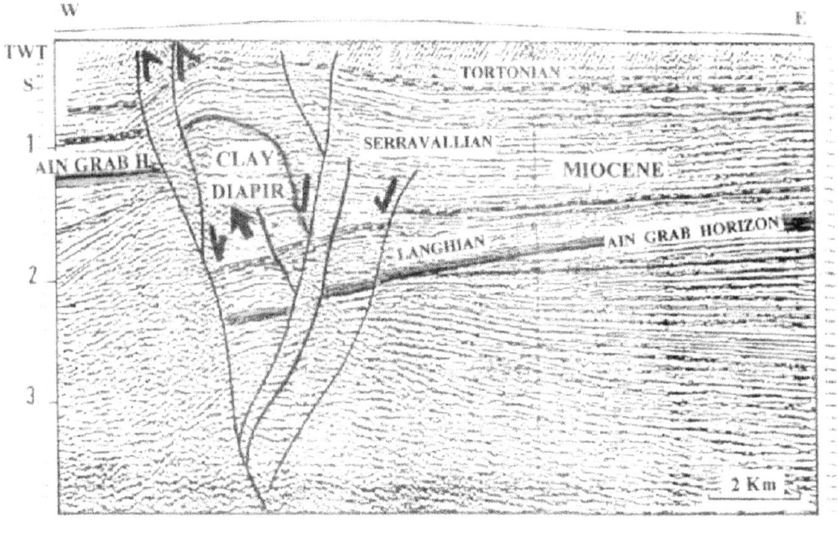

Figure 7. Seismic section of the North-South El Jem Flower fault corridor showing Miocene folded Half-Graben with decollement of Clay intrusive Diapir.

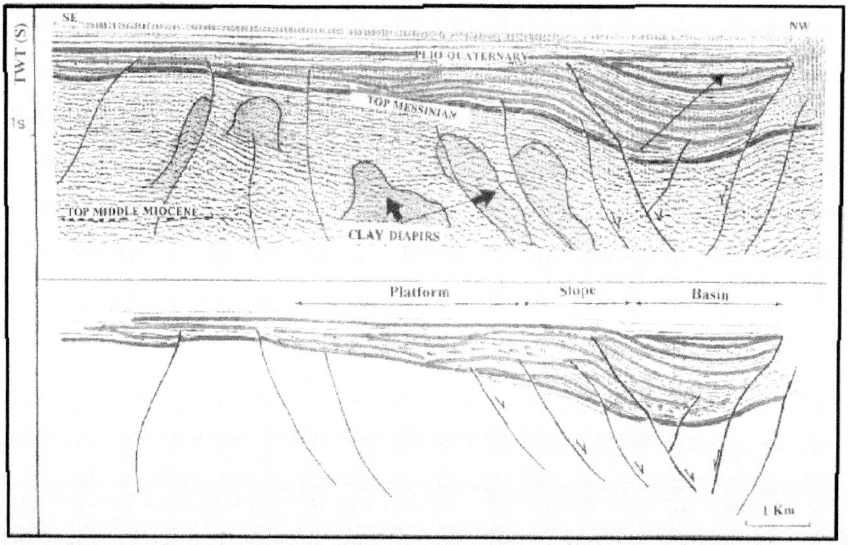

Figure 8. Seismic section of the N120 Hammamet-Maamoura Flower fault corridor showing Miocene Clay intrusive and Diapir structures and Pliocene to Quaternary platform to basin folded syncline Basin with lateral Depocenter Migration. Pliocene and Quaternary Downlap progradational and Onlap/Toplap aggradational and retrogradational sequence deposits take place on the slope basin border.

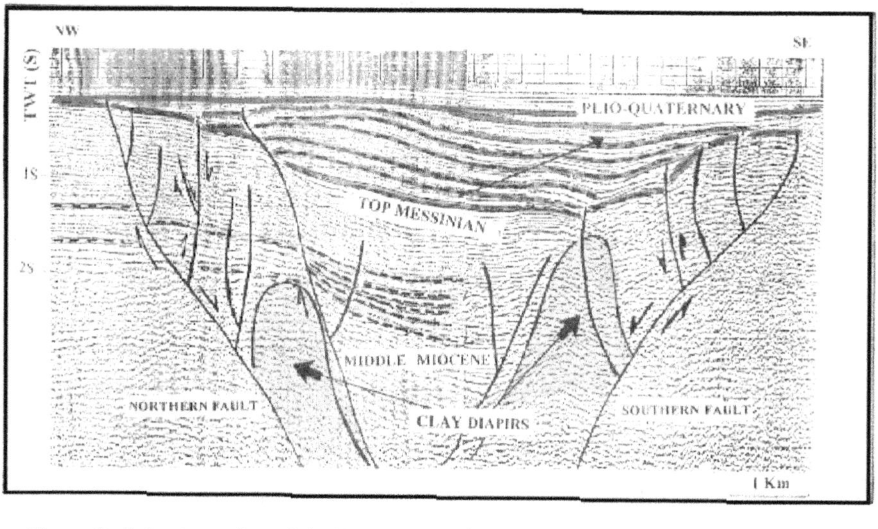

Figure 9 . Seismic section of the East-west Kuriat Pliocene and Quaternary Flower fault syncline basin showing Miocene Clay Diapirs and pinch outs and lateral Depocenter Migration

During the Neogene periods, the border faults of these corridors had been reactivated in transtensive and transpressive movements by the Northeast-Southwest distensive Langhian-Serravallian Miocene and Pliocene regional tectonic stress and Tortonian and Late Messinian Upper Miocene and Quaternary Northwest-Southeast to North-South compressive strains (Bédir, 1995; Bédir et al., 1996; Bédir et al., 1997; Ben Ayed, 1986; Ben Salem, 1992; Gaaloul et al., 1997). The impact of this tectonic reactivation of the faults combined with the eustatic sea level changes were very important for preparing the great accumulation rates of undercompacted clay and shale thick deposits in the syndepositional opened grabens and compressive synclines. The orientation of the distensive Miocene and Pliocene stress allows for the opening of the North-South tectonic corridors of Grombalia-Enfidha-Zéramdine-El Jem. The latter had permitted much greater accumulation rates of undercompacted clay and shale that would be activated by transtensional and transpressional fault plays inducing claykinesis and intrusive movements during Upper Miocene, Pliocene and Quaternary times. For this reason, clear and major spectacular mud diapirs and intrusive structures are present along this famous and major Grombalia-Enfidha-Zéramdine-El Jem flower fault corridor (Figs. 5, 6 and 7). On the other hand, the mud diapirs along the East-West corridors are more or less frequent and accentuated (Fig. 9).

Similarly to the Triassic halokinetic mechanisms highlighted in the central Atlassic Mesozoic and Cainozoic basins and structures (Bédir, 1995; Bédir et al., 2001; Boukadi, 1994; Boukadi and Bédir, 1996), the occurrence of clay diapirism seems to be favored in the fault corridor interferences known as a releasing strike-slip accommodation zones, as is the case of the Boumerdès area that constitutes the corner (Fig. 2) between the North-South Zéramdine-El Jem fault corridor and the East-West Boumerdès-Mahdia one (Bédir et al., 1992a).

4. CLAY DIAPIRIC AND INTRUSIVE STRUCTURES

Diapiric and intrusive seismic structures seem to be strongly associated to the faults limiting the graben, syncline and platform zones (Figs. 5, 6, 7 and 8). They are from Miocene sequences. They appear along the seismic deep-fault system and had been recognized in small outcrop scale in the Sahel in the Upper Miocene sequences showing metric to decametric lenticular intrusive shales along kilometric strike-slip faults (Fig. 10) (Bédir, 1995; Bédir et al., 1996; Bédir et al., 1997 ; Gaaloul, 1995) and in the North Atlassic Miocene deposits of Medjerda zone (Bédir et al., 1997). These structures consist of metric and decametric Upper Miocene Serravallian and Tortonian shales and clays intruded vertically and laterally across sand and sandstone beds. They had been also described in the Sahel by Gaaloul (1995) as micro-sand and clay volcanoes.

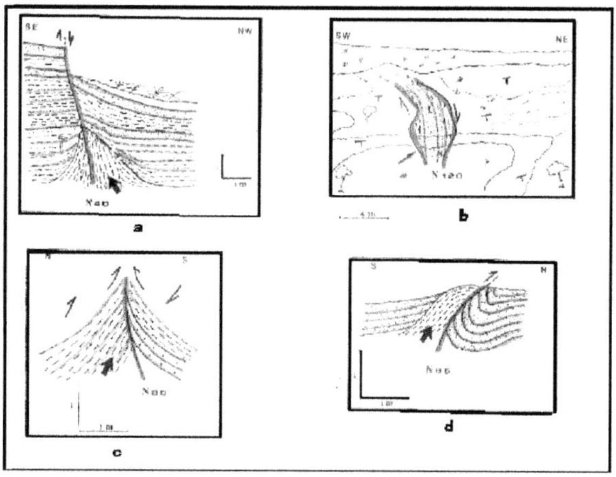

Figure 10. Serravallian and Tortonian Upper Miocene outcrops in the Zéramdine and Boumerdès Sahel localities (see Figure 2 for location) showing metric and decametric diapiric clays intruded in sandstone beds along normal (a) and reverse (d) faults and lenticular intrusive clay and sandstone layers along strike slip faults (b and c).

The mud diapirs and clay-kinesis of Miocene had been well studied in the Caraibes accretionary transform zone of Barbados where they constitute spectacular mud volcanoes (Griboulard, 1991). In the south-western margin of the gulf of Mexico, undercompacted Neogene shales and salt intrusive structures are associated to listric growth faults bordering syncline structures with trapped sediments (Shaub, 1983). Clay diapirs are also recognized in the western Mediterranean sea in the subsurface offshore Neogene deposits of Spain (Comas et al., 1992).

5. SEQUENCE AND BASIN STRUCTURING

The hypothesis of the presence of diapirs and claykinesis activity along the Neogene deposits in Tunisia is also shown by the synsedimentary structuring of sequence deposits and basin geometry and distribution. Miocene deposits are made up of seven to ten third order sequences (Bédir, 1995; Bédir et al., 1996; Gaaloul et al., 1997) and Pliocene and Quaternary deposits consist of six to eight Seismic sequences (S1 to S8) (Figs. 5 and 7). These sequences are recognized from the Neogene outcrops and from the sequence stratigraphic analyses of Neogene seismic horizons calibrated to the petroleum wells of the Offshore and Onshore zones in the gulf of Hammamet and the Sahel area (Bédir and Tlig, 1992b; Bédir, 1995; Bédir et al., 1997; Damak-Derbel and Zaghbib-Turki, 1995; M'ridekh, 1994; M'ridekh et al., 1995). These sequences were dated by the calibration of their system tracts and their limit boundaries to chrono-eustatic cycles of the global chart (Haq et al., 1987).

In the flanks of clay diapirs and basin borders, these sequences fossilize the tectonic subsidence activity by the distribution of their system tract structures. They are composed of high-angle downlap prograding system tracts at the base, overlapped by aggradational and retrogradational onlap and toplap deposits (Figs. 5 and 7). These sequences show structural and erosional unconformities and pinch outs in the flanks of diapiric structures (Figs. 5, 6, 7 and 9). The transtentional and transpressional border fault reactivation during Miocene, Pliocene and Quaternary tectonic stress combined to the great thickness and the ductil nature of Neogene deposits in this area induced vertical and lateral clay-kinesis movements that provoke the depocenter migrations and inversions along the fault corridors (Figs. 5, 6, 7 and 8). These basin migrations represent additional proof of the existence of clay-kinesis movements. This geodynamic processing recalls what we was illustrated for the Triassic halokinesis movements in the Mesozoïc basin migrations in the Atlassic and oriental regions of Tunisia (Bédir, 1995; Bédir et al., 2001).

6. BASIN MODELLING

The recognition of subsurface Neogene clay and mud diapiric and intrusive structures in the area of the gulf of Hammamet and the Sahel leads us to propose a new structural, sequence stratigraphic and basin model concerning the strike-slip fault corridor basin system (Fig. 11).

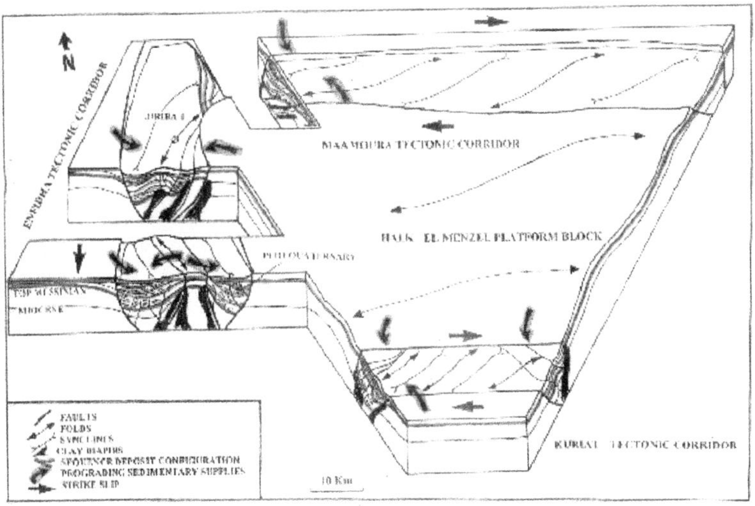

Figure 11. Basin Modelling in the gulf of Hammamet area showing the Neogene Clay Intrusives and Diapirs distribution associated to the transtensive and transpressive strike-slip Flower fault systems inducing Rim Syncline Depocenters, lateral Migrations and sequence deposit organization.

The block diagram in Figure 11 shows that the distribution of the clay and mud diapirs in the gulf of Hammamet block is well-guided by the bordering faults of grabens and syncline corridors. The structuring of East-West right lateral and North-South left lateral en echelon folds and synclines inside and outside the tectonic corridors indicates the strike-slip nature of the North-South and East-West flower faults. Depocenter and basin space lateral migrations along the fault corridors confirm the lateral and vertical movements of wrench faults and the ascention of clay intrusive diapirs. Syncline depocenters are clearly induced here by clay-kinesis dynamics and are very comparable to the rim syncline structures (Figs. 5, 6, 8 and 11) of the Triassic halokinetic structures. The directions of the sediment supplies of Pliocene and Quaternary deposits are given by the prograding system tracts. They are produced by the tilted limitroph platforms around the grabens and synclines and are from different pathways depending on the tilting orientation of the blocks (Fig. 11). This Pliocene and Quaternary infilling sediment dynamic model around thr platform, graben, fold and syncline syndepositional structures shows the straight control of strike-slip fault movement corridors and the fluctuating sea-level changes that guide the lowstand and the highstand sequence deposit system tracts. This model is very similar to the silicoclastic Miocene sedimentary model highlighted in the Onshore Sahel area of Mahdia and El Jem (Bédir, 1995; Bédir et al., 1996).

7. PETROLEUM IMPLICATIONS

Some Miocene oil fields in the gulf of Hammammet like Tazarka and Byrsa seem to be placed near such clay diapirs (Fig. 12).

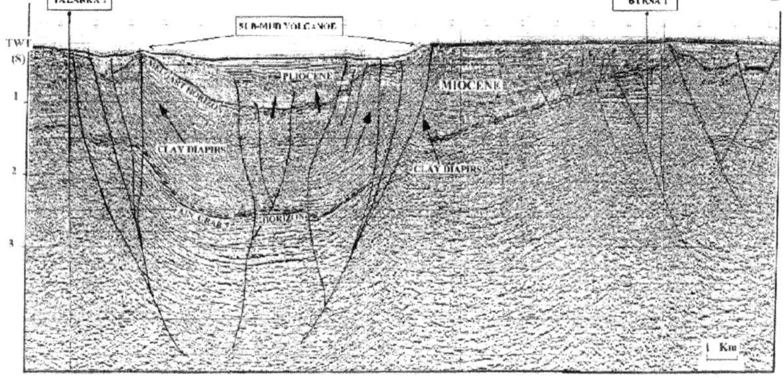

Figure 12 . Seismic section showing Tazarka and Byrsa oil field structures in the gulf of Hammamet associated to Flower fault structures and Clay Diapir and intrusives along graben, syncline and anticline Neogene structures.

As a matter of fact, these anticline structures are limited by North-South and East-West fault corridors bordering syncline structures and grabens. The discovery of Neogene mud diapirs in the Eastern margin of Tunisia constitutes an interesting new way of investigation for petroleum exploration. The petroleum implications of the diapiric intruded structures along the listric borders of flower faults concern the flanks and the tops of the domes where structural and stratigraphic unconformities and erosional surfaces around the rim synclines occur. Neogene sequence deposits contain reservoir turbidite sandstones and clay seals. Along these diapirs, the fracturation and the fluid migrations are very important with undercompacted shales rapidly deposited during the Miocene and Pliocene periods. Many oil fields are associated to such mud diapirs in the world like those of the Mexico gulf and the Venezuela and Caraibe areas. In Tunisia, the border fault corridors and the graben and platform borders are not yet explored and probably represent the future potential objectives to be pursued. In particular, these clay diapirs and their rim syncline flanks may represent new petroleum targets of Miocene traps and will probably allow us to understand their dynamics.

REFERENCES

1. Bédir, M., 1985. *Dynamique des bassins sédimentaires du Sahel de Mahdia d'après les données de surface et de subsurface. Seismo-stratigraphie, Evolution tectonique et sédimentologie.* Mémoire de Diplôme d'Etudes Approfondies, Faculté des Sciences de Tunis, Université de Tunis II, Tunis.
2. Bédir, M. and Bobier, CL., 1987. Les grabens de Mahdia et Sidi Chérif (Tunisie orientale). Dynamique des fossés oligo-miocènes induits au toit d'anticlinaux crétacés-éocènes par les jeux au Néogène de décrochements Est-Ouest et Nord-Sud, *Bulletin de la Société Géologique de France,* **8**, t. III, n° 6, 1143-1151.
3. Bédir, M., Zargouni, F., Tlig, S. and CL. Bobier, 1992a. Subsurface geodynamics and petroleum geology of transform margin basins in the Sahel of Mahdia and El Jem (eastern Tunisia), *Amer. Assoc. Petro. Geology,* **76**, n° 9, 1417-1442.
4. Bédir, M. and Tlig, S., 1992b. *Structuration des paraséquences miocènes le long des relais de coulissements de subsurface de la marge orientale de la Tunisie: Implications pétrolières,* 3ème Jour. Explo. Petro. E.T.A.P. , Tunis.
5. Bédir, M., 1993a. *Stratigraphie sismo-séquentielle et contrôle tectono-eustatique des dépôts miocènes de la marge orientale de la Tunisie*, Intérim Colloquium du RCMNS, Tunis, 28 Sept. 1993.
6. Bédir, M., 1993b. *Les bassins pliocènes et quaternaires des gouttières synclinales des plis messiniens de subsurface de la Tunisie nord orientale,* 14th Regional Meeting of Sedimentology, Marrakech, 27-29 April 1993.
7. Bédir, M. and Tlig, S., 1993c. *Les séquences turbiditiques miocènes du jebel Abderrahmane au Cap Bon en Tunisie Nord orientale,* 14th Regional Meeting of Sedimentology, Marrakech, 27-29 April 1993.
8. Bédir, M., 1995. *Mécanismes Géodynamiques des bassins associés aux couloirs de*

coulissements de la marge Atlasique de la Tunisie, Seismo-stratigraphie, Seismo-tectonique et implications pétrolières, Thèse de Doctorat en Sciences, Université de Tunis II, Tunis.

9. Bédir, M., Tlig, S., Bobier, Cl. and Issaoui, N., 1996. Sequence Stratigraphy, Basin dynamics and Petroleum geology of Miocene from the eastern Tunisia , American Association of Petroleum Geologist Bulletin (AAPG). U.S.A., **80**, N° 1, 63-81.
10. Bédir, M., El Manaa, S. and M'ridekh, A., 1997. *Subsurface Neogene Mud Diapirism and Basin Migration related to Strike-slip tectonics in the Eastern Margin of Tunisia,* Mud rocks at the basin scale Meeting, Londres , 28-29 Janvier 1997. Londres.
11. Bédir, M., Boukadi, N., Tlig, S., Ben Timzal, F., Zitouni, L., Alouani, R., Slimane, F., Bobier, CL. and Zargouni, F. , 2001. Subsurface Mesozoïc basins in the central Atlas of Tunisia: Tectonics, Sequence deposit distribution and Hydrocarbon potential, *American Association of Petroleum Geologist Bulletin. AAPG,* USA, **85**, N°. 5, 885-907.
12. Ben Ayed, N. , 1986. *Evolution tectonique de l'avant pays de la chaîne alpine de Tunisie du début du Mésozoïque à l'actuel,* Doctorat Es -Sciences, Université Paris Sud Orsay., Paris.
13. Ben Salem, H., 1992. *Contribution à la connaissance de la géologie du Cap Bon : stratigraphie, tectonique et sédimentologie,* Thèse de Spécialité, Université de Tunis II, Tunis.
14. Biély, A., Rakus, M., Robinson, P. and Salaj, J., 1972. Essai de corrélation des formations Miocènes du Sud de la Dorsale tunisienne, *Notes du Service Géologique de Tunisie*, n° 38, 73-92.
15. Boukadi, N., 1994. *Structuration de l'Atlas de Tunisie: signification géométrique et cinématique des noeuds et des zones d'interférences structurales au contact de grands couloirs tectoniques.* Thèse de Doctorat Es-sciences, Université de Tunis II, Tunis.
16. Boukadi, N. and Bédir, M., 1996. L'halocinèse en Tunisie: contexte tectonique et chronologie des événements, *Comptes Rendus, Académie des Sciences de Paris*, **322**, série II a, 587-594.
17. Burollet, P.F. ,1956. Contribution à l'étude stratigraphique de la Tunisie centrale, *Annales Mines et Géologie*, n° 18, 352 .
18. Colleuil, B., 1976. *Etude stratigraphique et néotectonique des formations néogènes et quaternaires de la région de Nabeul-Hammamet (Cap Bon, Tunisie)*, Diplôme d'Etudes Supérieures, Nice, France.
19. Comas, M. C., Garcia-Duenas, V. and Jurado, M. J., 1992. Neogene tectonic evolution of the Alboran sea from MCS data , *Geomarine Letters*, **12** ,157-164.
20. Damak-Derbel, F. ,1993. *Biostratigraphie et paléoenvironnement du Pliocène de la région Nabeul-Hammamet (Cap Bon, Tunisie)*, Thèse de troisième cycle, Université de Tunis II, Tunis.
21. Damak-Derbel, F. and Zaghbib-Turki, D., 1995. *Faunal characteristics of depositional sequences in Pliocene of Nabeul-Hammamet region (North-Eastern Tunisia)*, Réunion spécialisée, Faune et Flore et Stratigraphie séquentielle, Museum d'Histoire Naturelle, Paris.
22. EL Manaa, S., Bédir, M. and Zargouni, F., 1997. *Stratigraphie sismique et séquentielle des séries miocènes à travers les blocs tectoniques de la marge orientale de la Tunisie*, Colloque Marges Téthysiennes d'Afrique du Nord. Soc. Géol. France. Déc. 1997, Paris.
23. Gaaloul, N., 1995. *Les environnements siliciclastiques du Néogène du Sahel de la Tunisie : Palynologie et Biosedimentologie* , Thèse de Spécialité, Université de Tunis II, Tunis.
24. Gaaloul, N., Bédir, M., El Manaa S. and Razgallah, S., 1997. *Evolution des bassins Miocènes sur la marge orientale de la Tunisie,* Séance spécialisée de la Société Géologique de France, Les marges téthysiennes d'Afrique du Nord, p. 22.
25. Griboulard, R., Bobier, CL., Faugères, J.C. and Vernette, G. , 1991. Clay diapiric structures

within the strike slip margin of the southern Barbados prism, *Tectonophysics*, **192**, 383-400.

26. Haq, B.U., Hardenbol, J. and Vail, J.R. , 1987. Chronology of fluctuating sea-level since the Triassic , *Science*, **235**, 1156-1167.

27. M'ridekh, A.,1994. *Structuration tectonique et séquentielle des séries plio-quaternaires de subsurface du golfe d'Hammamet (Tunisie nord-orientale)* , Diplôme d'Etude Approfondies, Univ. Tunis II, Tunis.

28. M'ridekh, M., Bédir, M. and Tlig, S., 1995. *Contrôle tectonique et organisation séquentielle des séries plio-quaternaires du golfe de Hammamet (Tunisie Nord-orientale)*, 3ème Congrès National des Sciences de la Terre, Tunis, p. 123.

29. Shaub, F. J., 1983. Growth faults on the southwestern margin of the Gulf of Mexico, in *Seismic expression of Structural Styles: a Picture and Work Atlas*, Bally, A.W. (ed.) A.A.P.G., *Studies in Geology*, Series 15, V 2, p. 2. 3. 3-4, Rice Univ., Houston, Texas.

SEISMIC SIGNATURE OF GAS HYDRATE AND MUD VOLCANOES OF THE SOUTH AFRICAN CONTINENTAL MARGIN

Zvi Ben-Avraham[1], Moshe Reshef[2] and George Smith[3]

1 Department of Geophysics and Planetary Sciences, Tel Aviv University, Tel Aviv 69978, Israel, and Department of Geological Sciences, University of Cape Town, Rondebosch 7700, South Africa

2 Department of Geophysics and Planetary Sciences, Tel Aviv University, Tel Aviv 69978, Israel

3 Department of Geological Sciences, University of Cape Town, Rondebosch 7700, South Africa

Abstract: Widespread occurrence of bottom-simulating reflectors (BSRs) has been detected in multichannel seismic profiles on the upper continental slope in the southern periphery of the Orange River delta. Another remarkable feature in the area is the occurrence of a large number of mud volcanoes. The gas hydrate in this region may consist of a mixture of microbial and thermogenic gas, whereas much of the gas flowing through the mud volcanoes probably originated from deep-seated Aptian source shales. The mud volcanoes form a distinct lineament subparallel to the bathymetric contours in this area. They are of different sizes and different stages of development. In several locations, the volcanoes penetrate the seafloor while in others they are buried. The exposed size of these volcanoes is typically few hundred meters in diameter with a height of 10–40 m. Faults play a key role in the formation of mud volcanoes on this part of the southwest African continental margin.

Key words: gas hydrate, BSR, mud volcanoes, Southwest Africa

G. Martinelli and B. Panahi (eds.), Mud Volcanoes, Geodynamics and Seismicity, 17–27

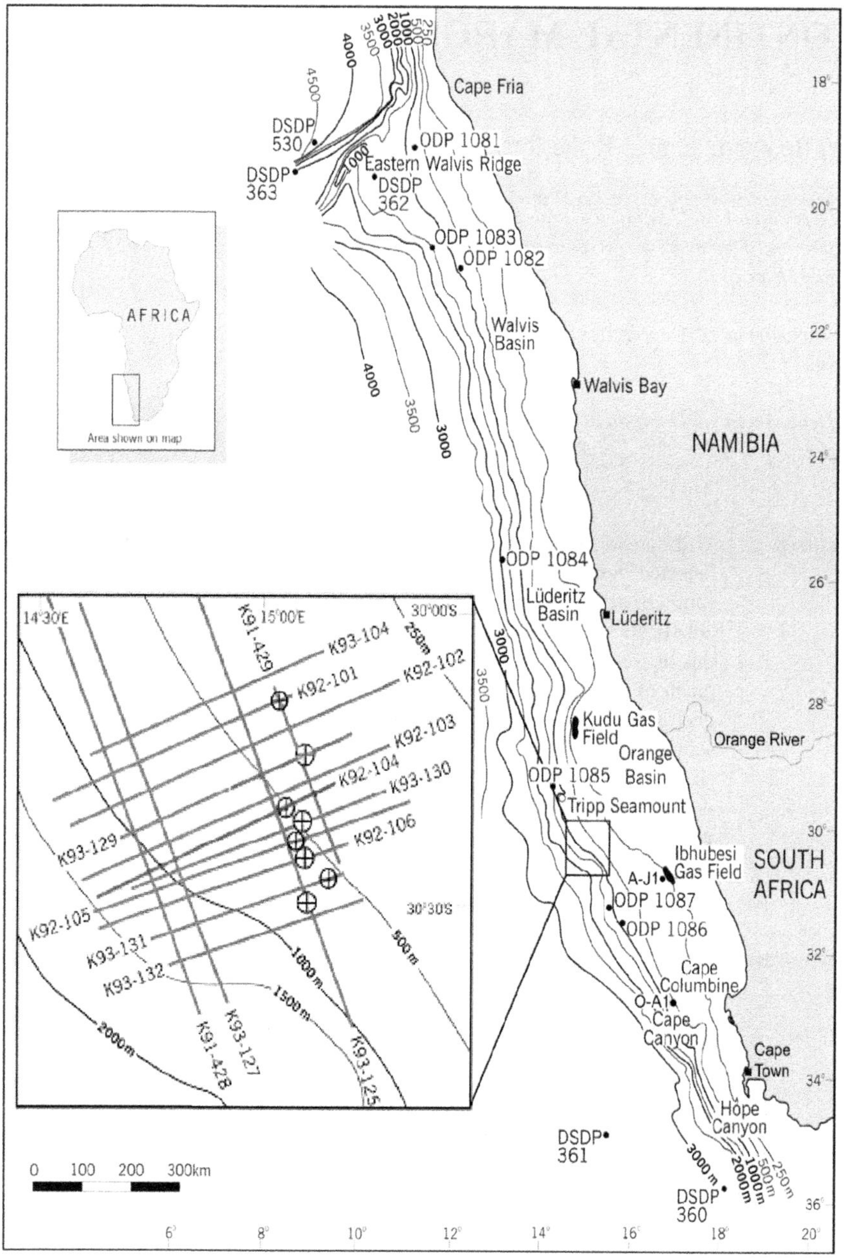

Figure 1. Location map showing main physiographic units of west coast of South Africa and position of ODP and DSDP drilling sites in this area. Position of analyzed seismic lines and outcropping mud volcanoes (marked with circles with +) are indicated on enlargement shown in lower left part of main figure.

1. INTRODUCTION

The physiography of the southwest African continental margin south of the Walvis Ridge can be divided into several provinces. The continental shelf is fairly uniform, ~110–160 km wide, from the southern edge of the Walvis Ridge (lat 21°S) south to 28°S. From this latitude south to Cape Columbine (~33°S), the continental shelf is much wider, ~170–230 km, reflecting the seafloor expression of the underlying Orange Basin. The present-day so-called Orange River delta is at the northern end of the Orange Basin.

In this area flows the Angola-Benguela current system with its associated upwelling, one of the five or six great upwelling regions in the world (Shipboard Scientific Party, 1998).

The current system extends over large parts of the western margin of South Africa and generates >180 $g \cdot m^{-2} \cdot yr^{-1}$ of carbon productivity. The western margin, therefore, is characterized by organic-rich sediments.

Numerous seismic reflection profiles have been recorded offshore South Africa by Soekor (Pty) over the last several decades. Examination of these profiles reveals indications of BSRs and diapirs, which are probably mud volcanoes, on the upper continental slope in the southern periphery of the Orange River delta (Fig.1). In the vicinity of this area, the A-K (Ibhubesi) gas field was discovered. The natural gas is trapped in fluvial channel-fill sandstones of Albian age. The gas is thought to have migrated from underlying Aptian source shales, mainly via faults (Jungslager, 1999).

Seismic lines from this region (Fig. 1) were examined, and a few were reprocessed. In this paper, seismic evidence for the presence of gas hydrate and mud volcanoes south of the Walvis Ridge are presented for the first time, and their implications are discussed.

2. SEISMIC DATA PROCESSING

Approximately 20 seismic lines were analyzed in the area marked by a rectangle in Figure 1. The data are two-dimensional marine lines that were shot and processed for Soekor (Pty) between 1991 and 1994. The data quality is generally good, and the available sections went through a standard marine processing scheme involving geometry, single-trace processing (gain, filters, deconvolution), multiple-suppression, Dip move-out (DMO) stack, and poststack time migration.

In order to examine the possible existence of a BSR in the area, several lines were selected for reprocessing. Emphasis was given to the wavelet processing and the imaging of the upper part of the sections.

The initial processing phase was mainly dedicated to deconvolution and filter tests in order to verify the reverse-polarity feature of the assumed BSR event. In areas where the event could be detected, its polarity was opposite to the polarity of the water-bottom reflector. A close look at the relationship between the BSR and the seafloor reflector is represented in Figures 2 and 3. Stacking velocity analysis, before and after DMO, did not reveal a major change in velocity values across the BSR event. Owing to the depth of the seafloor along the processed lines, water-bottom multiples did not interfere with the BSR. Multiple suppression was applied in the final processing sequence. Poststack time migration, followed by time and space filtering, generated the final images. Few iterations of migration were required to achieve optimal collapse of the diffractions generated by the faults.

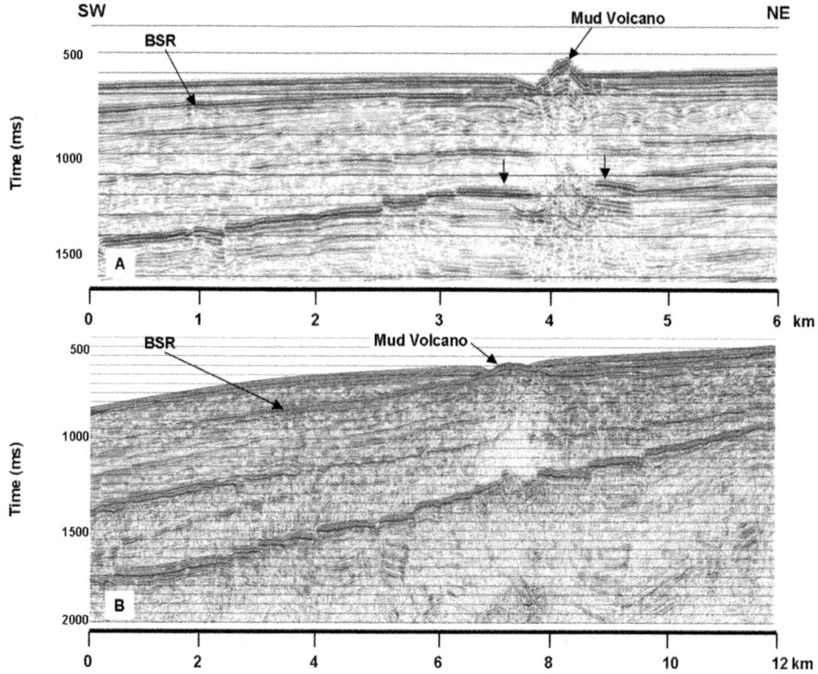

Figure 2. Mud volcanoes and BSRs on west margin of South Africa. A: BSR reaches seafloor near location of a mud volcano that crops out under shallower water. B: A mud volcano crops out under deeper water; BSR can be traced for another 2–3 km to right, toward shallow water. Note vertical offset of layers across mud volcano (for example, reflector that is marked by arrows in A), which suggests that mud volcano rises along a deep-seated fault.

3. THE OCCURRENCE OF A BSR AND MUD VOLCANOES

Our interpretation of a BSR in the area is based on the appearance of three of its known characteristics:

1. occurrence only in areas where water depth exceeds ~420 m
2. reverse polarity relative to the seafloor reflector
3. thickening of the layer containing the hydrate gas, as seafloor depth above increases (Fig. 3).

Although the BSR was identified over the entire area, it is not continuous along the seismic lines. Figure 3 shows the final processing result of one of the lines. If the BSR is followed along this line, from the shallow-water side toward the deeper water, it is evident that the BSR disappears between km 10 and km 6. A similar phenomenon was noticed along the other lines.

A relatively large number of apparent mud volcanoes was observed on the seismic sections. They form a distinct lineament subparallel to the bathymetric contours in this area (Fig. 1). The mud volcanoes are of different sizes and different stages of development (Fig. 4). They occur near the points where the BSRs approach the seafloor, ~420–450 m water depth. At locations where a mud volcano penetrates the gas hydrate layer, the BSR seems to be truncated by the diapir at the point in which the diapir cuts through the sea floor (Fig. 2).

4. DISCUSSION

The extensive occurrence of BSRs and mud volcanoes recorded by seismic reflection data on the continental slope off the Orange River delta suggests the existence of widespread gas hydrate layers in this area. So far the diapiric features and the suggested gas hydrate layers have not been sampled. However, the presence of BSRs, as suggested by the results of the seismic processing, provides for the first time good evidence for the presence of gas hydrates south of the Walvis Ridge. Gas hydrates have also been identified on the conjugate margin of the Atlantic, in the Pelotas Basin off Brazil (Sad et al., 1998).

The BSRs outcrop at a water depth of ~420 m. This depth agrees with the stability depth suggested for gas hydrates in phase diagrams (e.g., Kvenvolden and Lorenson, 2001). According to these diagrams, gas hydrates should be stable below 420 m where bottom-water temperatures are ~7 °C. Estimates of bottom-water temperatures in the upper continental slope of southwestern Africa are in this range (Dingle and Nelson, 1993). The situation here is similar to other locations where gas hydrate crops out on the seafloor, for example offshore central Oregon at a water depth of 600 m (Tréhu et al., 1995) and in the Gulf of Mexico at a water depth of 540 m (MacDonald et al., 1994).

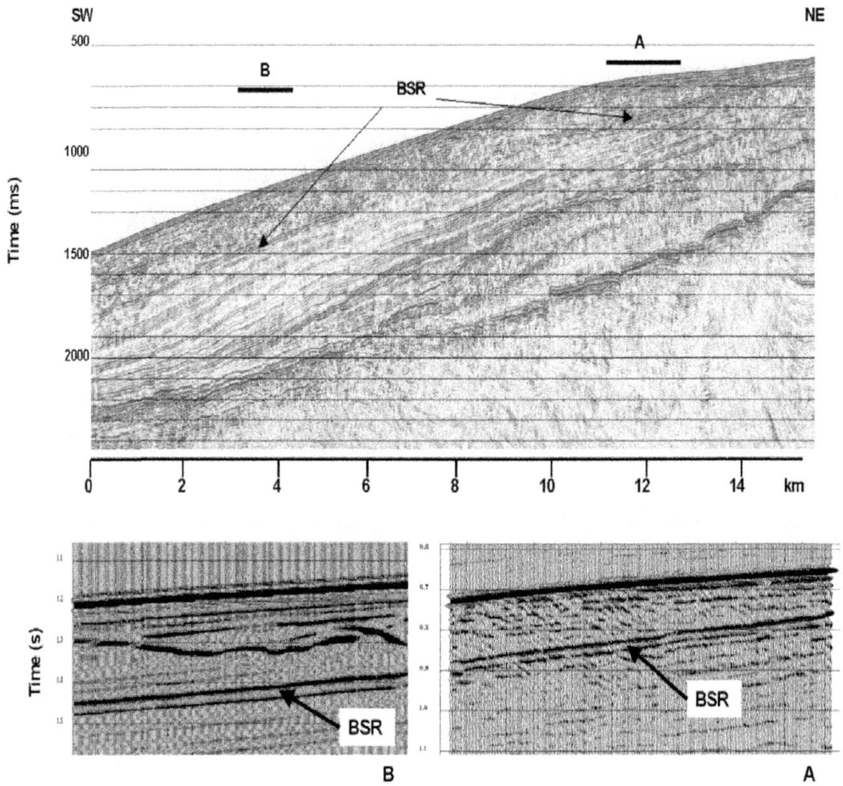

Figure 3. Discontinuity of BSR reflector. Notice close-to-vertical faults that can be traced almost to seafloor between km 6 and km 10. Reverse polarity (relative to water-bottom reflector) of BSR is represented in enlargements A and B. Also notice increase in distance of BSR from seafloor, as water column above thickens. See Figure 1 for location of profile.

5. SOURCE OF METHANE

The hydrates in this region may consist of a mixture of microbial and thermogenic gas. Ocean Drilling Project sites in the vicinity of the study area have detected a large concentration of gas in the subsurface. Moderately high amounts of methane and CO_2 were found in sediments from Site 1085 and Site 1087 (Fig. 1). High C_1/C_2 ratios and the absence of major contributions of higher-molecular-weight hydrocarbon gases indicate that the gas at these locations is microbial in origin (Meyers and Shipboard Scientific Party, 1998). It has been suggested that marine organic-rich sediments are deposited under upwelling zones, which are zones of high primary biologic productivity (Calvert et al., 1996). The data from the Deep Sea Drilling Project suggest that the sediments in the upwelling zone of the Angola-Benguela current system are a major source for gas of microbial origin in this area.

On the other hand, the close proximity of the probable gas hydrates to the relatively large gas fields of Kudu and Ibhubesi suggests that part of the gas within the hydrates, especially closer to shore, could be thermal in origin. In this case, hydrocarbon gases from deeper, thermally mature sediments may have migrated into the gas-rich zone, a situation similar to other places, such as the Exmouth Plateau, where a thermogenic source exists in underlying Jurassic rocks (Meyers and Snowdon, 1993). At shallower water, east of the study area, gas seepage on the seafloor is associated with active faults (Jungslager, 1999).

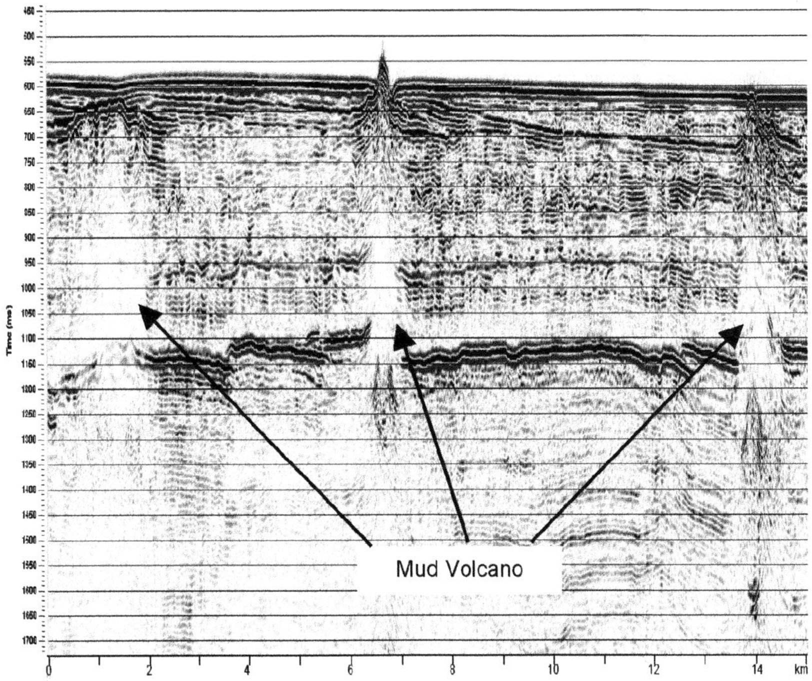

Figure 4. Mud volcanoes at different stages of development along a seismic line parallel to the shore line. Note the association of the volcanoes with the deep faults.

6. FAULT ACTIVITY AND METHANE DISTRIBUTION

A remarkable feature that has a distinct appearance in all the processed sections is the large zone of almost vertical faults. After the migration of the seismic data, the faults can be easily traced from the deeper part of the sections, all the way to the BSR (Figs. 2 and 3). In previous studies elsewhere, it has been suggested that a BSR is present usually where a free gas exists under the gas hydrates (e.g., Tréhu et al., 1995). In the areas where the BSR disappears (Fig. 3), it seems that the faults extend almost to the seafloor, indicating neotectonic activity. The faults may have caused an escape of a free gas to the seafloor, thus preventing the appearance of BSRs in these locations. Faulting activity, therefore, plays an important role in controlling the existence of BSRs in this region.

A closed examination of the BSRs shows that even where they are continuous, normal faults slightly displace them and penetrate into the supposed gas hydrate layer. Normal faults penetrating the BSR were also observed on the Blake Ridge in the western North Atlantic (Rowe and Gettrust, 1993). Periodic rupture along these faults may provide an avenue for migration of large quantities of methane from deeper reservoirs upward through the sediment until the methane enters the gas-hydrate stability zone, where it forms methane hydrate, inhibiting methane from migrating upward and causing it to pool at the base of the hydrate-bearing layer (Rowe and Gettrust, 1993). Where the BSR disappears in the study area, the faults are much larger and extend, as already mentioned, all the way to the seafloor.

7. MUD VOLCANOES

The area is also characterized by the occurrence of mud volcanoes. At this stage it is difficult to determine whether these features are indeed mud volcanoes, associated with gas flow, or shale diapirs. However, the relatively shallow water and the close proximity to large gas fields suggest that these features are indeed mud volcanoes. In several locations, the volcanoes penetrate the seafloor (Fig.1). The exposed size of these volcanoes is typically few hundred meters in diameter with a height of 10–40 m (Fig 2).

Faults also play a key role in the formation of mud volcanoes on this part of the southwest African continental margin. Almost every mud volcano in this area is associated with a deep-seated fault (Fig. 5). The faulting can be seen on both the shore-perpendicular and shore-parallel seismic sections. In a review of the worldwide distribution of mud volcanoes, Milkov (2000) described various mechanisms that have been proposed for the formation of submarine mud volcanoes. One of the key factors is the occurrence of faults.

Milkov (2000) also suggested that there are two key reasons for the formation of submarine mud volcanoes, high sedimentation rate and lateral tectonic compression. Indeed, most mud volcanoes were observed at collisional plate boundaries, such as in front of the Barbados accretionary wedge (Sumner and Westbrook, 2001), offshore north Panama (Reed et al., 1990), at the interaction of the Hellenic and Cyprean arcs in the eastern

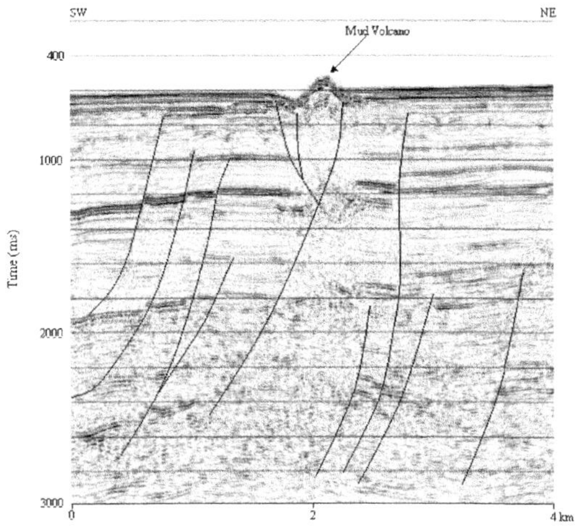

Figure 5. Deep faults in the vicinity of a mud volcano. The image clearly shows that the strong reflector, at about 1200ms, is faulted right under the marked mud volcano. Deeper reflectors, masked by the surface multiples, are probably broken by the same fault.

Mediterranean (Woodside et al., 1998), and offshore central Oregon (Tréhu et al., 1995). Most of these areas are regions of overpressure (Reed et al., 1990). However, the southwest African margin is a passive margin with a different tectonic setting. Nevertheless, the study area, which is located in the distal part of the Orange River delta, is also characterized by overpressure that results in active fluid expulsion as shown by existence of the mud volcanoes, pockmarks, and possibly cold water corals thriving on methane gas seeps (Jungslager, 1999).

The close proximity of the mud volcanoes to the zone where the BSRs crop out on the seafloor deserves attention. The seismic records strongly suggest that much of the gas in the mud volcanoes is originating from a deeper level than that of the gas hydrate. Faulting, again, could be the responsible factor for this unique situation. In other areas where gas hydrates are associated with mud volcanoes, such as the situation with the Haakon Mosby mud volcano in the Norwegian Sea, or the mud volcanoes in the Anaximander Mountains region in the Eastern Mediterranean (Woodside et al., 1998), the mud volcanoes are characterized by a concentric zonal distribution of gas hydrates. It was argued that the gas hydrates in these cases were probably formed from the gas that emanates from the central part of the mud volcano and is transported in solution by diffusion (Milkov, 2000). However, the occurrence of gas hydrates on the southwest African margin may not be directly linked to the activity of mud volcanos, but the gas seepage from these volcanoes may be a mixture of microbial gas from the gas hydrates and thermal gas from deeper in the section where mud volcanoes originate.

ACKNOWLEDGMENTS

This study was supported by the University of Cape Town and the Petroleum Agency of South Africa.

REFERENCES

1. Calvert, S.E., Bustin, R.M., and Ingall, E.D., 1996. Influence of water column anoxia and sediment supply on the burial and preservation of organic carbon in marine shales, *Geochimica et Cosmochimica Acta*, **60**, 1577–1593.
2. Dingle, R.V., and Nelson, G., 1993. Sea-bottom temperatures, salinity and dissolved oxygen on the continental margin off south-western Africa, *South African Journal for Marine Science*,**13**, 33–49.
3. Jungslager, E.H.A., 1999. Petroleum habitats of the Atlantic margin of South Africa, in *The oil and gas habitats of the South Atlantic*, Cameron, N.R., et al. (ed.), Geological Society [London] Special Publications 153, 153–168.
4. Kvenvolden, K.A., and Lorenson, T., 2001. The global occurrence of natural gas hydrate, in, *Natural gas hydrates: Occurrence, distribution and detection*, Paull, C.K., and Dillon, W.P.,(eds), American Geophysical Union Geophysical Monograph 124, 3–18.
5. MacDonald, I.R., Guinasso, Jr., N.L., Sassen, R., Brooks, J.M., Lee, L., and Scott, K.T., 1994. Gas hydrate that breaches the see floor on the continental slope of the Gulf of Mexico, Geology, **22**, 699-702.
6. Meyers, P.A., and Shipboard Scientific Party, 1998. Microbial gases in sediments from the southwest African margin, in, *Proceedings, Ocean Drilling Program, Initial reports*, Wefer, G., Berger, W.H., et al. (eds.) Volume 175, College Station, Texas, Ocean Drilling Program, 555–560.
7. Meyers, P.A., and Snowdon, L.R., 1993, Sources and migration of methane-rich gas in sedimentary rocks on the Exmouth Plateau: Northwest Australian continental margin, in *Biogeochemistry of global change* , Oremland, R.S., (ed.), Chapman and Hall, New York, 434–446.
8. Milkov, A.V., 2000, Worldwide distribution of submarine mud volcanoes and associated gas hydrates, *Marine Geology*, **167**, 29–42.
9. Reed, D.L., Silver, E.A., Tagudin, J.E., Shipley, T.H., and Vrolijk, P., 1990, Relations between mud volcanoes, thrust deformation, slope sedimentation, and gas hydrate, offshore north Panama, *Marine and Petroleum Geology*, **7**, 44–54.
10. Rowe, M.M., and Gettrust, J.F., 1993. Faulted structure of the bottom simulating reflector on the Blake Ridge, western North Atlantic, *Geology*, **21**, 833–836.
11. Sad, A.R.E., Silveira, D.P., Silva, S.R.P., Maciel, R.R., and Machado, M.A.P., 1998. Marine gas hydrates along the Brazilian margin, in *America Association of Petroleum Geologist International Conference* , Mello, M.R., and Yilmaz, P.O., (eds.), Rio de Janeiro, Extended Abstracts, 146–147. AAPG, Tulsa, USA.
12. Shipboard Scientific Party, 1998. Introduction: Background, scientific objectives, and principal results for Leg 175 (Benguela current and Angola-Benguela upwelling systems), in *Proceedings, Ocean Drilling Program, Initial reports*, Wefer, G., Berger, W.H., et al.,(eds.) Volume 175, College Station, Texas, Ocean Drilling Program, 7–23.
13. Sumner, R.H., and Westbrook, G.K., 2001. Mud diapirism in front of the Barbados

accretionary wedge: The influence of fracture zones and North America–South America plate motions, *Marine and Petroleum Geology*, **18**, 591–613.
14. Tréhu, A.M., Lin, G., Maxwell, E., and Goldfinger, C., 1995. A seismic reflection profile across the Cascadia subduction zone offshore central Oregon: New constraints on methane distribution and crustal structure, *Journal of Geophysical Research*, **100**, 15,101–15,116.
15. Woodside, J.M., Ivanov, M.K., Limonov, A.F., and Shipboard Scientists of the ANAXIPROBE Expeditions, 1998. Shallow gas and gas hydrates in the Anaximander Mountains region, eastern Mediterranean Sea, in *Gas Hydrates: Relevance to World Margin Stability and Climate Change*, Henriet, J.-P., and Mienert, J., (eds.), Geological Society, London, Special Publication, v. 137, 177-193.

GLOBAL DISTRIBUTION OF MUD VOLCANOES AND THEIR SIGNIFICANCE IN PETROLEUM EXPLORATION AS A SOURCE OF METHANE IN THE ATMOSPHERE AND HYDROSPHERE AND AS A GEOHAZARD

Alexei V. Milkov
Exploration and Production Technology Group, BP America, 501 Westlake Blvd, Houston, TX 77079 USA

Abstract: Mud volcanoes occur worldwide in areas of rapid sedimentation, lateral tectonic compression, and geologically recent magmatic activity. The total number of individual mud volcanoes on the Earth exceeds 2,000 and this number is growing as the exploration of deep oceans continues. Sediments and fluids expelled from mud volcanoes provide useful information on the geology and petroleum potential of deep sedimentary basins. Mud volcanoes are considered to be a minor but yet not fully recognized and properly quantified source of greenhouse gases (mainly methane) in the atmosphere. A significant (but still uncertain) amount of methane may escape into the ocean and affect the size and characteristics of the ocean carbon pool. Finally, mud volcanoes represent a recognized geohazard that affects life forms and petroleum exploitation. This paper reviews the results of recent studies into worldwide mud volcanism.

Key words: mud volcanoes, petroleum systems, methane, atmosphere, geohazards

1. INTRODUCTION

Onshore mud volcanism has attracted the attention of geologists for over two centuries (Goubkin and Fedorov, 1938; Yakubov et al., 1971; Higgins and Saunders, 1973; Hedberg, 1980; Barber et al., 1986; Rakhmanov, 1987; Kopf, 2002). A wide distribution of offshore mud volcanism has only been discovered in the past few decades as a result of an extensive exploration of deep oceans (Milkov, 2000). New mud volcanoes are being discovered every year as ocean exploration continues (Holland et al., 2003). This paper provides a concise review of the global distribution of mud volcanoes worldwide and highlights the main directions of mud volcano research.

2. GLOBAL DISTRIBUTION OF MUD VOLCANOES

Mud volcanoes occur worldwide and represent constructional features

G. Martinelli and B. Panahi (eds.), Mud Volcanoes, Geodynamics and Seismicity, 29–34

(diameter up to 10 km, relief up to 700 m) from which sediments and fluids (water, dissolved salts, gas, and oil) flow or erupt (Milkov, 2000; Kopf, 2002; Dimitrov, 2002). The main morphological features of mud volcanoes include the edifice created by mud volcano products, the crater which is the area of ongoing mud volcano activity (often populated by gryphons (vents) and salsas (small lakes)), and mud flows coming from the crater to the base of the edifice.

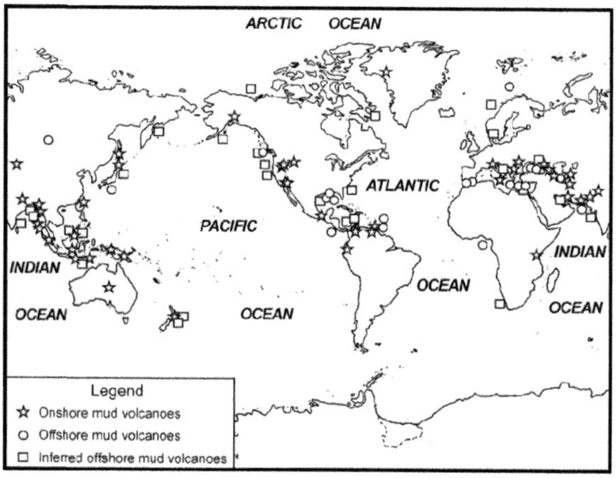

Figure 1. Worldwide distribution of MVs (modified from Milkov, 2000).

Mud volcanoes are documented in 44 onshore and 21 offshore areas, and there is indirect geological and geophysical evidence of offshore mud volcanoes in 25 areas (Fig. 1) Approximately 1,100 mud volcanoes are documented onshore and in shallow water on continental shelves (Dimitrov, 2002), and 1,000-100,000 mud volcanoes may exist on continental slopes and abyssal plains (Milkov, 2000). These features are most common in areas of rapid deposition and lateral tectonic compression with overpressure. Approximately 80% of all mud volcanoes may occur on convergent and transform continental margins.

3. DRIVING FORCES AND MECHANISMS IN THE FORMATION OF MUD VOLCANOES

There are many driving forces for mud volcanism identified in various areas. They can be divided into geological, tectonic, geochemical, and hydrogeological reasons, most of which are interrelated (Milkov, 2000). For example, if there is a lateral tectonic compression in the area, it is likely that abnormally high pressures and fluid flow would also be present. The reasons for mud volcanism can be summarized as two key reasons, namely the lateral tectonic compression at convergent and transform margins, and high sedimentation rates at divergent margins and in the abyssal parts of inland seas and lakes. All the other reasons listed in regional and local works (e.g., Goubkin and Fedorov, 1938; Yakubov et al., 1971; Higgins and Saunders, 1973; Hedberg, 1980; Barber et al., 1986; Rakhmanov, 1987) follow from these two key reasons.

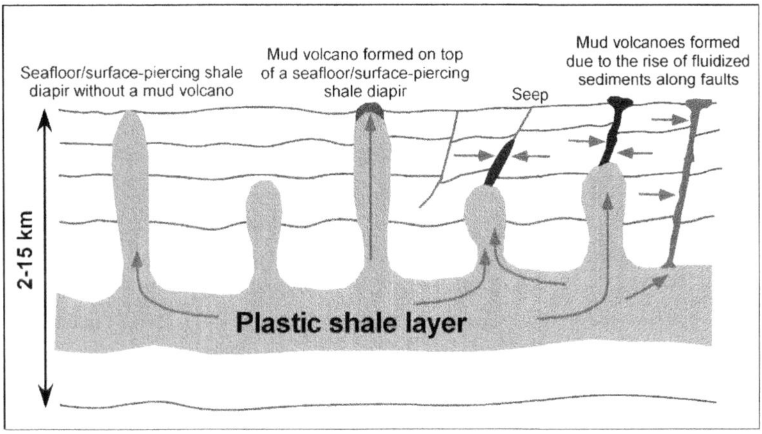

Figure 2. Picture showing submarine mud volcanoes formed by two basic mechanisms (modified from Milkov, 2000).

It is suggested that mud volcanoes are formed by two basic mechanisms (Fig. 2). The first mechanism is the formation of a mud volcano directly on top of a seafloor/surface-piercing shale diapir as a consequence of fluid migration along the body of the diapir. If the fluids do not migrate along the diapir, a mud volcano would not be formed and only a seafloor/surface-piercing shale diapir would occur. A mud volcano may or may not eventually develop on top of such a diapir. The second, and probably most common mechanism is the formation of a mud volcano as a result of the rise of fluidized mud along faults and fractures. In this case, sediments with a high fluid content reach the seafloor and form a mud volcanic edifice. This mud volcano may be connected to shale diapirs located at some depth beneath the seafloor. However, in many cases diapiric folds below mud volcanoes are not observed, and mud volcanic channels penetrate directly into the source layers. Fluid migration plays the primary role in both mechanisms (Fig. 2; Milkov, 2000).

4. THE SIGNIFICANCE OF MUD VOLCANOES IN PETROLEUM EXPLORATION

Mud volcanoes eject and erupt sediments and rocks of various shapes, composition and age that may come from a depth of 2-15 km. Therefore, sampling and studying of these products may help to better understand the deep lithosphere, especially in the frontier petroleum exploration areas. Basically, mud volcanoes may be considered as free deep wells.

The study of fluids expelled by the mud volcanoes may have even greater importance. Mud volcanoes are often associated with large actively producing petroleum basins (for example, Azerbaijan, the Gulf of Mexico, deepwater Nigeria) where they expel thermogenic hydrocarbons enriched in C_{2+} gases (Fig. 3). Therefore, mud volcanoes in frontier areas (for example, in the Gulf of Cadiz) may indicate an active petroleum system. However, there are some mud volcanoes that only emit bacterial methane, largely emit CO_2 or N_2, (Fig. 3) occur in areas with thin sedimentary cover (e.g., ~2 km) and in areas of recent magmatic activity (Milkov et al., 2003). These mud volcanoes may not indicate an active petroleum system. The expelled fluids should be carefully studied to evaluate the hydrocarbon potential of mud volcanic areas.

Gas hydrates associated with deep-water mud volcanoes may be considered as a potential energy resource (Hovland, 2000). However, many mud volcanoes are likely to provide only sub-economic gas hydrate resources because the volume of hydrate-bound gas in such accumulations may be insignificant (Milkov and Sassen, 2002).

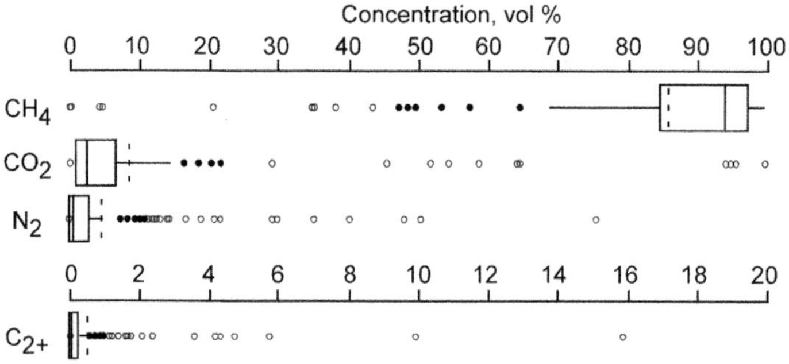

Figure 3. Box plots illustrating composition of gas flux from 161 onshore and offshore MVs. The 25th to 75th percentiles (interquartile range) of the data set values are represented by the box. The line inside the box corresponds to the median. The dashed line corresponds to the mean. The horizontal lines (whiskers) that extend to the right and left of the boxes stop at the last data point within a distance of 1.5 of the interquartile range. Closed circles correspond to the mild outliers (1.5-3 interquartile ranges). Open circles correspond to the extreme outliers (>3 interquartile ranges) (Milkov et al., 2003).

5. MUD VOLCANOES AS A SOURCE OF METHANE IN THE ATMOSPHERE AND HYDROSPHERE

Methane is the dominant gas in most mud volcanoes (Fig. 3). It is important to estimate how much gas escapes from mud volcanoes into the atmosphere and the ocean because methane is an important greenhouse gas. Mud volcanoes emit gases during eruptive and quiescent periods. Eruptions are rather catastrophic events often associated with flame, and therefore it is difficult to make direct measurements. It is much easier to estimate gas flux during quiescent periods. Such estimates vary from 100 m^3 to more than 10 million m^3 of gas per year, the estimates are log-normally distributed and the mean gas flux may be about 3-4 million m^3 (Milkov et al., 2003).

Approximately 600 Tg of methane is emitted into the atmosphere every year by all sources. Approximately 65 Tg of emitted methane is fossil (radiocarbon-free) and its sources are not well constrained. Using the mean gas flux from the individual mud volcanoes, it is possible to estimate the global gas flux from these features if their total number is known. Milkov et al. (2003) have suggested that mud volcanoes may emit ~6 Tg of gases (mainly methane) directly into the atmosphere. This amount accounts for ~1% of the total methane source, which is not very significant. However, almost all methane coming from mud volcanoes should be fossil. This source of fossil methane may be on the order of 10% of the total fossil sources. The gas flux from mud volcanoes to the ocean is much more badly constrained because the number of deep-water mud volcanoes is unknown. Assuming that there are 5,000 mud volcanoes in deep-water areas, Milkov et al. (2003) estimated that ~27 Tg of gases may escape into the ocean on continental slopes every year suggesting that the total hydrocarbon seepage from deep-water areas (currently estimated at 18-50 Tg per year) may be underestimated.

6. MUD VOLCANOES AS A GEOHAZARD

Eruptions of mud volcanoes often associated with flames (Aliyev et al., 2002) represent an apparent geohazard for surrounding life forms and constructions. From the perspective of the petroleum industry, mud volcanoes are a major geohazard. Mud volcanoes often occur in active petroleum-producing areas such as the Gulf of Mexico and deep-water Nigeria. The mud volcanoes are composed of relatively unconsolidated sediments, they produce numerous mud flows and may erupt, and they may host gas hydrates. These characteristics may lead to major slope instabilities. It is obvious that mud volcanoes should be avoided during the building of pipelines and other facilities.

7. CONCLUSIONS

Mud volcanoes occur both onshore and offshore, usually in areas having a high sedimentation rate, lateral tectonic compression and their combination. They are a result of diapirism and sediment/fluid migration along faults/

conduits. Mud volcanoes emit gases into the atmosphere and hydrosphere affecting the size and budget of carbon reservoirs. They represent a geohazard and may indicate active petroleum systems but the products (especially fluids) should be carefully studied.

REFERENCES

1. Aliyev, Ad. A., I.S. Guliyev, I.S., Belov I.S., 2002. Catalogue of recorded eruptions of mud volcanoes of Azerbaijan (for period of years 1810-2001). Publishing House "Nafta-Press", Baku.
2. Barber, A.J., Tjokrosapoetro, S., Charlton T.R., 1986. Mud volcanoes, shale diapirs, wrench fault and melanges in accretionary complexes, eastern Indonesia, *Bull. Am. Assoc. Pet. Geol.* **70**, 1729-1741.
3. Dimitrov, L., 2002. Mud volcanoes - the most important pathway for degassing deeply buried sediments, *Earth-Sci. Rev.* **59**, 49-76.
4. Goubkin, I.M., Fedorov S.F., 1938. Mud volcanoes of the Soviet Union and their connection with the genesis of petroleum fields in Crimean-Caucasus geologic province. USSR Academy of Science, Moscow (in Russian).
5. Hedberg, H.D., 1980. Methane generation and petroleum migration. In: Roberts III, W.H., Cordell, R.J. (eds.), *Problems of petroleum migration*, AAPG Studies in Geology 10, 179-206.
6. Higgins, G.E., Saunders J.B.,1973. Mud volcanoes - their nature and origin: contribution to the geology and paleobiology of the Carribbean and adjacent areas, *Verh. Naturforsch. Geschel. Basel*, **84**, 101-152.
7. Holland, C.,Etiope, G., Milkov, A.V., Michelozzi, E., Favali P., 2003. Mud volcanoes discovered offshore Sicily, *Mar. Geol.* **199**, 1-6.
8. Hovland, M., 2000. Are there commercial deposits of marine hydrates in ocean sediments ?, *Ener. Expl. Exploit.* **18**, 339-347.
9. Kopf, A.J., 2002. Significance of mud volcanism, *Rev. Geoph.* **40** (2), 1005, doi: 10.1029/2000RG000093.
10. Milkov, A.V., 2000. Worldwide distribution of submarine mud volcanoes and associated gas hydrates, *Mar. Geol.* **167**, 29-42.
11. Milkov, A.V., Sassen R., 2002. Economic geology of offshore gas hydrate accumulations and provinces, *Mar. Petrol. Geol.*, **19**, 1-11.
12. Milkov, A.V., Sassen, R., Apanasovich, T.V., Dadashev F.G., 2003. Global gas flux from mud volcanoes: a significant source of fossil methane in the atmosphere and the ocean, *Geoph. Res. Lett.*, **30** (2), 1037, doi: 10.1029/2002GL016358.
13. Rakhmanov, R.R., 1987. Mud volcanoes and their importance in forecasting of subsurface petroleum potential. Nedra , Moscow (in Russian).
14. Yakubov, A.A., Alizade, A.A., Zeinalov, M.M., 1971. Mud volcanoes of Azerbajan SSR. Atlas. Elm, Baku (in Russian).

STYLES AND PRODUCTIVITY OF MUD DIAPIRISM ALONG THE MIDDLE AMERICAN MARGIN

Part I: Margin Evolution, Segmentation, Dewatering and Mud Diapirism

Tobias Moerz[1], Achim Kopf[3], Warner Brueckmann[1,2], Noemi Fekete[1], Veit Huehnerbach[4], Douglas G. Masson[4], Daniel A. Hepp[1], Erwin Suess[1,2], and Wilhelm Weinrebe[1,2]

1 *Sonderforschungsbereich 574, Christian-Albrechts-University of Kiel, Wischhofstr. 1-3, 24148 Kiel, Germany*

2 *GEOMAR Research Centre, Wischhofstr. 1-3, 24148 Kiel, Germany*

3 *Research Centre Ocean Margins, University of Bremen, Klagenfurter Str., 28359 Bremen, Germany*

4 *Southampton Oceanography Centre, European Way, Southampton, SO14 3ZH, UK*

Abstract: Mud diapirism is a common phenomenon of accretionary convergent margins but less common in erosive margins. Fluid venting associated with mud diapirism is of importance for the dewatering of the forearc and the resulting devolatilisation of the entire subduction zone. The margin offshore Costa Rica is today interpreted as erosive and subdivided into two major structural domains on grounds of the roughness of the downgoing plate: a smooth domain in the north where normal oceanic crust originating at the East Pacific Rise, and a rough southern domain where the margin is uplifted and fractured by the collision of the Cocos Ridge and numerous adjacent volcanic seamounts. These structural differences are reflected in differences in the output at the volcanic arc, dewatering mechanisms, and the abundance and geometry of mud mounds in the forearc. Diapiric mud mound occurrences in the smooth domain are most abundant in the middle and upper slope and apparently do not correlate with the maximum of compactional water release of the incoming sedimentary sequence. We invoke rapid changes in sedimentation rate and addition of accommodation space due to extensional faulting of the wedge to explain the observed mound distribution.

Key words: Costa Rica margin, rough domain, smooth domain, mud diapirism, venting

1. INTRODUCTION

Onshore and offshore mud volcanism is a global phenomenon. Reports on

G. Martinelli and B. Panahi (eds.), Mud Volcanoes, Geodynamics and Seismicity, 35–48

onshore mud volcanism date back to the 12th century (Aliyev et al., 2002) and mud volcanism is scientifically studied for almost 200 years (e.g. Goad, 1816; Abich, 1857). Global compilations report ~1100 onshore and shallow marine (Dimitrov, 2002), and ~5000 estimated offshore deep water occurrences (Milkov et al., 2003). The number for deepwater occurrences is only a tentative estimate as surveys of new areas or improvements in the bathymetry and sidescan resolution will continuously provide additional information.

Mud volcanism has been described from a wide variety of tectonic settings, including passive continental margins, continental interiors, transforms and convergent plate boundaries (Hovland et al., 1997). Inferred from global distribution charts (Guliyev and Feizullayev, 1996; Kopf, 2002; Milkov, 2000; Milkov et al., 2003; Rhakmanov, 1987) all accretionary convergent systems show diapirism, whereas only a few occurrences are reported from non accretionary systems.

Mud diapirism is characterized by the extrusion of fluid-rich, fine-grained sediments through an overlying lithologic succession. Main driving forces are buoyancy contrasts and lateral or vertical stress on the formation. Seismicity and/or hydrocarbon generation may trigger the timing and amount of extruded material (Yassir, 1989). The location of mud uprise is often determined by confining structural elements or pre-existing structurally weak zones e.g. faults, that act also as dewatering pathways (Shipley et al., 1990).

The main reason to study mud diapirism in the forearc zone of convergent margins is their role in the dewatering process of the forearc region and hence their influence on the overall fluid and volatile budget of entire subduction zones. Offshore margin scale mud volcano flux budgets are difficult to obtain due to their inaccessibility and the sporadic nature of gas, fluid, and mud release. First order estimates regarding mud diapir flux rates have recently been attempted for various onshore and offshore regions (e.g. Etiope and Klusman, 2002; Henry et al., 1996; Kopf and Behrmann, 2000). In large accretionary complexes, like the Barbados or Mediterranean Ridges, diapir related fluid expulsion rates exceed those at the frontal part of the prism (Kopf et al., 2001). Direct emission of gas e.g. methane to the atmosphere from mud volcanism is of minor importance and is estimated to be ~1 % of the total global atmospheric methane sources (Milkov et al., 2003). This applies especially to mud volcanoes off the shelf in greater water depth where biochemical processes consume the methane before it can reach the atmosphere (Judd et al., 2002; Milkov et al., 2003). In the context of the entire subduction factory the release of fluid and gas in the forearc acts as a devolatilisation process and hence regulates the atmospheric and climatically relevant emission of the arc (Dia et al., 1999; Kopf et al., 2000).

Recently numerous new mounds have been mapped along the non-accretionary Pacific section of the Central American margin based on high resolution 30 kHz TOBI sidescan data and detailed swath bathymetry acquired during survey and mapping cruises (e.g. RV SONNE 144, 163, and RV METEOR M54-1; Bohrmann et al., 2002; Weinrebe and Flueh, 2002).

In the first part of the paper we summarize information on the margin evolution, segmentation, dewatering and diapirism and discuss new mound findings and their relation to the lower plate dewatering. In the accompanying second part we report on coring results of selected mounds and compare different mound styles in relation to their respective settings.

2. STRUCTURAL SETTING

2.1 Incoming plate

West of Costa Rica and southern Nicaragua, the late Oligocene to early Miocene oceanic crust of the Cocos Plate is rapidly subducted at the Middle America trench at about 90 mm/a (Barckhausen et al., 2001; DeMets et al., 1994). Geomorphologically the Pacific oceanic crust offshore Costa Rica and Nicaragua is divided into a smooth northern and a rough southern domain (von Huene et al., 2000; Fig. 1).

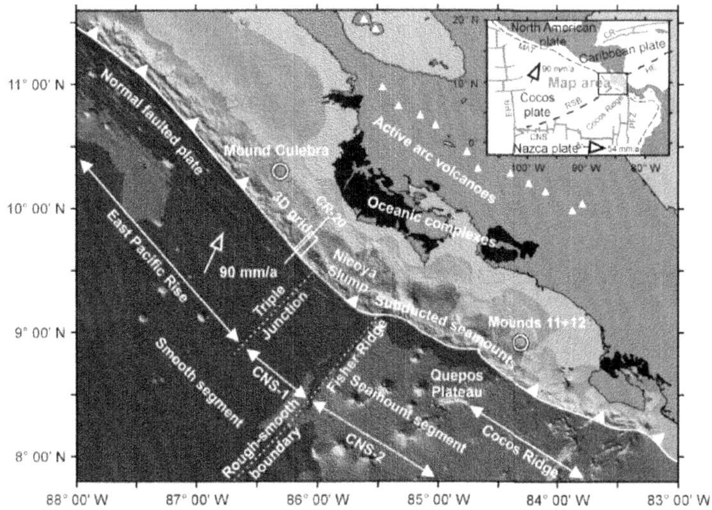

Figure 1. Geodynamic setting off Central America and locations of the two working areas Mound Culebra, Mound 11, and 12. The inset is giving a Plate scale view with the relative plate movement vectors (54 mm/a, after Mann, 1995; 85 mm/a, after DeMets et al., 1994). The margin bathymetry and segmentation nomenclature is after von Huene et al. (2000). Box off shore Nicoya Peninsula = 3D seismic grid of Shipley et al. (1992); Black signature on shore = Oceanic complexes; CNS = Cocos-Nazca Spreading Center; CR = Cayman Ridge; EPR = East Pacific Rise; HE = Hess Escarpment; MAT = Middle America Trench; PFZ = Panama Fracture Zone; RSB = Rough-Smooth Boundary.

The dominantly N-MORB type crust of the smooth northern domain is formed at the N-S trending East Pacific Rise spreading center (EPR) and in part by an abandoned precursor of the Cocos-Nazca spreading center CN-1 (Meschede et al., 1999; Fig. 1). At the trench in 5 km of water depth SW of Mound Culebra (Fig. 2) the incoming plate has a normal crustal thickness of ~7 km, is anomalous cold and the subduction angle is steep (~65°). Crustal ages at the trench are 24-25 Ma (Barckhausen et al., 2001; Harris and Wang, 2002; Protti et al., 1995; von Huene et al., 2000). Trench-parallel normal faults of the outer flexural bulge that increase in fracture width and offset to the north are prominent structural features (Ranero et al., 2000; von Huene et al., 2000). Seismic surveys demonstrate that the normal faults extent more than 10 km below the Moho (Ranero et al., 2003). The incoming sedimentary succession is ~400 m-thick and is dominated by diatomaceous and calcareous oozes open marine pelagic sediments as well as graded sand and silt layers and abundant ashes (Kimura et al., 1997; von Huene et al., 1985).

The rough southern oceanic plate domain relevant for Mound 11 and 12 originates at the E-W-trending active CN-2 Cocos-Nazca spreading center. At the shallow trench in 3 km of water depth, SW of Mounds 11 and 12 (Fig. 2), the anomalous 18 km-thick crust of the incoming plate is relatively warm and has a shallow subduction angle of ~40°. Crustal ages at the trench are 16-17 Ma (Barckhausen et al., 2001; von Huene et al., 2000). Important features of the southern domain are the colliding a seismic Cocos Ridge and a seamount dominated normal crust segment adjacent to the ridge (von Huene et al., 2000; Fig. 1). Both the Cocos Ridge and the adjacent seamounts represent anomalous thick, plume influenced crust generated above the Galapagos hotspot (Hauff et al., 2000; Hoernle et al., 2000).

2.2 Wedge and arc characteristics

A basic structural margin concept for both study areas consists of a small deformed prism behind the toe of the overriding plate and a landward thickening margin wedge with P-wave velocities between ~5 km/s at the coast and ~4 km/s at the lower slope (Fig. 3). This high velocity wedge is overlain by a slope sediment cover varying in thickness between 1 and 2 km (Ye et al., 1996). Ophiolites with variable degrees of alteration are exposed along the coast between the two study areas (Hauff et al., 2000; Fig. 1) Onshore/Offshore seismic profiles of the smooth domain image a trenchward continuation of the exposed ophiolithic material into the margin wedge (Christeson et al., 1999). It is suggested that between the study areas the entire high velocity margin wedge except the small toe consist of ophiolithic material overlain by a thin conglomeratic shallow marine facies (Kimura et al., 1997; Meschede et al., 1999; Vannucchi et al., 2001). Lateral seaward velocity decreases are attributed to an increase in fracture density and intensity of alteration (von Huene et al., 2000; Ye et al., 1996).

Geometry and morphological features of the oceanic plate shape the overlying margin wedge and even control compositional trends in volcanic output along the magmatic arc (von Huene et al., 2000). In the rough southern domain ridges and seamounds elevate, undermine, fracture and fold the overlying wedge. On the trailing flanks of subducting seamounts, uplifted and oversteepened sediments fail and leave trail like slump scars visible up to the shelf edge (von Huene et al., 2000). Especially the subduction of the Cocos Ridge severely affects the slope by a 2 km uplift and enhanced coupling between the upper and lower plate (Fig. 3).

In contrast, the regular slope morphology in the northern domain reflects the smooth character of the adjacent oceanic plate. Slope processes like sediment creeping and normal faulting are the dominant processes in this area and reflect the overall extensive stress regime (Meschede et al., 1999; Flueh et al., 2000; Ranero et al., 2000).

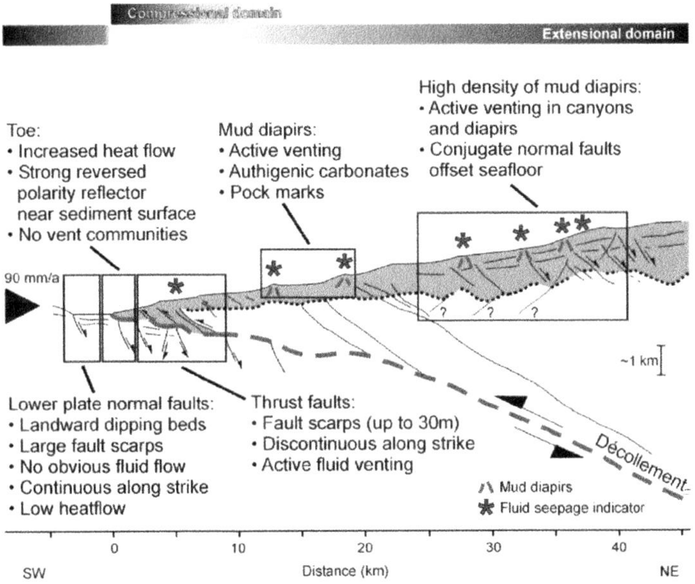

Figure 2. Schematic cross section of the smooth segment of the margin based on seismic line CR-20 off the Nicoya peninsula (McIntosh et al, 1993; Line location: Fig. 1). The fluid flow and venting summary is modified from McAdoo et al. (1996) to implement the new views of the margin (Ranero and von Huene, 2000). Offsets in the ophiolithic margin wedge in the mid- and upper slope originally seen as thrusts are now interpreted as normal faults according to the extensional margin model from Meschede et al. (1999).

Today the margin of Costa Rica is seen as non-accretionary and erosive with complete sediment subduction and largely extensional tectonic structures in the upper plate (Ranero and von Huene, 2000; Flueh et al., 2000; Moritz et al., 1999; Ranero et al., 2000; Vannucchi et al., 2001). ODP Leg 170 drilling recovered shallow 16.5 Ma marine nearshore deposits from a depth of 4000 m indicating rapid subsidence and tectonic erosion of the upper plate (Vannucchi et al., 2001). Current erosion rates of the upper plate differ between the smooth northern and the rough southern domain. This is illustrated by an offset in the alignment of active arc volcanoes, a back stepping of the Costa Rica coastline and a lower slope retreat of ~16 km coinciding with the rough-smooth boundary (von Huene et al., 2000). The cause for the observed differences in subduction style, forearc and arc behavior and the current rapid subsidence is seen in the proposed >5-7 Ma onset of the Cocos Ridge subduction and the adjacent plate sector studded with volcanic seamounts that influenced the entire margin beyond the rough-smooth boundary via long wavelength bulging (Vannucchi et al., 2003).

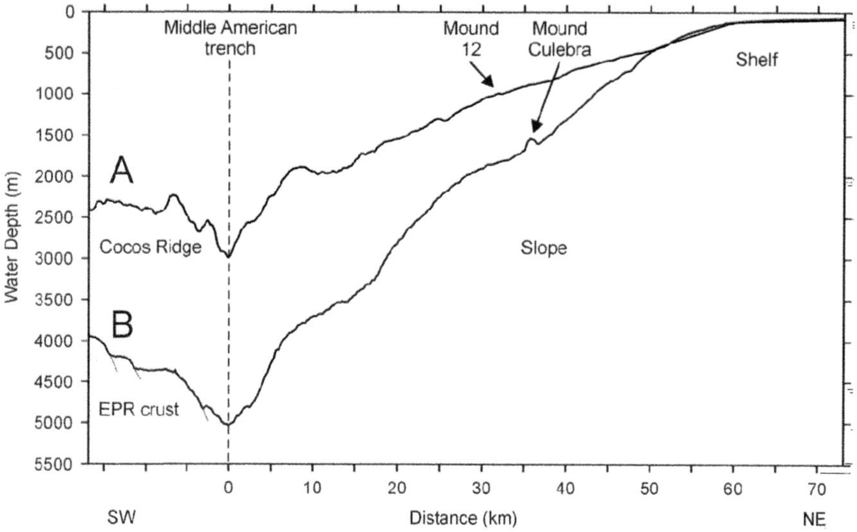

Figure 3. Trench normal bathymetric crossections through Mound 12 (rough domain) and Mound Culebra (smooth domain). Note the dramatic 2 km uplift of the margin prism due to the Cocos ridge subduction.

Differences in the lava chemistry along the arc correlate with fracture zone width and the magnitude of vertical fault displacement of normal faults of the incoming plate (von Huene et al., 2000). Patino et al. (2000) hold fracturing of the slab during flexure responsible for the higher slab signature and water content of the Nicaragua compared to the Costarican lavas. Ranero et al. (2001) and Ranero et al. (in press) reported on the deep penetration of these outer bulge normal faults and invoked deep serpentinisation into the upper mantle as a possible important process. The subduction input-output models of Ruepke et al. (2002) take up on this idea and demonstrate that differences in the degree of serpentinisation of the subducted crust may explain the observed differences in arc lava chemistry.

In contrast to conceptual models linking north-south structural differences to arc magmatism (REFS), such relationships have not been published for the dewatering and mound distribution in the forearc.

3. DEWATERING, MUD DIAPIRISM, HEAT FLOW AND GAS HYDRATES OF THE COSTA RICA FOREARC

The Costa Rica and Nicaragua margin has been investigated during numerous geophysical surveys and DSDP and ODP drilling with little groundtruthing related to mud diapirism.

The work of Shipley et al. (1990; 1992) and Stoffa et al. (1991) is based on a 3D reflection seismic box with ~33 m along profile resolution of the lower to middle slope and single profiles reaching from 800 m of water depth to the trench (Fig. 1). Together with Sea Beam bathymetrical information from first time mapping (Shipley et al., 1990), five mud diapirs and mud volcanoes were found within the 3D box (Fig. 1). A conduit and landward-dipping structures across the sediment cover to the base of the wedge below a mud diapir located in 3050 m water depth, ~13 km landward from the trench were resolved and interpreted as a possible dewatering pathway of underthrusted sediment. Calculations of fluid transfer from the subducted sediment based on compaction of the incoming sequence indicate that the highest gradients occur within the first 4 km off the trench (Shipley et al., 1990; McIntosh and Sen, 2000).. This however does not correlate with the observed abundance of mud diapirs which is greatest in the mid-and upper slope (Fig. 2). Shipley et al. (1990) therefore conclude that mud diapirs in the middle and upper slope are the result of sediment compaction within the wedge (assuming the wedge is accretionary) and only to a lesser extent linked to fluids from subducted sediments. Further processing of the 3D box data imaged a cross section of a low relief mud diapir in ~2500 m water depth ~ 22 km landward of the trench (Shipley et al., 1992). This diapir has a conduit diameter of ~1 km with a "Christmas tree" geometry and is well resolved down to 750 mbsf. The lower end of the resolved conduit seems to be linked to faults offsetting at the sediment apron/wedge boundary. Important findings are laterally extensive, trench parallel reflections interpreted as arcward dipping faults that commonly offset the margin wedge and sometimes extend to the seafloor surface. All well documented mud diapirs seem to be related to large offsets in the margin wedge. The margin model of Shipley et al. (1992) is still based on accretion and envisions a complex multi layer seaward structural propagation of the wedge.

ALVIN submersible dives explored the area of the 3D box in search of fluid venting expressions and for shallow sediment sample recovery (Kahn et al., 1996; McAdoo et al., 1996; McIntosh and Silver, 1996). Important findings were: a lack of vent communities in the first 3 km of the trench, active widespread venting where lower slope thrusts disrupt the seafloor, focused fluid expulsion at mud diapirs and some local vent communities associated with steeply landward and seaward dipping normal faults of the upper slope in 1000-1500 m water depth (McAdoo et al., 1996; Fig. 3). Again active venting on the surface is associated with highs or discontinuities of the apron/prism boundary (McIntosh and Silver, 1996). Interstitial water analyses of ALVIN push cores and various types of gravity cores and piston cores from mud diapirs and various other vent sites revealed some evidence for upward fluid flow from greater depth. The observed deviations from seawater are small, however strong admixture or downflow of seawater is suggested during fluid upward migration (Zuleger et al., 1996). Using boron and boron isotopes from ODP Leg 170 samples, Kopf et al. (2000) suggested effective fluid separation of the dewatering slope apron and underthrust sediments at the decollement. Saffer et al. (2000) showed that the average vertical permeability of the decollement and incoming strata is too low (by a factor of 100) to drain the expelled pore fluids of the downgoing slab sediments vertically and calls for either close spaced bedding parallel high permeability layers or vertical highly focused fluid flow in conduits and fractures. The absence of seeps in the toe and frontal part of the wedge and results from temperature driven fluid flow models of (Silver et al., 2000) support a strong horizontal dewatering component.

The study by Kopf et al. (2000) also emphasizes the importance of water release from gas hydrates in explaining the observed porewater compositions. Wide spread gas hydrate occurrences are known for the smooth domain from direct ODP and DSDP sampling and through margin-wide occurrences of BSRs (Pecher et al., 1998; 2001). In general, the BSR shallows toward the frontal prism where the highest heatflow values are observed within an otherwise anomalous cool margin wedge (Pecher et al., 2001; Ruppel and Kinoshita, 2000).

The mostly geophysical information from these surveys can be summarized as follows: the northern Costa Rica forearc is characterized by anomalous low heat flow decreasing from the frontal prism arcward and chemical pore water compositions that suggest mixing of apron pore water with fresher water from the underthrust sediment or gas hydrate dissolution (Silver, 2001). Fluid flow seems to be contained in three separated convective systems: the oceanic crust, subducted sediment and apron sediment (Saffer et al., 2000; Silver, 2001). Horizontal and vertical fluid and mass flux is highest but diffuse in the frontal prism and becomes more focused along structural pathways toward the arc. Controlling structures are thrusts reaching the decollement in the lower slope and normal faults in the middle and upper slope.

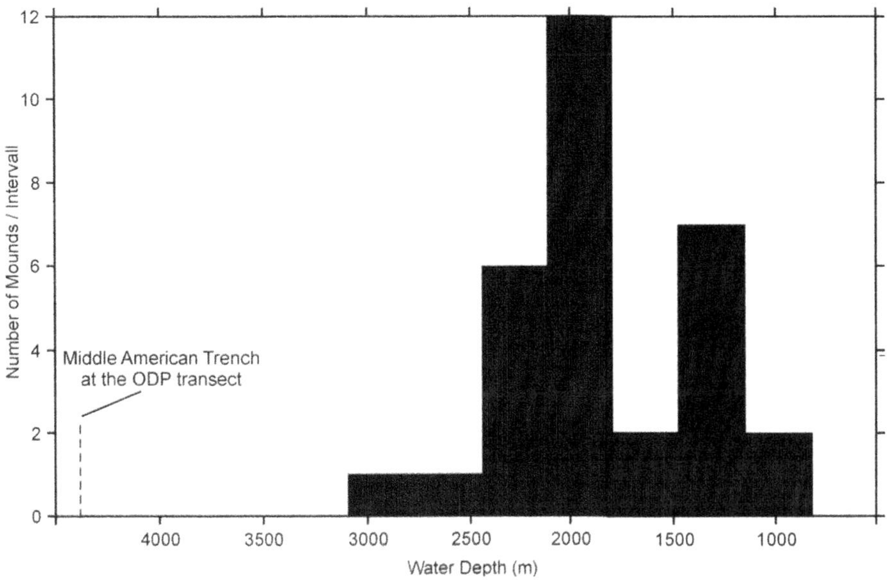

Figure 4. Depth distribution of selected, topographically well expressed mounds in the smooth margin domain of the Costa Rica margin. The class width is 325 m water depth.

On the base of detailed swath bathymetry and TOBI mapping numerous new mounds were anticipated. During SONNE 144 cruise (Bialas et al., 1999; Bohrmann et al., 2002) widespread fluid venting associated with vent fauna occurrences was observed during nine video sled surveys in the rough domain associated with landslide headwalls, seamount scars, various types of faults and one survey of a midslope mound of the smooth domain. The SONNE 163 cruise surveys (Weinrebe and Flueh, 2002) extended the video observations over thirteen new mounds and together with the extended TOBI mapping settled the ground for the geophysical cruise METEOR 54-1 and the sampling campaigns of METEOR M54-2 and M54-3 cruises in the same year (Soeding et al., 2003).

Recent efforts to identify all mound-like structures based on TOBI backscatter images, multi beam bathymetry data and the existing 3D seismic coverage revealed around 70 surface expressed mounds for the Costa Rica margin. With only 9 comparatively small mounds occuring in the rough domain of the margin (Sahling et al., in prep.). These numbers are preliminary and represent the current state of research and may change in the course of new or higher resolution surveys or by extending the data base to the shelf edge where significant data gaps remain.

Preliminary frequency versus depth distribution calculations using 31 topographically well expressed mounds choosen on the base of closed contours on a 10 m resolution bathymetric map have been performed for the smooth domain. The data show an increase of mound occurrences in the middle and upper slope with high mound abundances in the interval between 2500 and 1250 m water depth corresponding to ~25 to ~41 km distance from the trench (Fig. 4).

4. DISCUSSION AND CONCLUSION

The erosive, pacific margin of Costa Rica is segmented with regard to its morphology, arc chemistry and rate of erosion. Recent surveys show that this segmentation is also reflected in the dewatering style and occurrence of vent sites and the distribution of diapiric mounds. Mound formation and mound distributions seems to be therefore closely linked to the differences in margin evolution.

The comparably more stable tectonic environment of the smooth domain with steadier changes in geometry and larger structural features fosters a high abundance of larger mounds. Although much faster than for evaporitic diapirism, mud diapirism is generally a slow process and some diapiric features are long lived elements that are capable to stand up against background sedimentation and can resist erosion for millions of years (Robertson et al., 1996). This may also imply a need of relative geometric stability in the source region of a diapir that seems to be provided in the smooth northern domain.

The intense uplift, undermining and fracturing processes active in the rough domain seem to prevent the formation of larger mounds or may have destroyed preexisting mounds prior to the onset of the Cocos Ridge subduction. The observed variety of venting environments (Bohrmann et al., 2002) in the rough domain is coupled with complex and often smaller scale local tectonism allowing fluids to escape vertically without long lateral migration. Vertical tectonics due to Cocos Ridge and seamount subduction caused intense fracturing of the ophiolitic margin wedge and the slope sedimentary cover. Mounds need a confined environment with only discrete zones of weakness for them to form or erupt. Intense fracturing and multiphase extension and shortening due to positive features creates plenty of fluid pathways and no need for greater pressure buildups and lateral mass movements toward distal weak zones that subsequently become diapirs or domes.

The observed distribution of mounds in the smooth domain supports ideas of Shipley et al. (1992) that mound occurrences are decoupled from the compactional dewatering of the incoming sequence. Instead, the increasing number of mounds toward the middle and upper slope correlates with the increasing thickness of the slope apron (Fig. 2). This in turn suggests that diapirism is dominantly routed in the upper plate slope sediment apron. Extensional tectonics following the onset of subduction erosion lead to a local thickening of the slope apron (e.g. Meschede, 1999) with extreme temporal variations in the local sedimentation rate and consolidation history. This sets the base for locally underconsolidated sediment packages with a positive buoyancy that later in the slope development started to rise.

Slope sediment roots for the diapiric mounds doe not exclude the contribution of deep exotic fluids e.g. from the ophiolithic part of the margin wedge or the subducting sediment sequence. The incoming sedimentary succession for example may contribute by means of clay mineral dehydration (Dählmann and Lange, 2003) or relic pore water that got preserved in the gussets of the bulge horst and graben structures. This contribution however is not causal to the mound formation.

ACKNOWLEDGEMENTS

The ideas of the paper gained significantly from fruitful discussions with Heiko Sahling, Ernst Flueh, Tim Reston, Arnim Berhorst, Cesar Ranero, and Martin Meschede. Many thanks also to Jens Schneider and Dirk Schwertfeger who carried out the mound hunt in endless data matrixes and Heiko Sahling who made absolut numbers available for the SFB 574. This publication is contribution no. 32 of the Sonderforschungsbereich 574 "Volatiles and Fluids in Subduction Zones" at the Christian-Albrechts-University of Kiel.

REFERENCES

1. Abich, O.W.H., 1857. Über Schlammvulkane und ihre Bedeutung für die Geologie, *Ber. Ver. Dt. Naturwiss.*, **33**, 101.
2. Aliyev, A.A., I.S. Guliyev,I I.S., and Belov, I. S., 2002. *Catalogue of Recorded Eruptions of Mud Volcanoes of Azerbaijan (For Period of Years 1810-2001)*, Nafta, Baku.
3. Barckhausen, U., Ranero, C.R., von Huene, R., Cande, S.C., and Röser, H.A., 2001. Revised tectonic boundaries in the Cocos Plate off Costa Rica: Implications for the segmentation of the convergant margin and for plate tectonic models, *J. Geophys. Res.*, **106** (B9), 19,207-19,220.
4. Bialas, J., Flueh,E.R., and Bohrmann, G., 1999. FS SONNE, Cruise Report SO144/1&2. PAGANINI, Panama 1999, in *GEOMAR Rep.*, **94**, GEOMAR, Kiel.
5. Bohrmann, G., Heeschen, K., Jung, C., Weinrebe, W., Baranov, B., Cailleau, B., Heath, R., Hühnerbach, V., Hort, M., Masson, D., and Trummer, I., 2002. Widespread fluid expulsion along the seafloor of the Costa Rica convergent margin, *Terra Nova*, **14** (2), 69-79.
6. Christeson, G.L., McIntosh, K.D., and Shipley, T.H., 1999. Structure of the Costa Rica convergent margin, offshore Nicoya Peninsula, *J. Geophys. Res.*, **104** (B11), 25,443-25,468.
7. Dählmann, A., and de Lange, G.J., 2003. Fluid-sediment interactions at Eastern Mediterranean mud volcanoes: a stable isotope study from ODP Leg 160, *Earth Planetary Sci. Lett.*, **212**,

377-391.

8. DeMets, C., Gordon, R.G., Argus, D.F. and Stein, S., 1994. Effect of recent revision to the geomagnetic reversal timescale on estimates of current plate motions, *Geophys. Res. Lett.*, **21**, 2191-2194.
9. Dia, A.N., Castrec-Rouelle, M., Boulègue, J., and Comeau, P., 1999. Trinidad mud volcanoes: Where do the expelled fluids come from ?, *Geochim. Cosmochim. Acta*, **63**, 1023-1038.
10. Dimitrov, L.I., 2002. Mud volcanoes - the most important pathway for degassing deeply buried sediments, *Earth Sci. Rev.*, **59**, 49-76.
11. Etiope, G., and Klusman, R.W., 2002. Geologic emissions of methane to the atmosphere, *Chemosphere*, **49**, 777-789.
12. Flueh, E.R., Ranero, C.R. and von Huene, R., 2000. The Costa Rican Pacific margin: from accretion to erosion, *Zbl. Geol. Paläont, Part 1, 7/8*, 669-678.
13. Goad, S.T., 1816. Miscellaneous observations on the volcanic eruptions at the islands of Java and Sumbawa, with a particular account of the mud volcano at Grobogan, *J. Sci. Arts*, **1**, 245-258.
14. Guliyev, I.S., and Feizullayev, A.A., 1996. *All about mud volcanoes, Azerbaijan*, Geol. Inst. Azerbaijan, Baku.
15. Harris, R.N. and Wang, K., 2002. Thermal models of the Middle America Trench at the Nicoya Peninsula, Costa Rica, *Geophys, Res. Lett.*, **29** (21), doi 10.1029/2002GL015406.
16. Hauff, F., Hoernle, K., van den Bogaard, P., Alvarado, G., and Garbe-Schönberger, D., 2000. Age and geochemistry of basaltic complexes in western Costa Rica: Contributions to the geotectonic evolution of Central America, *Geochem., Geophys., Geosyst.*, **1**, 1-42.
17. Henry, P., Le Pichon,X., Lallemant,S., Lance, S., Martin, J.B., Foucher, J.-P., Fiala-Médioni, A., Rostek, F., Guilhaumou, N. Pranal, V., and Castrec, M., 1996. Fluid flow in and around a mud volcano field seaward of the Barbados accretionary wedge: Results from Manon cruise, *J. Geophys. Res.*, **101** (B9), 20,297-20,323.
18. Hoernle, K.,Werner, R., Morgan, J.P., Garbe-Schönberger, D., Bryce, J., and Mrazek, J., 2000. Existence of complex spatial zonation in the Galápagos plume for at least 14 m.y., *Geology*, **28** (5), 435-438.
19. Hovland, M., Hill, A., and Stokes, D., 1997. The structure and geomorphology of the Dashgil mud volcano, Azerbaijan, *Geomorphology*, **21**, 1-15.
20. Judd, A.G., Hovland, M., Dimitrov, L.I., Gar a Gil, S., and Jukes,V., 2002. The geological methane budget at continental margins and its influence on climate change, *Geofluids*, **2**, 109-126.
21. Kahn, L.M., Silver, E.A., Orange, D.L., Kochevar, R., and McAdoo, B.G., 1996. Surficial evidence of fluid expulsion from the Costa Rica accretionary prism, *Geophys. Res. Lett.*, **23** (B8), 887-890.
22. Kimura, G., Silver, E.A., Blum, P., et al., 1997. Costa Rica accretionary wedge, Sites 1039-1043, in *Proc. ODP, Init. Rpts.*, Ocean Drilling Program, College Station TX .
23. Kopf, A., 2002. Significance of Mud Volcanism, *Rev. Geophys.*, **40**, doi 10.1029/ 2000RG000093.
24. Kopf, A., and Behrmann, J.H., 2000. Extrusion dynamics of mud vulcanoes on the Mediterranean Ridge accretionary complex, in *Salt, Shale, and Igneous Diapirs in and around Europe*, Vendeville, B., Mart, Y., and Vigneresse, J.-L. (eds.) , Geol. Soc., London, 169-204.
25. Kopf, A., Deyhle, A., and Zuleger, E., 2000. Evidence for deep fluid circulation and gas hydrate dissociation using boron and boron isotopes of pore fluids in forearc sediments from Costa Rica (ODP Leg 170), *Mar. Geol.*, **167**, 1-28.

26. Kopf, A., Klaeschen, D., and Mascle, J., 2001. Extreme efficiency of mud volcanism in dewatering accretionary prisms, *Earth Planetary Sci. Lett.*, **189**, 295-313.
27. Mann, P., 1995. Geologic and tectonic development of the Caribbean plate boundary in southern Central America, *Geol. Soc. Am. Spec. Pap.*, **295**, 1-349.
28. McAdoo, B.G., Orange, D.L., Silver, E.A., McIntosh, K.D., Abbott, L., Galewsky, J., Kahn, L.M., and Protti, M., 1996. Seafloor structural observations, Costa Rica accretionary prism, *Geophys. Res. Lett.*, **23** (B8), 883-886.
29. McIntosh, K.D. and Sen, M.K., 2000. Geophysical evidence for dewatering and deformation processes in the ODP Leg 170 area offshore Costa Rica, *Earth Planetary Sci. Lett.*, **178**, 125-138.
30. McIntosh, K.D., and Silver, E.A., 1996. Using 3D Seismic reflection data to find fluid seeps from the Costa Rica accretionary prism, *Geophys. Res. Lett.*, **23** (8), 895-898.
31. Meschede, M., Zweigel, P., and Kiefer, E., 1999. Subsidence and extension at a convergent plate margin: evidence for subduction erosion off Costa Rica, *Terra Nova*, **11**, 112-117.
32. Milkov, A.V., 2000. Worldwide distribution of submarine mud volcanoes and associated gas hydrates, *Mar. Geol.*, **167**, 29-42.
33. Milkov, A.V., Sassen, R., Apanasovich, T.V., and Dadashev, F.G., 2003. Global gas flux from mud volcanoes: A significant source of fossil methane in the atmosphere and the ocean, *Geophys. Res. Lett.*, **30** (B2), doi 10.1029/2002GL016358.
34. Moritz, E., Bornholdt,S., Westphal, H., and Meschede, M., 2000. Sediment subduction lacking accretion at the Costa Rica convergent margin (ODP Leg 170), *Earth Planetary Sci. Lett.*, *174*, 301-312,.
35. Patino, L.C., Carr, M.J., and Feigenson, M.D., 2000. Local and regional variations in Central American arc lavas controlled by variations in subducted sediment input, *Contrib. Mineral. Petrol.*, **138**, 265-283.
36. Pecher, I.A., Kukowski, N., Huebscher, C., Greinert, J., Bialas, J., and GEOPEG Working Group, 2001. The link between bottom-simulating reflections and methane flux into the gas hydrate stability zone – new evidence from Lima Basin, Peru Margin, *Earth Planetary Sci. Lett.*, **185**, 343-354.
37. Pecher, I.A., Ranero, C.R., von Huene, R., Minshull, T.A., and Singh, S.C. , 1998. The nature and distribution of bottom simulating reflectors at the Costa Rican convergent margin, *Geophys. J. Int.*, **133**, 219-229.
38. Protti, M., Güendel, F., and McNally, K., 1995. Correlation between the age of the subducting Cocos plate and the geometry of the Wadati-Benioff zone under Nicaragua and Costa Rica, in *Geologic and Tectonic Development of the Caribbean Plate Boundary in Southern Central America*, Mann, P. (ed.) , GSA Special Papers, Boulder CO, 309-343.
39. Ranero, C.R., Morgan, J.P., McIntosh, K., and Reichert, C., 2003. Bending-related faulting and mantle serpentinization at the Middle America Trench, *Nature*, **425**, 367-373.
40. Ranero, C.R., Morgan,J.P., McIntosh, K.D., and Reichert, C., 2001. Flexural faulting and mantle serpentinization at the Middle America Trench, *Eos*, 1-82.
41. Ranero, C.R., and von Huene, R., 2000. Subduction erosion along the Middle America convergent margin, *Nature*, **404**, 748-752.
42. Ranero, C.R., Flueh, E.R., Duarte, M., Baca, D., and McIntosh, K.D., 2000. A cross section of the convergent Pacific margin of Nicaragua, *Tectonics*, **19** (2), 335-357.
43. Rhakmanov, R.R., 1987. *Mud Volcanoes and Their Importance in Forecasting of Subsurface Petroleum Potential (in Russian)*, Nedra, Moscow.
44. Robertson, A.H.F. and Shipboard Scientific Party, 1996. Mud volcanism on the Mediterranean Ridge: Initial results of Ocean Drilling Program Leg 160, *Geology*, **24** (3), 239-242.

45. Ruepke, L.H., Morgan, J.P., Hort,M., and Connolly, J.A.D., 2002. Are the regional variations in Central American arc lavas due to differing basaltic versus peridotitic slab sources of fluids ?, *Geology*, **30** (11), 1035-1038.
46. Ruppel, C., and Kinoshita ,M., 2000. Fluid, methane, and energy flux in an active margin gas hydrate province, offshore Costa Rica, *Earth Planetary Sci. Lett.*, **179**, 153-165.
47. Saffer, D.M., Silver, E.A., Fisher, A.T., Tobin,H., and Moran, K., 2000. Inferred pore pressures at the Costa Rica subduction zone: Implications fo dewatering processes, *Earth Planetary Sci. Lett.*, **177**, 193-207.
48. Shipley, T.H., McIntosh, K.D., Silver, E.A. and Stoffa, P.L., 1992. Three-Dimensional Seismic Imaging of the Costa Rica Accretionary Prism: Structural Diversity in a Small Volume of the Lower Slope, *J. Geophys. Res.*, **97** (B4), 4439-4459.
49. Shipley, T.H., Stoffa, P.L., and Dean, D.F., 1990. Underthrust Sediments, Fluid Migration Paths, and Mud Volcanoes Associated with the Accretionary Wedge off Costa Rica: Middle America Trench, *J. Geophys. Res.*, **95** (B6), 8743-8752.
50. Silver, E.A., Miriam, K., Andew, F., Morris, J., McIntosh, K., Saffer, D., 2000. Fluid flow paths in the Middle America Trench and Costa Rica margin, *Geology*, **28** (8), 679-682.
51. Silver, E.A., 2001. Leg 170: Synthesis of Fluid-Structural Relationships of the Pacific Margin of Costa Rica, in *Proc. ODP, Sci. Res.*, Ocean Drilling Program, College Station TX.
52. Sahling, H, Ranero, C.R., Soeding, E., Weinrebe, W., Huehnerbach, V. and Masson, D.G., Abundant indication for fluid venting connected to subduction erosion at the continental margin of Costa Rica and Nicaragua, *Marine Geology*, in prep.
53. Soeding, E., Wallmann, K., Suess, E., and Flueh, E., 2003. RV METEOR, Cruise Report M54/2+3. Fluids and Subduction, Costa Rica 2002, in *GEOMAR Rep., 111*, GEOMAR, Kiel, 2003.
54. Stoffa, P.L., Shipley, T.H., Kessinger, W., Dean, D.F., Elde, R., Silver, E.A., Reed, D., and Aguilar, A., 1991. Three-Dimensional Seismic Imaging of the Costa Rica Accretionary Prism: Field Program and Migration Examples, *J. Geophys. Res.*, **96** (B13), 21,693-21,712.
55. Vannucchi, P., Ranero, C.R., Galeotti, S., Straub, S.M., Scholl, D.W., and McDougall-Ried, K., 2003. Fast rates of subduction erosion along the Costa Rica Pacific margin: Implications for non-steady rates of crustal recycling at subduction zones, *J. Geophys. Res.*, **108**, n.B11, 2511.
56. Vannucchi, P., Scholl, D.W., Meschede, M., and McDougall-Ried, K., 2001. Tectonic erosion and consequent collapse of the Pacific margin of Costa Rica: combined implications from ODP Leg 170, seismic offshore data and regional geology of the Nicoya Peninsula, *Tectonics*, **20**, 649-668.
57. Von Huene, R., Ranero, C.R., Weinrebe, W., and Hinz, K., 2000. Quaternary convergent margin tectonics of Costa Rica, segmentation of the Cocos Plate, and Central American volcanism, *Tectonics*, **19** (2), 314-334.
58. Von Huene, R. et al., 1985. Site 565, in *Init. Repts. DSDP*, von Huene, R. and J. Aubouin, J. (eds.) ,US Govt. Printing Office, Washington DC, 21-77.
59. Weinrebe, W., and Flueh, E., 2002. FS/RV SONNE, Cruise Report SO163. Subduction 1, Costa Rica 2002, in *GEOMAR Rep., 106*, GEOMAR, Kiel.
60. Yassir, N.A., 1989. *Mud volcanos and the behaviour of overpressured clays and silts*, PhD thesis,Univ. Coll. London, London.
61. Ye, S., Bialas, J., Flueh, E.R., Stavenhagen, A.U., and . von Huene, R., 1996. Crustal structure of the Middle American Trench off Costa Rica from wide-angle seismic data, *Tectonics*, **15** (5), 1006-1021.
62. Zuleger, E., Gieskes, J.M., and You, C.-F., 1996. Interstitial water chemistry of sediments of the Costa Rica accretionary complex off the Nicoya Peninsula, *Geophys. Res. Lett.*, **23** (B8), 899-902.

STYLES AND PRODUCTIVITY OF MUD DIAPIRISM ALONG THE MIDDLE AMERICAN MARGIN

Part II: Mound Culebra and Mounds 11 and 12

Tobias Moerz[1], Naomi Fekete[1], Achim Kopf[3], Warner Brueckmann[1,2], Stefan Kreiter[4], Veit Huehnerbach[5], Douglas Masson[5], Daniel A. Hepp[1], Mark Schmidt[1,6], Steffen Kutterolf[1], Heiko Sahling[1], Friedrich Abegg[2], Volkhard Spiess[3], Erwin Suess[1,2], and Cesar R. Ranero[1,2]

1 *Sonderforschungsbereich 574, Christian-Albrechts-University of Kiel, Wischhofstr. 1-3, 24148 Kiel, Germany*

2 *GEOMAR, Wischhofstr. 1-3, 24148 Kiel, Germany*

3 *Research Center Ocean Margins, University of Bremen, Klagenfurter Str., 28359 Bremen, Germany*

4 *Technische Universität Berlin, Mueller-Breslau-Str., 10623 Berlin, Germany*

5 *Southampton Oceanography Centre, European Way, Southampton, SO14 3ZH, UK*

6 *Institute of Geosciences, Ludwig-Meyn-Str. 10, 24118 Kiel, Germany*
corresponding author: T. Moerz, e-mail: tmoerz@geomar.de, Fax: +49-431-600-2916

Abstract: We present sedimentological and structural data and conceptual models for the evolution of two types of diapiric mud moundss offshore Costa Rica. Dozens of exposed mud mounds are found in the smooth domain of the margin with Mound Culebra being the most prominent example. Mound Culebra is a fault controlled feature with steep flanks (~10-20°) and a lack of recent mud flows. The extruded material consists of overcompacted silty clay with signs of intense brittle deformation, brecciation, hydrofracturing and secondary perforation by closely spaced conduits. The southern rough domain is characterized by numerous local tectonic regimes linked to seamount subduction and the collision of the Cocos ridge all associated with diverse forms of venting. Diapirism seems to play only a minor role and Mound 11 and 12 are examples of fault controlled, low relief mud volcanoes with shallow-dipping flanks. Mud flow sequences, vent debris and the presence of gas hydrates in the shallow subsurface favor an episodic, gas driven eruption behavior.

Key words: Costa Rica margin, mud diapirism, mud volcanoes, mud flows, fluid conduits

1. INTRODUCTION

In this second part of the study we report on two types of mud diapir

G. Martinelli and B. Panahi (eds.), Mud Volcanoes, Geodynamics and Seismicity, 49–76

occurrences along a non-accretionary 520 km section of the Central American trench offshore Costa Rica and Nicaragua. The purpose of this paper is to characterize two different styles of mud volcanism based on three well sampled mounds. The findings will be discussed in the context of mud diapir conceptual models and differences in subduction styles and dewatering mechanisms in the corresponding margin segments.

The presented data have been collected during numerous geophysical surveys and mapping cruises (e.g. RV SONNE 144, 163 and RV METEOR M54-1; Bohrmann et al, 2002; Weinrebe and Flueh, 2002) and an intense coring and sampling program during RV METEOR M54-2 and M54-3 cruises (Soeding et al., 2003) along the Costa Rica and Nicaraguan Pacific margin. Main tools of investigation are sediment structural observations in conjunction with logging-, sediment index-, and geotechnical data from gravity cores. This point sampling is aided by high resolution bathymetry, 4 kHz digital Parasound sediment echosounder profiles and 30 kHz TOBI sidescan data. Specifically, this study compares a mud diapir occurrence off the northern Nicoya Peninsula (Mound Culebra) to an occurrence west of the Osa Peninsula (Mounds 11 and 12; Fig. 1). From their morphological and acoustic expression the described mud volcanoes/diapirs are representative for a number of other occurrences in their respective settings. The selection of the presented individual mounds is based on their complete sample coverage and overall sedimentphysical, geophysical and chemical data base.

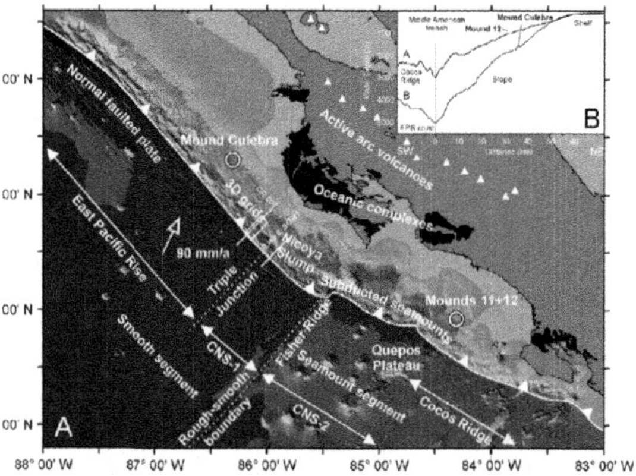

Figure 1. **A,** Geodynamic setting off Central America and locations of the two working areas Mound Culebra and Mound 11, 12. The inset is giving a Plate scale view with the relative plate movement vectors (85 mm/a after DeMets et al., 1994). The margin bathymetry and segmentation nomenclature is after von Huene et al., 2000). Box off shore Nicoya Peninsula = 3D seismic grid of Shipley et al. (1992); Black signature on shore = Oceanic complexes; CNS = Cocos-Nazca Spreading Center; CR = Cayman Ridge; EPR = East Pacific Rise; HE = Hess Escarpment; MAT = Middle America Trench; PFZ = Panama Fracture Zone; RSB = Rough-Smooth Boundary. **B**, trench normal bathymetric cross sections through Mound 12 (rough domain) and Mound Culebra (smooth domain).

2. MOUND CULEBRA, SMOOTH DOMAIN

2.1 Morphology and seafloor observations

The primary bathymetry of Mound Culebra and the adjacent slope has been obtained by RV SONNE using a Simrad EM120 swath mapping echosounder during SONNE cruise 163 (Weinrebe and Flueh, 2002). More detailed bathymetric information was added during RV METEOR cruise M54-1 with the Hydrosweep system. Mound Culebra is located on the upper slope, 36 km landward off the trench in ~1500 m of water depth (Fig. 1). The slope around Mound Culebra is depressed for several km and the feature is situated at the upslope end of this depression, SW of terminating Canyon systems with higher slope angles (Fig. 1, inset). The mound has an distorted oval shape with a SW-NE-oriented basal long axis of 1750 m and a NW-SE-oriented basal short axis of 1400 m. (Fig. 2, 3 and 4). The long axis has a normal orientation to the slope. Based on the SIMRAD bathymetry, Mound Culebra's crest resembles a saddle with two elevated ends which are separated by depressions. The mound crest in 1508 m water depth (86°18.3'W/10°17.8'N) rises over the surrounding seafloor by about 115 m. The base of Mound Culebra covers an area of 2.08 km^2 and the volume of exposed material above the average surrounding seafloor is 0.073 km^3. The NW and SE slopes of the mound have steep uniform angles of 19.3°. Whereas the slope of the SW face, toward the trench, is in average 14° with a steeper lower part of 19° and a shallower upper part of 10°.

The slope surrounding Mound Culebra is covered by turbidites and the local seafloor morphology is modeled by small-scale slumps and trenchward dipping normal faults with offsets up to ten meters (Fig. 3). Based on sediment coring along the slope and the acoustic facies, the prominent sediment accumulation in the lee of Mound Culebra is interpreted as a protected overbank deposit of turbidity currents focused in the gullies NW and SE of the mound. Lee in this case is not related to the main bottom water current direction, which is directed toward the NW (Mau et al., in prep.), but rather addresses the leeward side of turbidity current prevalent direction. Mound Culebra is located above a NW-SE-trending trench parallel growth fault that is continuous along strike for 10th of km. The surface expression of the fault is best seen in SOC TOBI (Towed Ocean Bottom Instrument) images (Fig. 2) and a normal offset along an exposed fault face with soft sediment infill on the downthrown block is suggested by the acoustic illumination from the SW (Fig. 2B). In the bathymetry the fault intersection with the seafloor is expressed as a small-scale flattening of the slope. The mound itself is intersected by the fault expressed in two large and several smaller scars across the crest. The distinct lateral distortion of the oval shape of the mound along the scars suggests a small strike slip component in addition to the normal movement of the fault (Fig. 2C).

Mapping with the TOBI side scan sonar system (Weinrebe and Flueh, 2002) yielded areas with high backscattering for the crest and upper flanks of the mound (Fig. 4) coinciding in part with strong reflectivity in the Parasound (Fig. 3). Subsequent video sled surveys revealed the presence of authigenic carbonates and typical cold vent faunal communities (vesicomyid clams, pogonophoran tubeworms and mytilid mussels) partially covering the mound crest and upper flanks (Mau et al., in prep.). No indicators of recent, surface expressed mud flow activity were observed. These observations questioned the classification of the mound being a mud volcano or diapir,

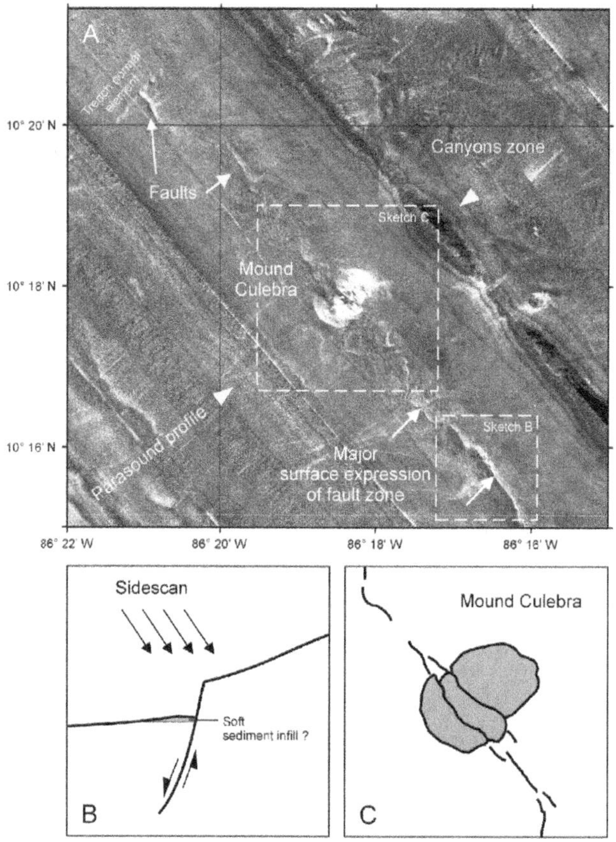

Figure 2. Regional TOBI side scan map of Mound Culebra and surroundings. **A**, displays the slope canyons to the NE, the surface expression of the trench parallel fault and a trench normal structural element in the NW corner of the map. **B**, Interpretation of the surface expressed fault scare educed from the sidescan illumination direction. **C**, Close up of at least two major fault scars bisecting the mound. White triangles indicate the approximate location of the Parasound Profile of Fig. 3.

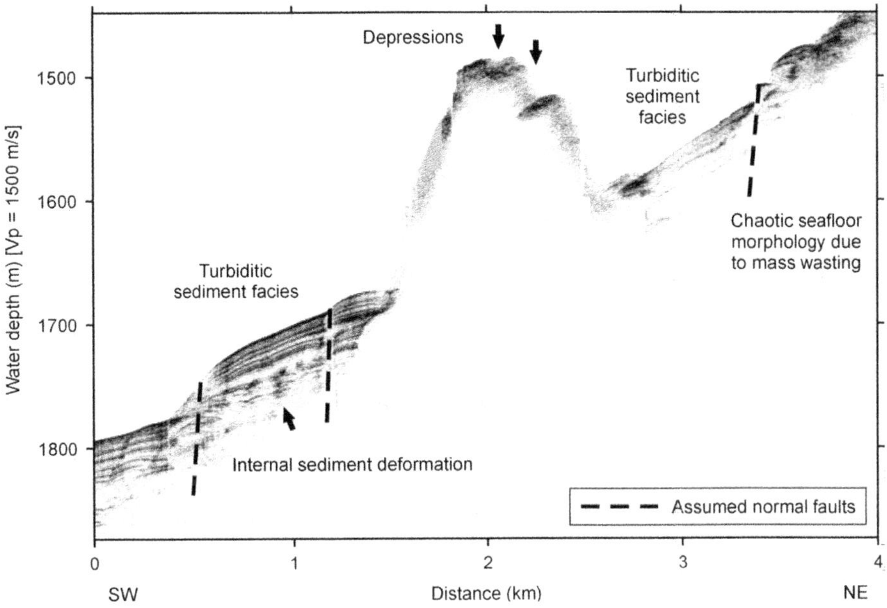

Figure 3. Digital SW-NE, 4 kHz parasound image of Mound Culebra and the adjacent slope. The slope is characterized by turbidite deposition and mass wasting. The thick sediment pile SW of the mound is interpreted as turbiditic overbank deposit in the flow lee of the Mound, onlapping on the mound flank. The location of the profile is indicated with arrows in Fig. 2.

based on the interpretation of similar mounds in the area (Shipley et al., 1992) and raised suggestions that the mound could be made up entirely of chemoherm-like carbonates and carbonate cemented sediment (Weinrebe and Flueh, 2002). Detailed sampling with gravity cores and information on the penetration or tipping of heatflow probes revealed that massive carbonate occurrences are limited in spatial and subsurface extent (Fig. 4). Massive carbonates are restricted to the depth interval between the mound crest and the ~1550 m depth contour. Even within this depth interval long sediment cores have been recovered (e.g. M54-18, 19, 27; Fig. 4). Elevated heatflow (~1.6 times the background of 36 mW/m^2; Grevemeyer et al., 2004) and methane anomalies in water samples that are up to one order of magnitude higher than the background values of 1-2 nmol/L, indicate that Mound Culebra is an active vent site (Mau et al., in prep.) and of extrusive origine.

2.2 Core observations and sediment physical properties

Of the 12 gravity cores and 5 short push cores taken at Mound Culebra and its immediate surrounding, 6 gravity cores will be discussed in greater detail to illustrate important structural and sedimentological characteristics. Logging data on whole cores has been obtained with the GEOTEK Multi Sensor Core Logger (MSCL). The system allows continuous profiling of gamma density, magnetic susceptibility, P-wave velocity and attenuation, core temperature and core liner thickness. Gamma densities have been cross-checked with laboratory-derived index property measurement on discrete, undisturbed samples (water content (*wc*), volumetric wet bulk density, pycnometer-derived grain density (*gd*) and recalculated wet bulk densities from *wc* and *gd*). Shear strength measurements using a laboratory shear vane and a fall cone apparatus together with spectral color scan information compliment the obtained physical property inventory.

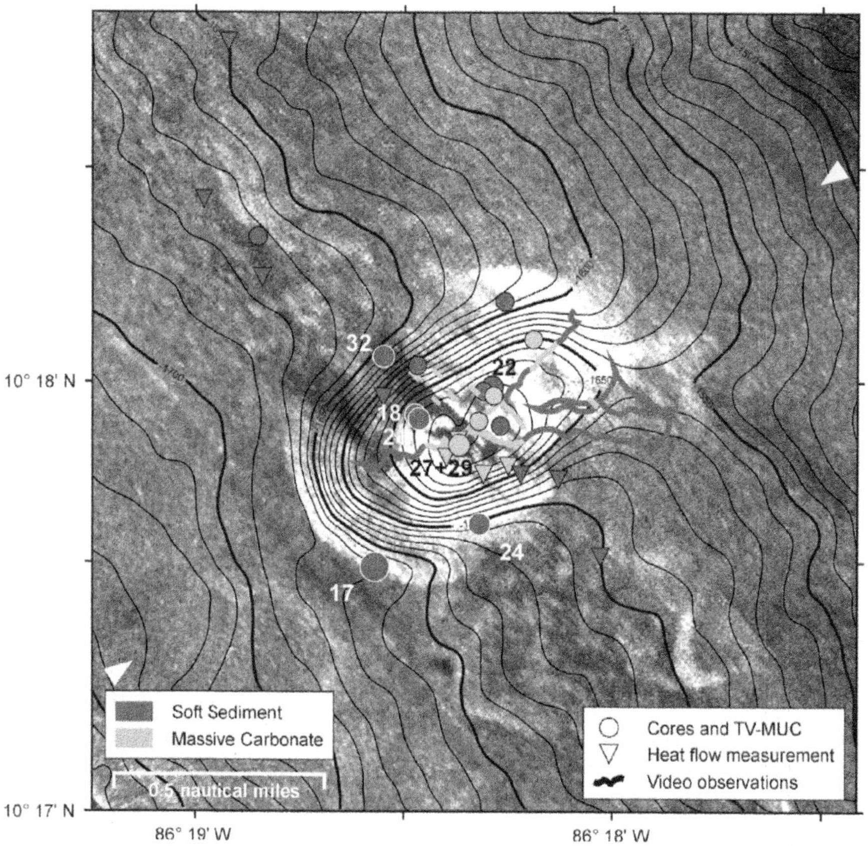

Figure 4. TOBI side scan image and bathymetry of Mound Culebra. Sample stations and seafloor observations are displayed and used to show the surficial extend of massive carbonate and sediment. Massive Carbonates are restricted to the < 1500 mbsf contour. For a best fit the side scan data has been shifted compared to the bathymetry by 75.6 m to the N and 41.5 m to the E. The applied shift is well within the navigational uncertainities of the TOBI system that was deployed without a short baseline navigation system.

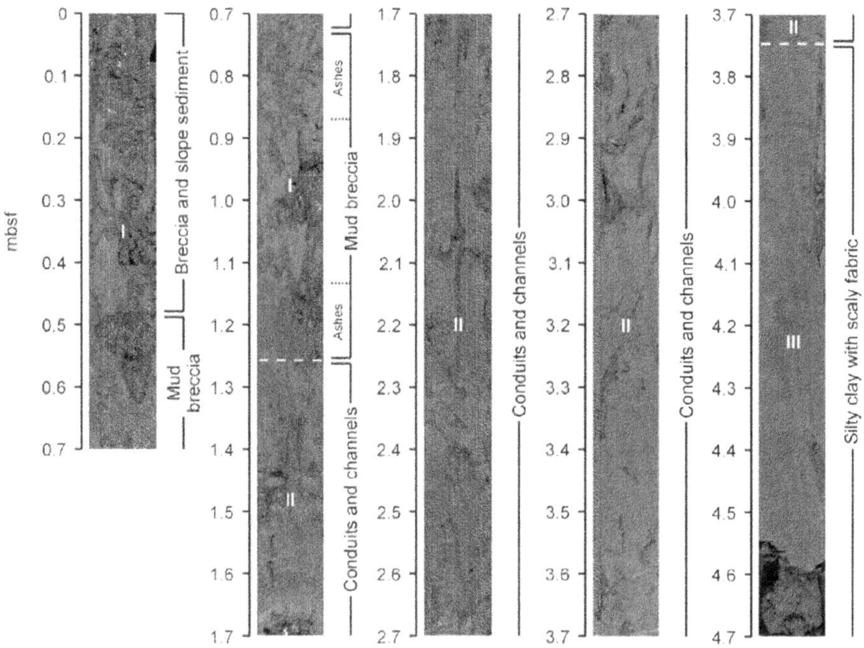

Figure 5. Image of core M54-21, Mound Culebra. The 4.70 m long gravity core from the central depression of the mound (Fig. 2 and 4) is characterized by intensely deformed sediment, hydrofracturing and piercing of numerous fluid conduits. Units I-III are indicated to allow a direct comparison with the sediment physical core log M54-18 (Fig. 6).

Coring stations M54-18 and -21 are only a couple meters apart from each other and located within the central depression of the mound, which is seen as a dark, shaded, NW-SE trending zone in the side scan image (Fig. 4). The recovered lithologies (5.5 and 4.7m) are very similar in the two cores and silty clay is the dominant constituent (Fig. 5).

The upper 50 cm of core M54-21 consists of hemipelagic sediment and brecciated, indurated silty clay (Fig. 5). The interval from 50 to 125 cmbsf (centimeter below seafloor) is entirely made up of brecciated silty clay with cm to dm long individual clay clasts. Two mafic ash layers are found within this interval. The ash layers can be traced across the core width through several neighboring clasts. This indicates little relative movement between the clasts for this interval. The fractures and void between clay clasts is filled with water, low viscous mud slurry and sand-size clay aggregates. This upper part of the core has a low sediment stength with values < 20 kPa (Fig. 6) Open and filled pipe-like channels, or conduits, are found in the interval 125-374 cmbsf. Channel orientation is commonly near vertical (Fig. 5), but part of the conduits also occur in an *en echelon*-like orientation with angles of 20-25 degrees off the core axis (Fig. 7G, H).

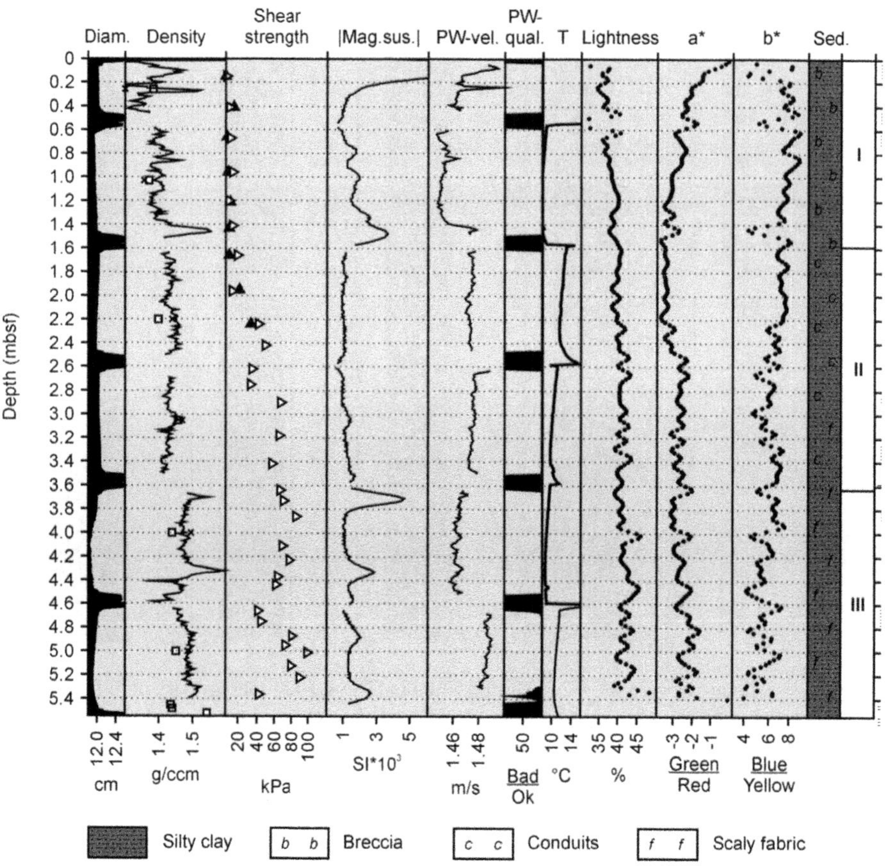

Figure 6. Logging data and physical property collection of gravity core M54-18 taken at the same location as M54-21 (Fig. 4). Shear strength values in the lower part of the core (unit III) exceeds 100 kPa. Shear strength column; filled triangles: rotary shear vane, open triangles: miniature fall cone test. Density column (wet bulk density); line: MSCL-data, square box: density from wet equal volume samples, cross: recalculated wet bulk density from grain densities.

The conduits have diameters between ~5 and 10 mm and individual conduits can be traced for more than 220 cm (Fig. 7F, L, O). Parts of core M54-18 and -21 haven been analyzed with a medical X-ray computer tomography (CT) to image internal details. Most of the conduits found have near round to slightly elliptic cross-sections (Fig. 7N), a few resemble intruded and expanded cracks (Fig. 7F, N) or interconnections between originally separated conduits (Fig. 7F). A tectonic control upon the location and orientation of the channels is evident, as channels follow fault plane intersections and micro-brecciated sections of the structured clays (Fig. 7E, F, N).

Channel fillings display a great variability:

- some of the conduits in core M54-18 (3.10-3.70 mbsf) exposed after splitting the core in halves were filled with clear water (Fig. 7C) and showed dark stains along the channel walls (manganese oxide ?);
- other conduits are completely filled with coarser, fine sand-size material, but similar in water content to the host sediment;
- the third category of conduits is filled with a mobile low-viscosity mud slurry that extrudes once the core is cut (Fig. 8F);
- multiphase fillings are observed in the CT- images. Coarse fillings of the first generation are forming a passive outer ring along the channel walls, which in turn is filled by an inner, second generation cylinder of mud slurry or water of low viscosity (Fig. 7N);
- mixed infills are observed in conjunction with altered basaltic fragments that seem to block some of the smaller conduits or form coarse (>1 mm diameter) constituents in the fine sand fillings (Fig. 7B, D, N).

In some instances filled and clogged conduits are reactivated and new microcracks emanate from them (Fig. 7M). This seems to be a true original fabric and no coring artifact since carbonate mineralization covers the walls of newly opened fractures (e.g. Fig. 7M). CT images from core interval M54-18 2.2-2.6 (Fig. 7L) show a chaotic clast supported mud breccia and some intact and some collapsed conduits that seem to terminate in-between the clasts.

Below 3.7 mbsf in cores M54-18 and -21, the number of conduits decreases and sediment from there on downward to the core catcher is well indurated and characterized by abundant stress-induced brittle deformation features. Undrained shear strength in this section exceeds 100 kPa and bulk wet density increases to 1.54 g/cm^3. The stiff, overcompacted and tectonically deformed clay body is intensely structured by closely spaced joints and small-scale faulting of variable orientation. The orientation of the faults and joint planes can not be explained by a single deformational regime and a multidirectional stress history which could imply rotation of the ascending material is inferred. The micro fault planes or fractures are either smooth or covered by well expressed slickensides indicating considerable net movement (Fig. 7J, K). The described clay fabric is common to the margins of mud diapirs and is referred to as 'scaly fabric' by various authors (Brown, 1990).

Coatings of carbonate, manganese or salt on some cleavages and fault planes have been observed on broken-up pieces of the M54-18 core catcher or imaged as thin bright traces in most computer tomography (CT) cross sections (Fig. 7I, M, N). Mineral precipitation that can heal and seal frac-

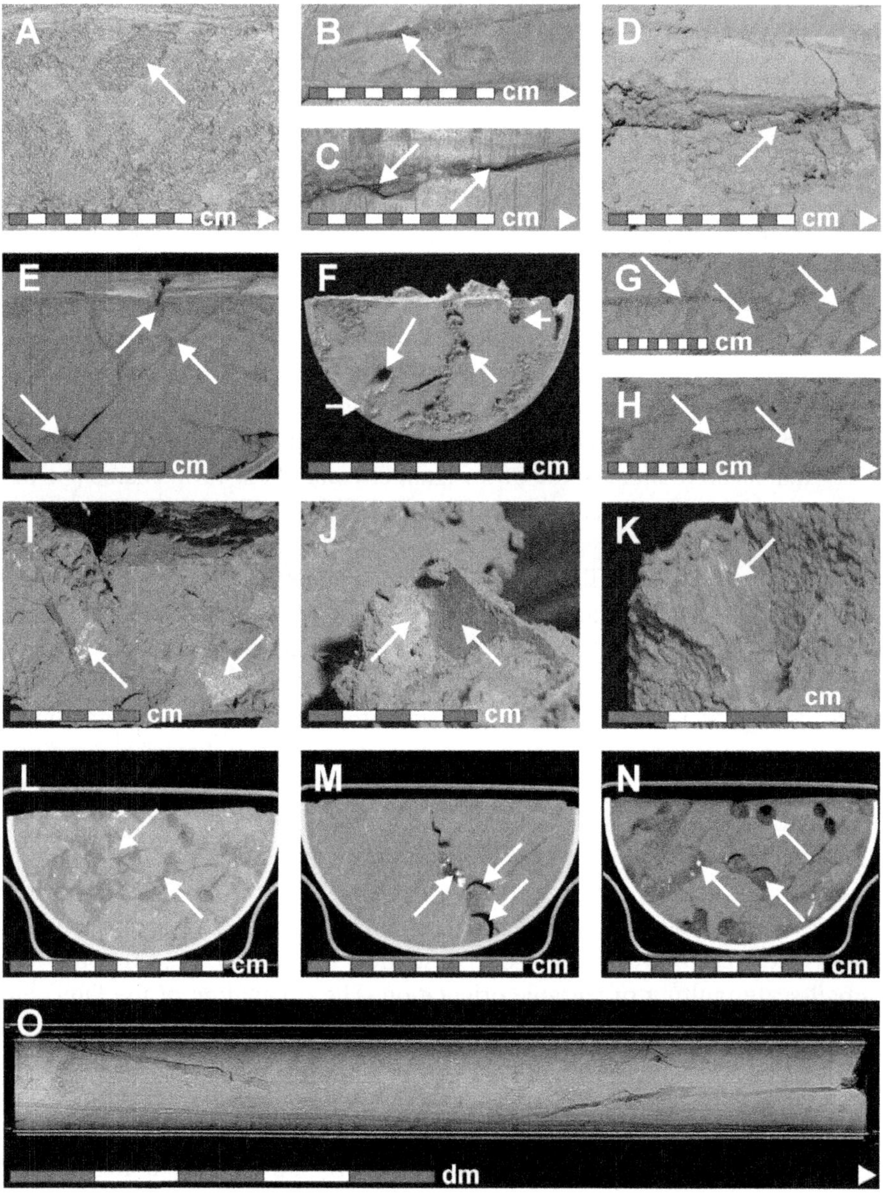

Figure 7. Structural details from photographs and x-ray computer tomography of cores M54-18 and -21, see text for details. **A**, M 54-18, 0.54-0.69 mbsf. **B**, M 54-18, 2.57-2.72 mbsf. **C**, M 54-18, 2.80-2.95 mbsf. **D**, M 54-18, 3.55-3.70 mbsf. **E**, M 54-18, 3.55 mbsf, Cross section, **F**, M 54-21, 1.60 mbsf. **G**, M 54-21, 2.09-2.39 mbsf. **H**, M 54-21, 2.40-2.70 mbsf, **I-K**, M 54-18, 5.30 mbsf, core catcher. **L**, M54-18, 2.32 mbsf. **M**, M 54-18, 2.71 mbsf. **N**, M 54-21, 1.80 mbsf. **O**, M 54-18, 2.50-3.50 mbsf.

A summary plot of all core logging data, index properties, color- and shear strength values of core M54-18 is given in Fig. 6. Important trends are the downcore increase of shear strength and density related to structural variations (breccia, conduits abundance, and dense scaly clays). The excursions of the magnetic susceptibility values in the upper part of the core are related to the entrainment of slope turbidites and ashes (0-1.4 mbsf). In the lower part the excursions reflect mafic volcanic clasts (3.6-5.4 mbsf).

In general conventional gravity coring was unsuccessful in areas of the mound crest with massive authigenic carbonate coverage. Cores M54-27 and -29 however, recovered carbonate and sediment and might illustrate the mound buildup and structural relations of these otherwise unsampled areas (Fig. 4). Core M54-27 is only 34 cm long and starts at the seafloor end with an unconsolidated hemipelagic sediment layer of ~5 cm containing large mytiloid shell fragments. A core diameter size white, authigenic carbonate piece with finger-like extensions and holes is found at the interface between the hemipelagic layer and the more consolidated silty clay of the remaining core (Fig. 8A, B). Dark, 'sulfidic' carbonates are found below 26 cmbsf starting at another lithological boundary. The dark carbonates and the surrounding sediments of this lowermost interval give off H_2S odor and vesicomyid clam shells are also abundant. A conduit with softer and coarser sediment can be traced from shortly below the white authigenic carbonated toward the dark colored interval (Fig. 8A, B). Together with carbonate recrystallization along fractures within the dark carbonates (Fig. 8B) we interpreted this as evidence for past or present advective fluid flow.

Abundant carbonate concretions are also found in core M54-22 (4.7 m length) and M54-29 (1.67 m length; Fig. 4). Size and quantity seem to decrease with depth. Many of the carbonate concretions in core M54-29 are embedded in grayish, soft marl aureoles and some concretions resemble lithified fluid conduits.

Cores taken from the base and flanks of Mound Culebra (M54-17, -24, -32; Fig. 4) show no clay clasts, scaly fabric or clear evidence for mud flows. The occurrence of basaltic volcanic clasts of up to 10 mm diameter within these cores however, suggests admixing of extruded material and hemipelagic sediments. Mud slurry expelled from the crest conduits and material from the crest region transported by mass wasting processes may contribute to the mound flank and basal deposits.

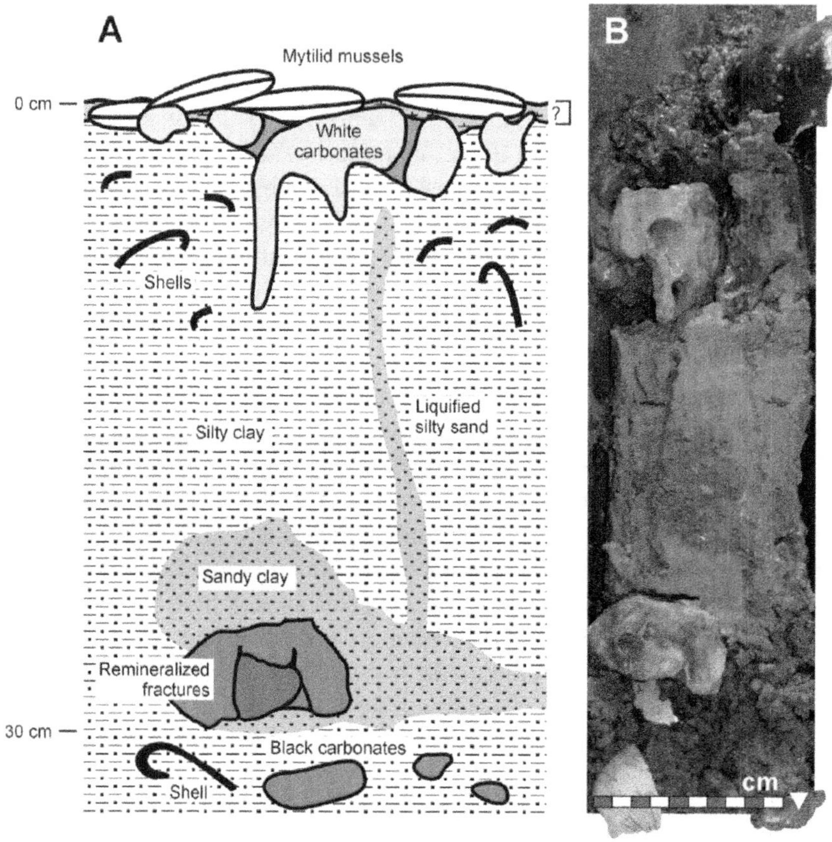

Figure 8 . Sketch and photograph of carbonate bearing core M54-27 (Fig. 4, core location). Structures like the sandy conduit are thought to provide fluids and methane for the above vent community and authigenic carbonate production.

3. MOUND 12 AND 11, ROUGH DOMAIN

3.1 Morphology and seafloor observations

Improved bathymetric charts of Mound 11 and 12 and the adjacent slope have been acquired during SONNE SO163 cruise (Weinrebe and Flueh, 2002) using the Simrad EM120 swath mapping system. Mound 11 and 12 and a third adjacent mound are located on the upper slope, ~32 km landward of the trench in ~1000 m of water depth (Fig. 1, inset). The slope around the Mounds is depressed for ~ 3 km and the two mounds are located at the

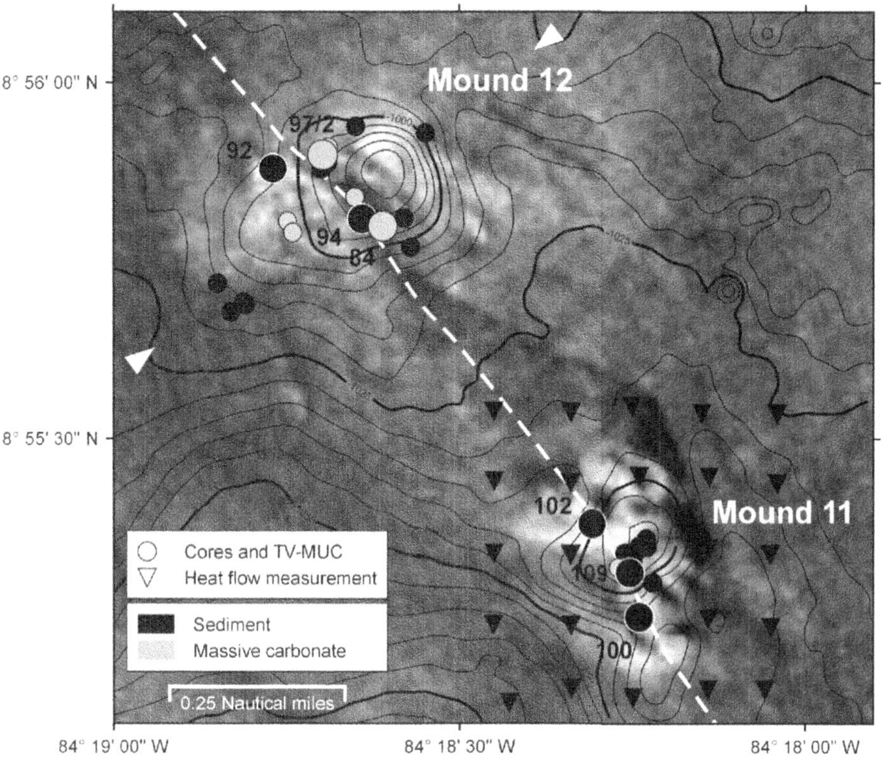

Figure 9. Detailed TOBI side scan image and bathymetry of Mound 11 and 12. Sample stations and seafloor observations are displayed and used to show the surficial extend of massive carbonate and sediment.

downslope end of this depression. The slope steepens trenchward (SW) of the mounds and is cut by prominent canyons. Mound 12, the more prominent of the three mounds, has an approximately equal dimensional, roundly shape in map view with a diameter of 800 m. Based on the SIMRAD bathymetry and a SW-NE-oriented Parasound profile, Mound 12 features a hunch-like higher pinnacle at its NE end and a low profile ridge toward the SW (Fig. 9, 10). The pinnacle in 977 m water depth (84°18.62'W/08°55.91'N) dominates the surrounding seafloor by about 38 m, whereas the ridge crest toward the SW in 1005 m water depth has a height of only 14 m. The base of Mound 12 covers an area of 0.65 km^2 and the volume of exposed material above the average surrounding seafloor is 0.0059 km^3 (i.e. approximately one order of magnitude smaller than that of Mound Culebra).

The steep northeastern slope of the pinnacle has an average slope angle of 10° compared to 6-7° of the NW and SE flanks. In contrast the more gentle SW-side of the pinnacle and the ridge have slopes of ~6°.

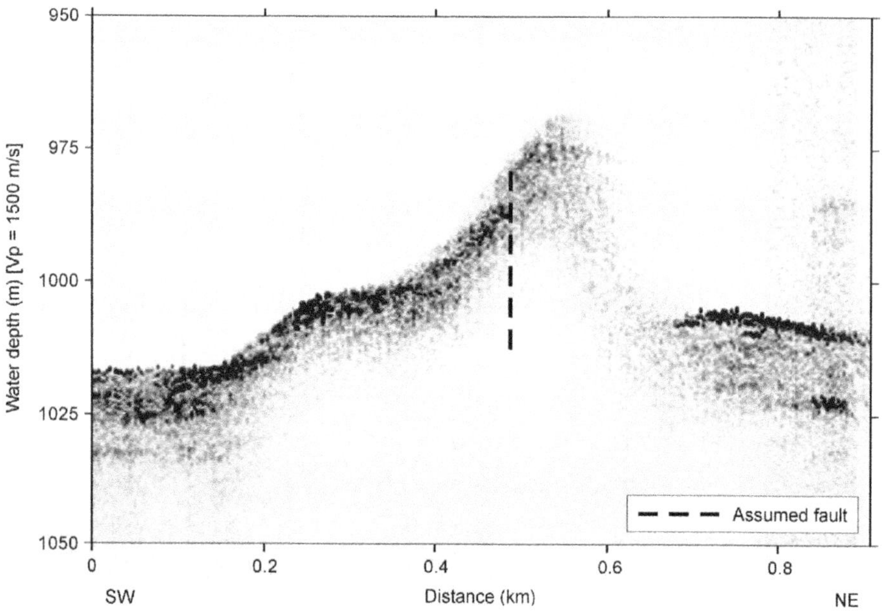

Figure 10. Digital SW-NE, 4 kHz parasound image of Mound 12. The approximate location of the profile is indicated with arrows in Fig. 9.

The lowest slope angles are encountered in trench parallel cross sections (NW-SE) through the ridge with ~4° to NW and ~3° to the SE.

Mound 12 and 11 as well as a third mound to the NW, not shown here, are aligned parallel to the trench. TOBI sidescan image combined with bathymetry data and changes in seafloor reflectivity in the parasound profile suggest a NW-SE-trending fault controlling the mound locations (Fig. 9, 10, 11). The fault is well expressed across Mound 12 in the side scan image as a depression (dark shade) where it separates the high mound pinnacle from the lower southwestern ridge. Mound 11 seems to be intersected by the fault in a similar way but the precise location of the fault trace is less clear from our geophysical data (Fig. 9). So far we have no good control on the fault orientation or movement sense.

Video sled surveys revealed the presence of authigenic carbonates and typical cold vent fauna communities (Mytilid mussels, pogonophoran tubeworms and bacterial mats) partially covering the pinnacle top and upper crest of the SW ridge flanks (Mau et al., in prep.). The occurrences coincide with TOBI high backscatter intensities (Fig. 9, 11) and in part with a highly reflective seafloor in the digital Parasound record (Fig. 10). Living vent fauna, elevated heat flow (~1.4 times the background of 42 mW/m^2; at Mound 11; Schmidt et al., submitted; Soeding et al., 2003); and the highest

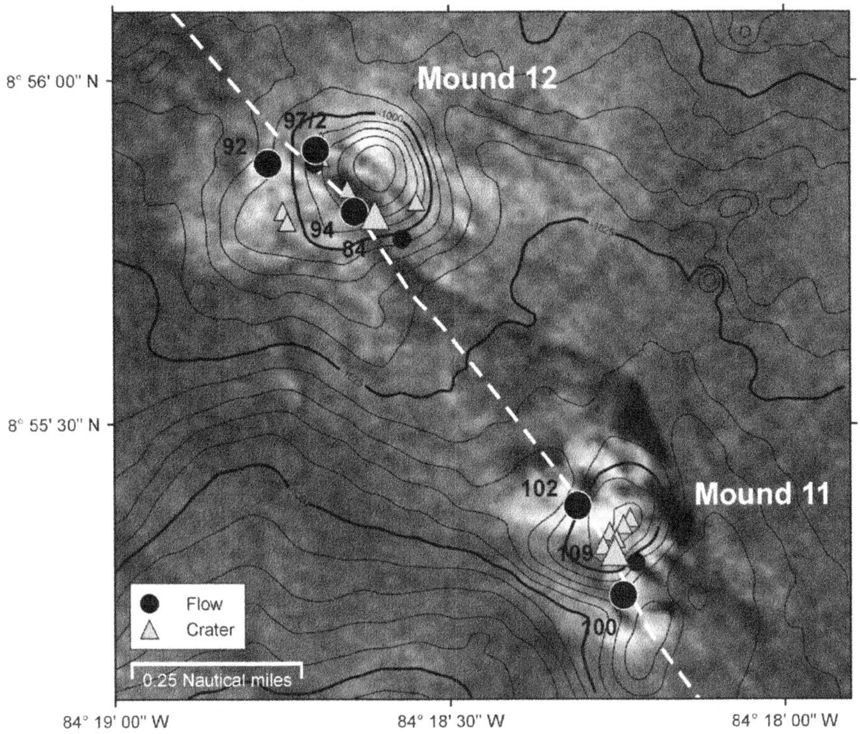

Figure 11. Attempt to distinguish mud flow occurrences from „crater" deposits that are characterized by vertical mass movement processes.

bottom water methane concentrations of all surveyed mounds (e.g. up to two orders of magnitude above the regional background of 1-2 nmol/L at Mound 12) indicate that Mound 11 and 12 are active vent sites (Mau et al., in prep.).

3.2 Core observations and sediment physical properties

Using the same approach as for Mound Culebra we collected all available direct sub-surface information from gravity cores, multi cores, TV grabs and the penetration or tipping of heat flow probes to map the area and subsurface extend of massive carbonates and sediment. The results are compiled on a combined bathymetric and TOBI image map (Fig. 9). Massive carbonates play no major role at Mound 11 and seem to be restricted in vertical extent on Mound 12 since long sediment-dominated cores have been obtained from various regions of the mound. Of the 19 gravity cores and 8 short push cores taken at various stations on and around Mounds 11 and 12, 5 gravity cores

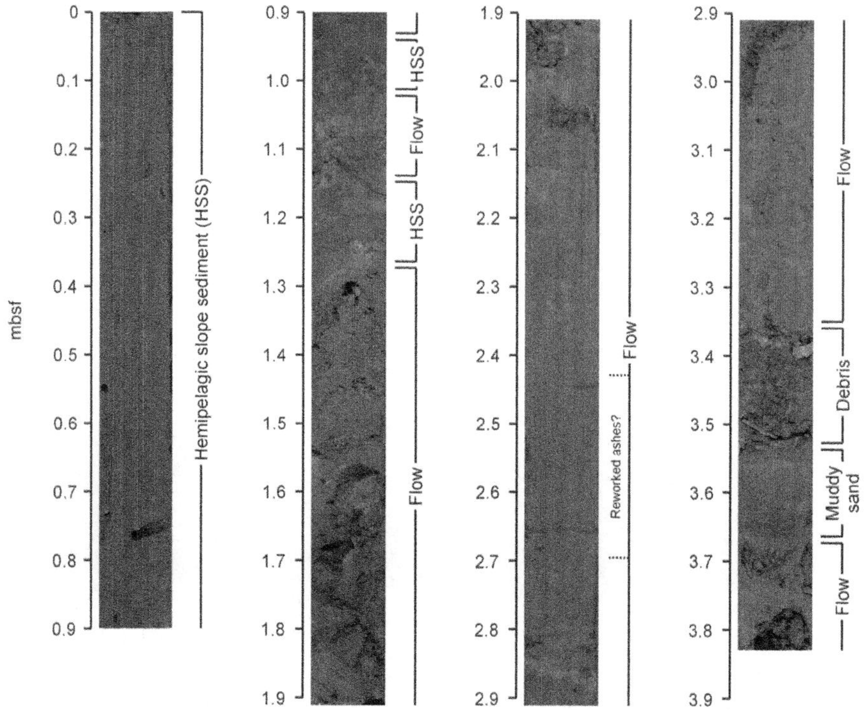

Figure 12. Image of core M54-97-2. The 3.83 m long gravity core from Mound 12 shows a succession of mud flows interstratified with hemipelagic sediment (for core location: Fig. 9, 11).

will be discussed in greater detail to illustrate important structural and sedimentological characteristics.

Core M54-97-2 is located ~ 180 m WNW of the pinnacle, close to the inferred location of the fault (Fig. 9, 11). The 3.83 m long sedimentary succession is characterized by alternating brownish and grayish units of varying thickness and abundant allochtonous and authigenic carbonate concretions and vent fauna debris. The upper 105 cm of core M54-97-2 consist of heavily bioturbated hemipelagic and turbidite-influenced sediment with a dominantly silty clay lithology. This uppermost unit is characterized by high magnetic susceptibility values, moderate wet bulk densities averaging ~1.49 g/cm^3, low lightness and comparably high a* values (Fig. 12, 13). The contact to the underlying 12 cm thick lighter colored chaotic unit (~1.0-1.12 mbsf) with a distinct lower magnetic susceptibility is rough but interpreted to be depositional. Except of a stratified slope layer (1.15-1.27 mbsf, Fig. 12, 13), this clay and carbonate rich, gray colored unit with low magnetic susceptibility values, and only slightly higher densities (~1.54 g/cm^3) continues down to 3.38 mbsf. We interpret this unit as a mud flow sequence containing exotic reworked material. The interval 3.38-3.54 mbsf contains unsorted and chaotically oriented mud supported carbonate and shell fragments interpreted as vent fauna debris. This debris flow deposits rest unconformably (= erosional surface) on a upward fining 13 cm-thick sand unit of turbiditic origin with high magnetic susceptibility values in the core log (Fig. 12, 13). The lowermost part of the core 3.67-3.83 is again part of a flow sequence.

A summary plot of all core logging data, index properties, color- and shear strength values of core M54-97-2 is given in Fig. 13. The low and near

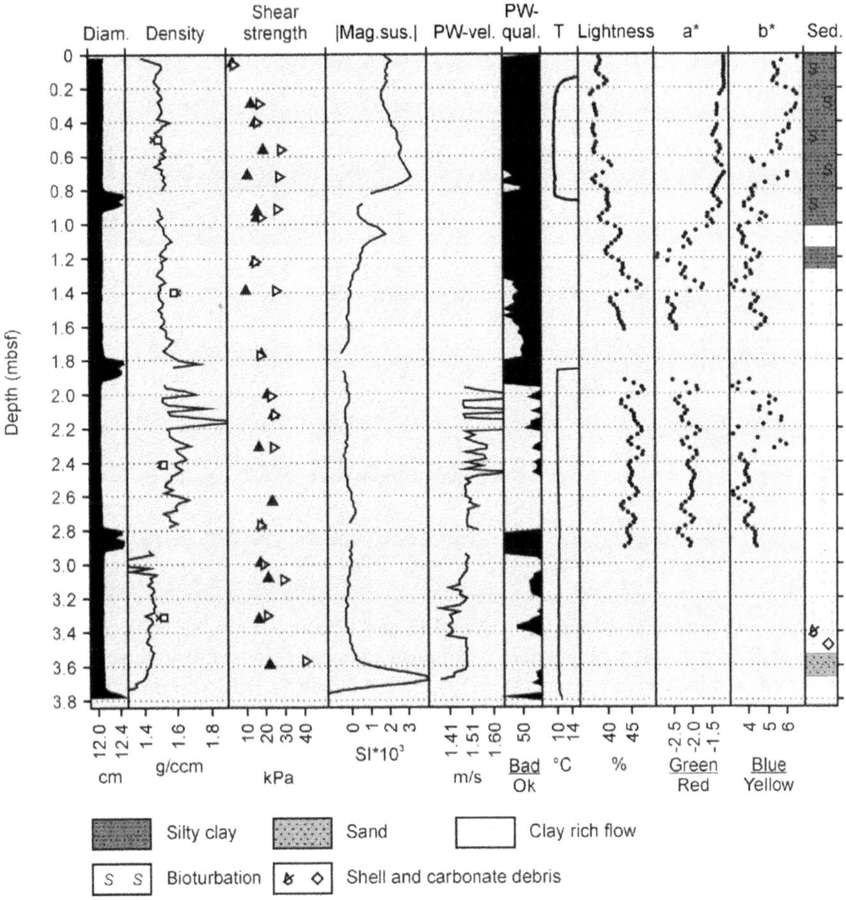

Figure 13. Logging data and physical property collection of gravity core M54-97/2. Hemipelagic and turbiditic core sections show high suszeptibilies and are easily distinguishable from mud flow units. Density column (wet bulk density); line: MSCL-data, square box: density from wet equal volume samples, cross: recalculated wet bulk density from grain densities. Shear strength column; filled triangles: rotary shear vane, open triangles: fall cone test.

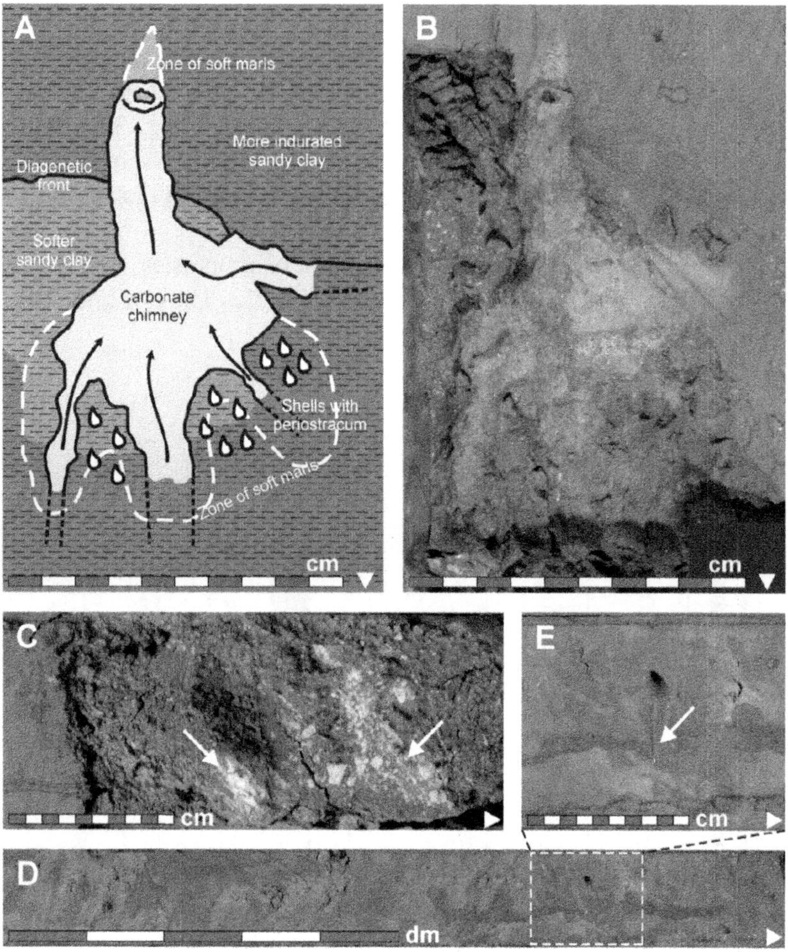

Figure 14 . Structural details from Mound 11 and 12 cores displaying vertical material ascent. **A,B**, sketch and photograph of a carbonate chimney (M54-84, 0.80 mbsf). **C**, rotated block of disseminated gas hydrate (M54-109, 3.70-4.00 mbsf). **D,E**, Section of vertical mud mixing and a conduit filled with low viscous mud (M 54-109, 0.53-1.53 mbsf).

uniform values for the undrained shear strength (~20 kPa) with no contrasts between extruded and hemipelagic slope sections are in sharp contrast to the high shear strength values (> 100 kPa) of extruded material from Mound Culebra core M54-18 (Fig. 6).

Other good indications for mud flows in combination with reworked vent fauna material have been found in cores M54-92, -94, -100 and -102 (Fig. 9, 11). In contrast, cores M54-84 and -109 are proximal to the mud source with no interbedded hemipelagic slope sediment, but abundant vertical structural elements like calcified chimneys (core M54-84; Fig. 14A, B) or vertical mud mixing (Mound 11, core M54-109; Fig. 14D, E). Core M54-

109 of Mound 11 is also a site of gas hydrate recovery from 2.18 mbsf and elevated free gas concentrations (Fig. 14C; Schmidt et al., submitted).

An approach to show the cored distribution of flows in contrast to proximal cores with indication for vertical material flux is shown in Figure 11. From the information available flows seem to originate from topographically higher areas on the NW and SE side of Mound 11 and 12 along the proposed fault intersection. No indication of flows have been found on the NE flank of Mound 12.

4. DISCUSSION AND CONCEPTUAL MOUND MODELS

4.1 Morphology

Mound Culebra (smooth domain) and Mound 12 (rough domain) show major differences in their morphological surface expression. In general the fluid content (i.e. viscosity) and the stress and strain history of the extruded material primarily control the surface morphology of extrusions (Brown, 1990). Using a simple first order model (Azizi, 1999), the relationship of shear strength and arising stable slope angles is explained. A slope segment of trapezoid shape will fail or deform when the ratio F of the material shear strength (τ_f) and the mobilizing shear forces (τ_{mob}) are <1.

$$F = \frac{\tau_f}{\tau_{mob}} \tag{1}$$

The acting shear force can be derived from the specific gravity of the segment, the slope angle β and the height of the segment (Fig. 17A).

$$\tau_{mob} = 1 \times h \times \cos\beta \times \gamma \times \sin\beta \tag{2}$$

Whereas τ_f is known and equal to the undrained shear strength determined shortly after retrieval of the sediment cores on deck.

$$\tau_f = c_u \tag{3}$$

F resolves now to

$$F = \frac{c_u}{2 \times h \times \gamma \times \frac{\sin 2\beta}{2}} \tag{4}$$

$F = 1$ at the point of failure and β can be expressed as

$$\beta = \frac{\arcsin\left(\frac{c_u}{2 \times h \times \gamma}\right)}{2} \tag{5}$$

Observed slope angles and undrained shear strength values of the two mounds together with the failure criterion (Eq. 5) are shown in Figure 15.

For a comparison we excluded measurements in diagenetically altered zones (M54-97) and measurements in conduit-rich sections (M54-18, -21). Slope angles are taken from the NW and SE of the mounds that are thought to be the least influenced by masswasting processes of the continental slope. Extruded material (clays with scaly fabric) of Mound Culebra is 5 times more shear resistant than values from mud flows of Mound 12. Looking at mound slope angles Mound Culebra has up to 4 times steeper slopes as compared to Mound 12. The resulting values for the height of the failed segment seems to be in a reasonable range for Mound Culebra. For a 115 m-high mound material failure and related slope adjustment below a 8-16 m-thick segments seems plausible. Failure likely occurs along clast boundaries or fluid channels where the shear strength is reduced by 60-90% compared to adjacent mud clasts. Measured shear strength values on extruded structurally contiguous clasts of Mound Culebra are probably lower than shortly after emplacement due to fabric relaxation. In the case of Mound 12 with a height of 30 m and observed mud flow thickness of ~2 m, *h* values between 6 and 16 m are unlikely. Here the extruded and flowed out material is completely remodeled and an initially much higher water content was lost during compaction after flow deposition. The measured undrained shear strength values of the paleoflow, now covered by 1.1 m hemipelagic sediment are therefore today much higher than during the flow event. Realistic heights of the failed segment should be in the range of the flow thickness of around 2 to 3 m resulting in shear strength values ~5 kPa during the mud flow event.

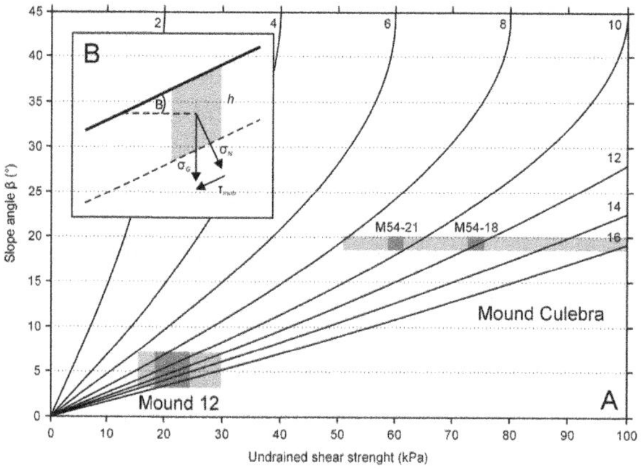

Figure 15. Comparison of slope angles and undrained shear strength values from Mound 12 and Mound Culebra with the slope failure criterion (Eq. 5). Curved lines indicate the stability limit for various thicknesses of the failing slope segment. For a given segment height *h* the slope fails toward higher slope angles and lower shear strength. White boxes indicate the range of observed values in the extruded material, outliers and conduit values are excluded. Dark boxes indicate average values of extruded material (M54-97: flow unit, M54-18, -21: unit III below 3.5 mbsf).

The narrow range of observed slope angles for Mound Culebra point toward a uniform mode of material ascend at least for the exposed cone. Wider ranges of observed slope angles for Mound 12 and especially the differences between the pinnacle (~10°) and the ridge point toward a multiphase extrusion history with the possible initial extrusion of higher viscous mud breccia followed by a mud flow dominated phase.

4.2 Mound Culebra model

The occurrence of mud breccia and deformed clays with only an insignificant hemipelagic cover in the central depressions of Mound Culebra are interpreted as the most active parts of the mound in terms of material uprise. Despite abundant indicators for fluid ascent and flow (conduits) no vent fauna or authigenic carbonates have formed yet. Overcompacted and deformed clays with scaly fabric, and clay breccias may be the main constituents and products of the mound in absence of flows. High shear strength values in combination with high slope angles support this assumption. Brown (1990) stresses the importance of this fabric in terms of preferential fluid pathways.

So far we do not have a stratigraphic control on the source depth of the ascending material. Basic, conservative isostatic equilibrium calculations following a model of Murton and Biggs (2003) indicate source depths of 600-800 m for the mud breccia of Mound Culebra. These numbers are in agreement with preliminary results from an ongoing carbon maturity study on mud breccia of Mound Culebra. Both findings are comparable to diapir source depths imaged by Shipley et al. (1992) for the lower and mid slope.

Fluids low viscous mud slurry and basaltic clasts however may have a different, deeper origin. Based on core observations it is likely that fluid and mud ascent at Mound Culebra are temporarily decoupled. Fluid and low viscous conduits develop after an inferred bulk uprise of the mud breccia. This suggest a hierarchical and episodic scenario. Initial emplacement of new mud breccia during the peak of activity is followed by hydrofracturing of succeeding fluids and terminates in stable conduits which maintain venting for longer time periods. Setting the stage for vent fauna growth authigenic carbonate formation and eventually self sealing of the conduits due to precipitation. This implies that areas now covered by vent fauna and authigenic carbonate have underlying conduit systems (core M54-27, Fig. 10). Comparable, highly advective conduits of 1 cm^2 effective diameter are described from Barbados diatremes by Henry et al. (1996). Slightly larger advective vent structures are found on top of mud diapirs from the Mediterranean Ridge (Cita et al., 1989). Local circulatory cells as proposed by Henry et al. (1996) may also play a role and affect the fluid composition since the observed structural pathways in combination with the morphological obstruction of the mound into the bottom water flow regime may induce secondary hydraulic gradients that temporarily sustain shallow seawater circulation. These effects, however, are not responsible for the primary formation of the observed fluid conduits and may be disrupted by deep-seated venting.

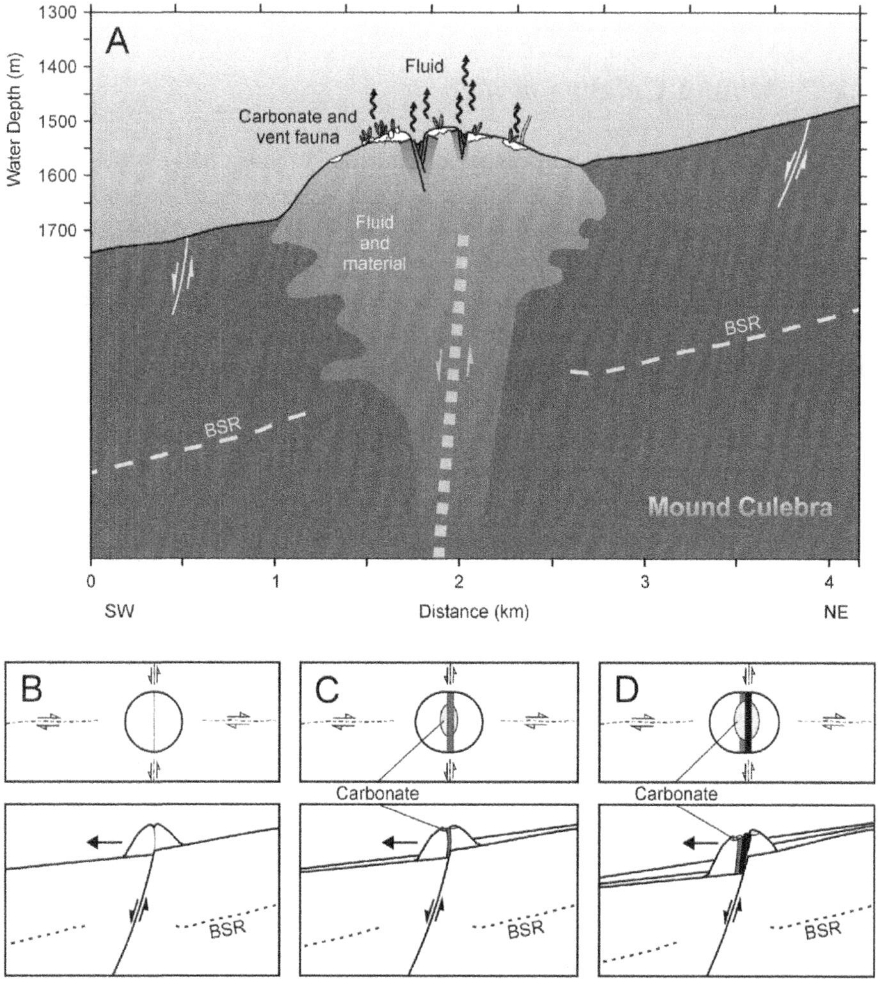

Figure 16. Conceptual mound model of Mound Culebra. **A**, schematic cross-section through the fault controlled breccia cone of the diapir. The approximate location of the profile is indicated with arrows in Fig. 9. **B-D**, Comics to illustrate tectonic controlled sheet like intrusion of mud breccia and related lateral growth of the mound.

Gas hydrates do not seem to play a major role below the central part of Mound Culebra despite a geophysically well documented regional gas hydrate occurrence (Pecher et al., 2001). Preservation of dedicated structural fabric in the sub mm scale like slickensides on micro-faults are inconsistent with hydrate growth in the pore space or along fractures.

In summary, Mound Culebra is interpreted as a diapiric breccia dome controlled by a underlying deep reaching extensional growth fault or faults with a likely seaward dip. A deep penetration of the faults at least into the margin wedge is inferred from sparse basaltic fragments in the extruded breccia. Successive movements along the faults created locally additional vertical accommodation space for slope sediment (Line 8, SO-81, Fig. 1; Meschede, 1999). The tectonic, episodic trigger on the additional accommodation space caused extreme temporal variations in the local sedimentation rate and consolidation history. This sets the base for locally underconsolidated sediment packages with a positive buoyancy that later in the slope development started to rise, preferably along the tectonically existing pathways.

The along strike positioning of the mounds in the smooth domain is likely controlled by trench normal tectonic elements. The surface expression of such an element is seen in TOBI data NW of Mound Culebra (Fig. 2A). Faults normal to the trench may originate from long wavelength deformation and folding of the entire margin wedge e.g. related to the uplift and caused by the Cocos Ridge subduction (Vannucchi et al., in press), or due differences in extension following subduction erosion.

The geodynamic evolution of the exposed part of the mud diapir may have stated initially as a cone like structure growing rapidly due to its buoyancy (Fig. 16B). Subsequent movements along the normal fault causes segmentation of the cone and a temporal pressure release of the mud breccia column. This pressure release initiates rise of mud breccia that subsequently fills the emerged fracture (Fig. 16C). Extreme pressure sensitivity is also known from the Azerbaijan mud volcanoes. Eventually as a result of repeated extensional faulting and infilling of rising material the mound will get an oval shape in map view with the long axis normal to the strike of the fault (Fig. 16D). Besides explaining the map shape of the mound the proposed model has several other implications. The model predicts an absolute seaward movement of the SW half of the mound and relative landward movement of the mound center. A slightly listric shape of the fault would cause a landward tilt of the down-thrown part of the mound. The proposed conceptual model implies also a relative landward shift of the most actively venting part of the mound which may be combined with a increasing calcification of the tops of older mud breccia intrusion (Fig. 16D). Small-scale fracture-like depressions north of the main depression may indicate a landward shift in activity. Those small depressions might in future widen up and could be infilled by a rising mud breccia body. In case of an erroneous inferred dip direction of the normal fault the described movements would be reversed.

4.3 Mound 12 model

Based on the widespread occurrence of mud flows in cores Mound 11 and 12 are interpreted as mud volcanoes. Again the mound location and mud uprise is controlled by a trench parallel fault of uncertain orientation and movement sense. Cores proximal to the inferred fault show vertical mixing of different mud bodies and fluid conduits and chimneys. The described flow sequence of core M54-97 with interfingered slope sediment implies frequent mud eruptions and phases of quiescence when only venting might occur. The described vent fauna and carbonate debris on the base of a > 2 m thick mud flow unit gives important hints regarding eruption mechanism and driving forces. Constituents of the debris once formed an active vent community on a dormant though actively venting mound. A high energy event caused brecciated and chaotic relocation of shells and carbonate pieces, shortly afterward (no hemipelagic sediment in-between) a mud flow emerged and covered the debris. We suggest a violent, gas-driven mechanism to explain the observed sedimentary record.

Gas hydrates occur in shallow depth (~ 2 m) below Mound 11 and likely also below Mound 12, although they had not been recovered when sampling the latter. Based on a model by Reed et al. (1990), all material extruded must either pass through this zone of hydrate or be derived from within it. Gas hydrates may create temporal barriers for uprising fluids and muds. Rupture of a hydrate seal within a mud volcano may be caused by increased temperatures initiated by the release of warmer pore fluids, tectonic movements, or likely a combination of both. Once the hydrate seal is broken, pore fluid and mud pressure will suddenly decrease leading to the rapid expansion or even release of free methane (Reed et al., 1990). Hydrate formation and decomposition in a moving body of muddy sediments of Mound 11 and 12 may also be held responsible for the complete destruction of primary tectonic or sedimentary fabrics of the extruded material, then leading to a tooth past appearance. Hydrate growth and decay cycles within the ascending mud column may possible aid in the mixing process of solid and fluid phase and leads to low viscosity, water-rich flows as inferred from morphological considerations and shear strength measurements. Again, there is no stratigraphic control on the source depth of the rising material at Mounds 11 and 12. Using the same approach as for Mound Culebra we derive a depth of ~230 mbsf. This depth falls within the regional hydrate stability field of around 275 mbsf in 100 m of water depth (Pecher et al., 2001). The derived depth value may not be considered the source depth of fluids or solids, but rather the mixing depth of solid and fluid phases leading to the buoyancy contrasts that initiated the mud ascent.

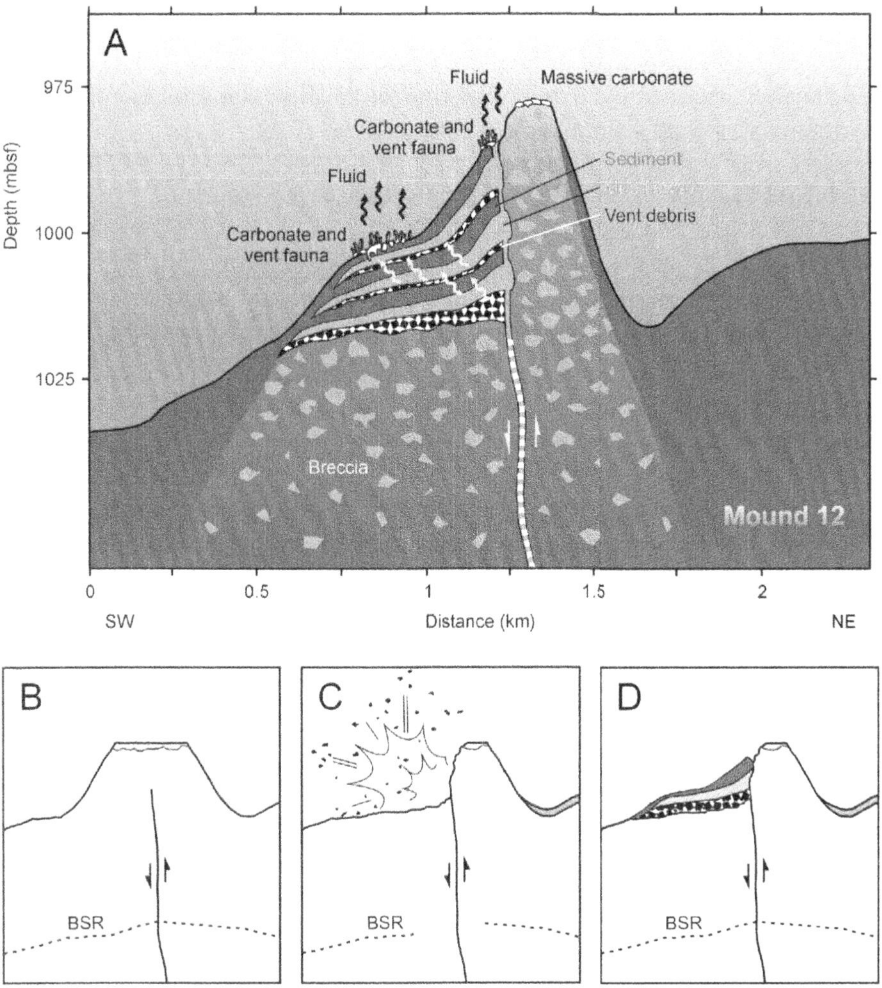

Figure 17. Conceptual mound model of Mound 12. **A**, Schematic cross-section through the fault controlled mud volcano and the older NE pinnacle. **B-D**, failure of the seaward half, formation of a basal carbonate and shell debris eruption layer that subsequently gets covered by mud flows and hemipelagic sediments.

Based on data and observations we suggest the following mound model (Fig. 17). Mound 12 started off as a fault controlled diapiric cone with slope angles of around 10° made up of mud breccia or high viscous breccia rich flows (Fig. 17B). Tectonic movements or an initial violent eruption either dislocated or removed part of the seaward side of the original mound (Fig. 17C). A succession of mud flow events and phases of hemipelagic sedimentation triggered by sealing and rupture of a hydrate seal form on the remnants of the old cone. Vent fauna and authigenic carbonates are stripped from the region above the feeding part when a new mud flow cycle follows a quiescent period. Some of the venting found on the lower SW flank of Mound 12 may be sustained by slow dewatering of the flow deposits themselves or hydraulic short cuts from the main conduit (Fig. 17A).

5. SUMMARY AND CONCLUSIONS

Using structural observations, logging and single sample physical property data in combination with side scan and bathymetrical information we described and characterized two different styles of mud extrusion along the Middle American margin. All described mounds are fault-controlled, and massive carbonates observed on both mound types are limited to the crestal area of the mounds. They do not represent a significant volumetric constituent of the mounds nor explain the mound's origin.

The Culebra style diapirism is characteristic for the smooth northern domain resulting in larger mounds with steep flanks consisting of overcompacted mud breccia with scaly fabrics and a high density of secondary fluid conduits. This observation attests effective decoupling of mud and fluid migration during the quiescent periods. The productivity of the cm scale diameter and meter long conduits is highly variable and ranges from clear water or mud slurry to exotic clasts. Due to relaxation the measured shear strength values on extruded material is probably now lower than shortly after extrusion. Our conceptual model explains the oval shape of the mound and slightly differing slope angles by successive mud breccia emplacement through the underlying normal fault. The height of the mound is isostatically controlled and source depth estimates are in the range of 600-700 mbsf. Diapirism is hence routed in the upper plate slope sediment apron and likely caused by extreme temporal variations in the local sedimentation rate and consolidation history during extensional tectonics following the onset of subduction erosion (e.g. Meschede, 1999).

The Mound 11 and 12 style mud volcanism is characteristic for the rough southern domain resulting in comparably small mounds with low-angle flanks made up of mud flow successions alternating with slope sediments. Observed shear strength values of flow sequences indicate normal consolidation after flow deposition and point toward initially low viscosity flows. Fast buried vent debris and the presence of gas hydrates in the shallow subsurface favor a cyclic and gas driven eruption behavior model. The extruded material lacks a coherent relationship between clay matrix, fluids and clasts so that mixing processes at the depth of the gas hydrate stability zone is inferred.

The presented study indicates that structural differences along the margin strongly control the forearc dewatering and style of diapirism.

ACKNOWLEDGEMENTS

We gratefully acknowledge the support of captain H. Papenhagen and the crew of RV Meteor during Leg M54-2. Many thanks also to the scientist of precursor cruises (SO144, SO163, and M54-1) that set the stage and data base for the presented data. We especially acknowledge the enthusiasm and persistence of Klaus Wallmann in coring the Mounds. The ideas of the paper gained significantly from many fruitful discussions with SFB members Susan Mau, Ernst Flueh, and Christian Hensen. Emmanuel Soeding helped in extracting bathymetric data and provided essential location maps with representations of all stations. This publication is contribution no. 36 of the Sonderforschungsbereich 574 "Volatiles and Fluids in Subduction Zones" at the University of Kiel.

REFERENCES

1. Azizi, F., 1999. *Applied Analyses in Geotechnics*, E & FN Spon, London, New York, 1999.
2. Bohrmann, G., Heeschen, K., Jung, C., Weinrebe, Baranov, W.B., Cailleau,B., Heath, R., Hühnerbach, V., Hort, M., Masson, D., and Trummer, I., 2002. Widespread fluid expulsion along the seafloor of the Costa Rica convergent margin, *Terra Nova*, **14**, 69-79.
3. Brown, K.M., 1990. The Nature and Hydrogeologic Significance of Mud Diapirs and Diatremes for Accretionary Systems, *J. Geophys. Res.*, **96**, 8969-8982.
4. Cita, M.B.,Camerlenghi, A., Erba, E., F.M. McCoy, F.M., Castradory, D., Cazzani, A., Guasti, G., Gianbastiani, M., Lucci, R., Nolli, V., Pezzi, G., Redaelli, M., Rizzi, E., Torricelli, S. and Violanti, D., 1989. Discovery of mud diapirism on the Mediterranean Ridge. A preliminary report, *Boll. Soc. Geol. It.*, **108**, 537-543.
5. DeMets, C., Gordon, R.G., Argus, D.F., and Stein, S., 1994. Effect of recent revision to the geomagnetic reversal timescale on estimates of current plate motions, *Geophys. Res. Lett.*, **21**, 2191-2194.
6. Grevemeyer, I., Kopf, A. J., Kaul, N., Fekete, N., Villinger, H., Heesemann, M., Gennerich, M. Müller, H.-H., Wallmann, K., and Weinrebe, W., 2004. Fluid flow through active mud dome Mound Culebra offshore Nicoya Peninsula, Costa Rica: evidence from heat flow surveying, *Mar. Geol.*, **207**, 145-157.
7. Henry, P., Le Pichon, X., Lallemant, S., Lance, S., Martin, J.B., Foucher, J.P., Fiala-Médioni, A., Rostek, F., Guilhaumou, N., Pranal, V., and Castrec,M., 1996. Fluid flow in and around a mud volcano field seaward of the Barbados accretionary wedge: Results from Manon cruise, *J. Geophys. Res.*, **101**, 20,297-20,323.
8. Mau, S., Sahling, H., Rehder, G., Soeding,E., Masson, D., and. Suess, E. Seafloor observations and methane plumes of mud extrusions along the erosive convergent margin of Costa Rica, *Mar. Geol.*, in prep.
9. Meschede, M., Zweigel, P., Frisch, W., and Völker, D., 1999. Mélange formation by subduction erosion: the case of the Osa mélange in southern Costa Rica, *Terra Nova*, **11**, 141-148.
10. Murton, B.J. and Biggs, J., 2003. Numerical modelling of mud volcanoes and their flows

using constraints from the Gulf of Cadiz, *Mar. Geol.*, **195**, 223-236.

11. Pecher, I.A., Kukowski, N., Huebscher, C., Greinert, J., Bialas, J., and GEOPEG Working Group, 2001. The link between bottom-simulating reflections and methane flux into the gas hydrate stability zone – new evidence from Lima Basin, Peru Margin, *Earth Planetary Sci. Lett.*, **185**, 343-354.
12. Reed, D.L., Silver, E.A, Tagudin, J.E., Shipley, T.H., and Vrolijk, P., 1990. Relations between mud volcanoes, thrust deformation, slope sedimentation, and gas hydrate, offshore north Panama, *Mar. Petrol. Geol.*, **7**, 44-54.
13. Shipley, T.H., McIntosh, K.D., Silver, E.A., and Stoffa, P.L., 1992. Three-Dimensional Seismic Imaging of the Costa Rica Accretionary Prism: Structural Diversity in a Small Volume of the Lower Slope, *J. Geophys. Res.*, **97**, 4439-4459.
14. Schmidt, M., Hensen, C., Mörz, T., Müller, C., Grevemeyer, I., Kaul, N., Wallmann, K., Mau, S., and Brückmann, W. Shallow surface methane hydrate accumulation related to mud volcanism at „Mound 11" (Costa Rica forearc), *Earth Planetary Sci. Lett.*, submitted.
15. Soeding, E., Wallmann, K., Suess, E., and Flueh, E., 2003. RV METEOR, Cruise Report M54/2+3. Fluids and Subduction, Costa Rica 2002, in *GEOMAR Rep., 111*, GEOMAR, Kiel.
16. Vannucchi, P., Ranero, C.R., Galeotti, S., Straub, S.M., Scholl, D.W., and McDougall-Ried, K., 2003. Fast rates of subduction erosion along the Costa Rica Pacific margin: Implications for non-steady rates of crustal recycling at subduction zones, *J. Geophys. Res.*, 108, n. B11, 2511.
17. Von Huene, R., Ranero, C.R., Weinrebe, W., and Hinz, K., 2000. Quaternary convergent margin tectonics of Costa Rica, segmentation of the Cocos Plate, and Central American volcanism, *Tectonics*, **19**, 314-334.
18. Weinrebe, W., and Flueh, E., 2002. FS/RV SONNE, Cruise Report SO163. Subduction 1, Costa Rica 2002, in *GEOMAR Rep., 106*, GEOMAR, Kiel.

Chapter 2
GEODYNAMIC IMPLICATIONS OF MUD VOLCANISM

MUD VOLCANOES AND SEISMICITY IN ROMANIA

Calin Baciu[1], Giuseppe Etiope[2]

1 Babes-Bolyai University, Department of Geology & Institute for Interdisciplinary Experimental Research, M. Kogalniceanu str. 1, 3400 Cluj-Napoca, Romania

2 Istituto Nazionale di Geofisica e Vulcanologia, via Vigna Murata 605, 00143 Rome, Italy

Abstract: Due to the occurrence of mud volcanoes and active seismic areas in close proximity in Romania, good conditions can be found for studying the relationship between the two phenomena. The most important earthquakes are generated within the Vrancea zone, related to a lithospheric fragment sunken into the astenosphere. Strong seismic shocks can be accompanied by an intensification of the mud volcanic activity, proving the sensitivity of the fluid reservoirs to strain variations in the shallow crust. Natural gas seeps and mud volcanoes are distributed over wide areas in Romania, and the monitoring of their activity in some selected points should give a better knowledge of the deformation field related to earthquakes. Mud volcano activity is strongly influenced by external factors, which alter the results of the measurements by introducing a high level of noise, which can be avoided by implementing a complex monitoring system.

Key words: mud volcanoes, seismicity, Romania

1. INTRODUCTION

The relationship between mud volcanism and seismicity is generally accepted (e.g., Aliyev, Guliev & Panahi, 2000), although most of the studies are descriptive and specific geophysical and/or geochemical monitoring is rare or at an early stage. While it is clear that seismicity can enhance degassing processes (Gold & Soter, 1985) and mud volcanic activity (e.g., Aliyev, Guliyev & Belov, 2002), there is very little evidence and reliable, systematic investigations as to the reverse process, that is mud volcanic activity as a seismic precursor. In order to fully understand this link, the first step lies in the knowledge, for a given mud volcanic area, of the main seismic

G. Martinelli and B. Panahi (eds.), Mud Volcanoes, Geodynamics and Seismicity, 77–87

parameters (hypocenter depth, focal mechanism, released energy), the deep tectonic structure and fluid reservoirs; indeed, it is essential to define the spatial correlation between seismogenetic structures and the fluid-bearing faults.

In Romania both seismicity and fluid reservoir characteristics are well known due to seismic monitoring, on the one hand, and the oil and gas exploration and exploitation activities, on the other. Therefore, it is possible to draw a realistic picture of the geodynamic and deep fluid circulation settings. Accordingly, theoretical limitations for understanding the mud volcanism-seismicity link are summarised and possible work prospects for detailed investigations and monitoring activities are defined.

2. GENERAL OUTLOOK ON THE GEOTECTONICS OF ROMANIA

Most of the authors accept the presence on the Romanian territory of parts of three major geotectonic units (fig. 1): the East European Plate (EEP), the Moesian micro-plate (MP) and the Intra-Alpine micro-plate (IAP).

The EEP is a very stable and thick lithospheric segment, consolidated during the Precambrian. Its south-western part (the Moldavian Platform) is making the basement of north-eastern Romania. EEP and IAP are separated by a compression limit – the Tornquist-Teisseyre zone. To the south, the EEP is in contact with the Paleozoic area of the Moesian micro-plate. The limit between the two units is an active dextral strike-slip fault, the Peceneaga-Camena Fault. A contact of the same nature – the Trans-Getica Fault – represents the limit between MP and IAP. The triple junction of the plates seems to be in the Vrancea region, characterized by active geodynamic conditions.

The folded units delimited in Romania, were generally formed during Alpine orogenesis. A distinct segment of the alpine domain is the North-Dobrudja Orogene, situated between the Moldavian and the Moesian platforms. Its outcrop area on land is very limited, but extends on the shelf of the Black Sea, continuing with the alpine Crimean orogene. The most important orogene, the Carpathian belt, covers more than half of Romania's surface, and is superposed upon the IAP.

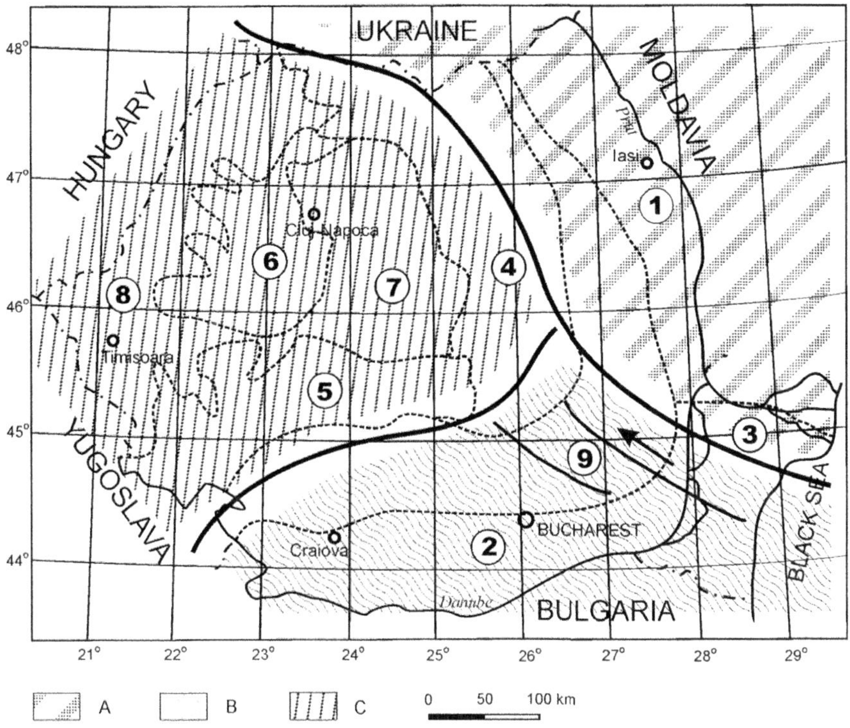

Figure 1. Main structural units in Romania. A – East European Plate; B – Moesian microplate; C – Intra-Alpine microplate; 1 – Moldavian Platform; 2 – Moesian Platform; 3 – North-Dobrudja Orogene; 4 – Eastern Carpathians; 5 – Southern Carpathians; 6 – Apuseni Mountains; 7 – Transylvanian Depression; 8 – Pannonian Depression; 9 – Carpathian Foredeep.

Its structure is very complicated, being one of the most complex segments of the European alpine belt. The mountainous chain consists of three branches: Eastern Carpathians, Southern Carpathians and Apuseni Mountains. The first two units are parts of the deformed European continental margin. The northern part of the Apuseni Mountains belongs to the Fore-Apulia micro-continent (Sandulescu, 2003). The southern segment of the Apuseni Mountains, and also a part of the Transylvanian Depression basement and some units of the Eastern Carpathians (Pienides), containing ophiolitic complexes and flysch deposits, is associated to the Main Tethyan Suture Zone. The post-tectonic low areas of the Transylvanian and Pannonian Depressions are filled with Tertiary sediments. Two compression stages which contributed to the formation of the Carpathian realm were distinguished: Cretaceous and Miocene (Sandulescu, 1984).

In its external part, the Carpathian chain is bordered by thick molassic deposits belonging to the foredeep.

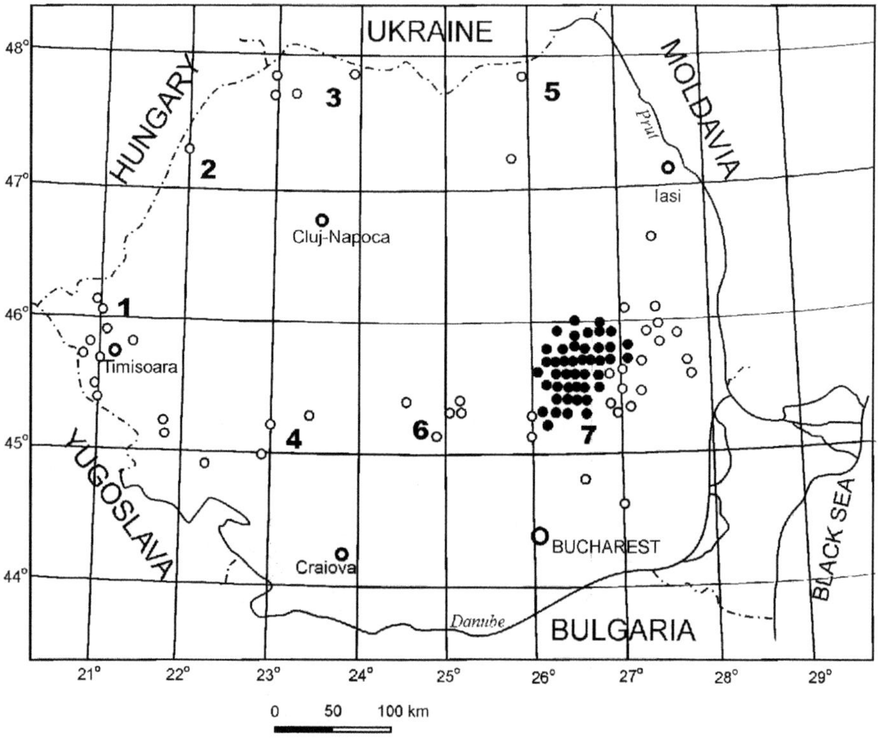

Figure 2. Epicentre distribution in Romania, M > 4, time period 1901-1979 (after Balan et al, 1982). Open circles – normal depth; black dots – intermediate depth; 1 – Banat; 2 – Crisana; 3 – Maramures; 4 – Oltenia; 5 – Bucovina; 6 – Fagaras; 7 – Vrancea.

3. SEISMIC AREAS IN ROMANIA

Romania can be considered a territory with a moderate seismic rate, characterized by the presence of more areas of normal depth earthquakes and one limited zone with intermediate depth hypocenters (fig. 2).

3.1 Normal depth seismogenic zones

The normal earthquakes are produced by intracrustal foci (generally 5 to 30 km deep). Some of them can liberate important amounts of energy, like in the Fagaras seismic zone, where magnitudes up to $M_{max} = 6.5$ were reached, with a frequency of one event per century. For the other zones, the magnitude does not exceed $M_{max} = 5.5$. Very often, the normal earthquakes are manifested on limited areas, but in some of the cases, with high intensities in the epicentral zone. Such events were recorded in Banat, Crisana, Maramures, northern and western Oltenia, Bucovina, etc.. Generally, they are located on deep tectonic lines. As an example, the Maramures seismic zone should be related to the contact between EEP and IAP. In western Romania, the seismic zones Banat and Crisana are supposed to be connected to a limit between two sectors (Pannonian and Transylvanian) of the Intra-alpine micro-plate. The energy liberated by the above mentioned normal-depth seismic zones accounts for less than 3% of the total amount calculated for Romania. The remaining energy is generated by Vrancea intermediate depth hypocenters.

3.2 Vrancea zone

The most important earthquake-prone area in Romania is Vrancea, where intermediate-depth seismic events, with moment magnitudes up to 7.7 were recorded. This is a very distinct seismic area, in regard to its isolation and seismic patterns, being comparable only to Bucaramanga (Columbia – South America) and Hindu-Kush area, in Central Asia (Gutenberg & Richter, 1965).

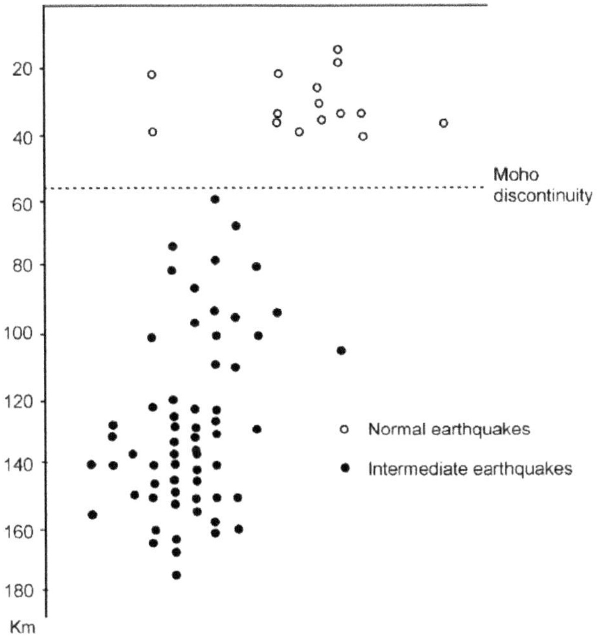

Figure 3. Hypocentre distribution on a NW-SE vertical section in Vrancea (after Constantinescu & Enescu, 1985).

Some arc-like features, deduced from its general geological structure, can express similarities to the Calabrian-Sicilian Arc from the Tyrrhenian Sea (Morelli et al, 1975). The Vrancea region is one of the most active seismic zones in Europe. Four major events, with moment magnitudes higher than $M_w = 6.9$ were recorded in the period 1940-1990. According to the seismic interpretations, the majority of the hypocenters of the intermediate earthquakes are confined within a very steep or almost vertical columns, with a 30×70 km section and extending from 70-80 km to about 180-200 km in depth. The hypocentral volume corresponds to a high velocity body, which extends down to 400 km, as shown by seismic tomography. However, the seismic shocks rarely occur below 200 km depth. The intermediate-depth events occur due to active faulting within the high velocity sunken body.

Above this volume, between 20 and 40 km in depth, an intracrustal seismic zone is developing. The two hypocentral domains are separated by a seismic gap, which was evidenced in the interval 45 to 70-80 km (fig. 3). The normal-depth seismic area is shifted towards the east relative to the intermediate zone, and only generates low energy earthquakes ($M_{max} < 5.25$).

The theories that seek to explain the origin of the intermediate earthquakes involve two geodynamic models (Diaconescu et al, 2001). The first admits the subduction of a remnant oceanic slab, and the second is based on the hypothesis of delamination of a thickened lithospheric root. Most of the subduction theories accept the detachment and lateral migration of an oceanic slab, but also the roll-back of the slab, orthogonal to the direction of inferred subduction, can be admitted (Knapp et al, 2001). The last remains of the oceanic plate were consumed about 8-10 Ma ago, when subduction ceased consequent to the collision of the two continental plates.

The strong earthquakes trigger a "fault reactivation process", which migrates from the epicentral zone as far as several hundreds of kilometres. Consequent to the major 1977 event, the seismic activity migrated with velocities in the order of $km \cdot min^{-1}$ in the foredeep sediments, $km \cdot h^{-1}$ within the Carpathian belt and $km \cdot day^{-1}$ towards Bucharest (Balan et al, 1982).

In Vrancea region, the fault plane solutions indicate that the compressive stress axis orientation alternates between a SE – NW direction and a SW – NE direction. During the stress manifestation in the first direction, seismic events with $M_{max} > 6.0$ were recorded. The second direction is characterized by weaker shocks, with $M_{max} < 6.0$. On a statistical basis, it was inferred that each period of stress orientation had a duration of 17 to 21 years (Constantinescu & Enescu, 1985). The historical study of the seismic events evidenced a general rate of three major events per century, with $M_{max} > 7.0$, during the last 600 years.

The specific focal depth of the earthquakes originating in Vrancea makes the observation of some of the most conclusive forerunners, like electromagnetic, geomagnetic and geoelectric phenomena, velocity variations of the longitudinal and transversal waves, etc. very difficult (Constantinescu & Enescu, 1985; Enescu, Enescu & Constantin, 1999).

3.3 Geophysical phenomena associated to earthquakes

In the period of time close to the important seismic events, different types of associated phenomena were observed: changes in the atmospheric state transduced by electrical and light phenomena and changes related to the geological environment. The most complete available observation database is related to the major 1977 event. In very few cases, the analysis was based on data directly collected in the field by specialists. Generally, the information were obtained on the base of questionnaires filled in by citizens. Here we will discuss the geological phenomena alone. A first category is related to groundwater. Usually, the observations describe the water table variations. In some locations, the groundwater erupted from the soil at elevations from 0.5 up to 5 m. In the same areas, increases in water turbidity from domestic wells were observed.

In many cases, oil well yield variations were reported. The fissures related to the earthquake temporarily opened up some reservoirs of methane or carbon dioxide. Finally, changes in the mud volcano activities were noticed. This aspect will be discussed in more detail later on.

4. DISTRIBUTION AND CHARACTERISTICS OF MUD VOLCANOES IN ROMANIA

Extensive field research conducted during the 20th century lead to the identification of about 1,000 points of oil and natural gas seepages in Romania (fig. 4). Out of the total number of points, 73% represent oil springs and 27% are gas emanations. The seepages are generally related to low age formations: 68% to Neogene, 26% to Paleogene, and 6% to Cretaceous (Paraschiv, 1984). From a tectonic point of view, 98.5% of the seepages occur into the Carpathian folded domain and only 1.5% were found on the platforms. At the same time, 55% of the known hydrocarbon-bearing structures belong to the first domain and 45% to the second. This difference is mainly related to the tectonic stability of the platforms. It seems that the appropriate conditions for the existence of hydrocarbon seepages are fulfilled only in areas with active tectonics. The presence of plicative and/or disjunctive elements, namely faulted anticlines, domes, salt diapirs, faults, is an important prerequisite for the existence of mud volcanoes and other forms of gas emission. Paraschiv (1984) shown that the majority of the hydrocarbon seeps in Romania occur in areas affected by recent positive crust movement.

The most impressive mud volcanoes in Romania are located in the Berca-Arbanasi hydrocarbon-producing area. This structure roughly consists of an anticlinal fold, 20 km long, affected by a system of longitudinal faults, intersected by transversal fractures. Four mud volcanoes fields are active: Paclele Mari, Paclele Mici, Fierbatori and Beciu.

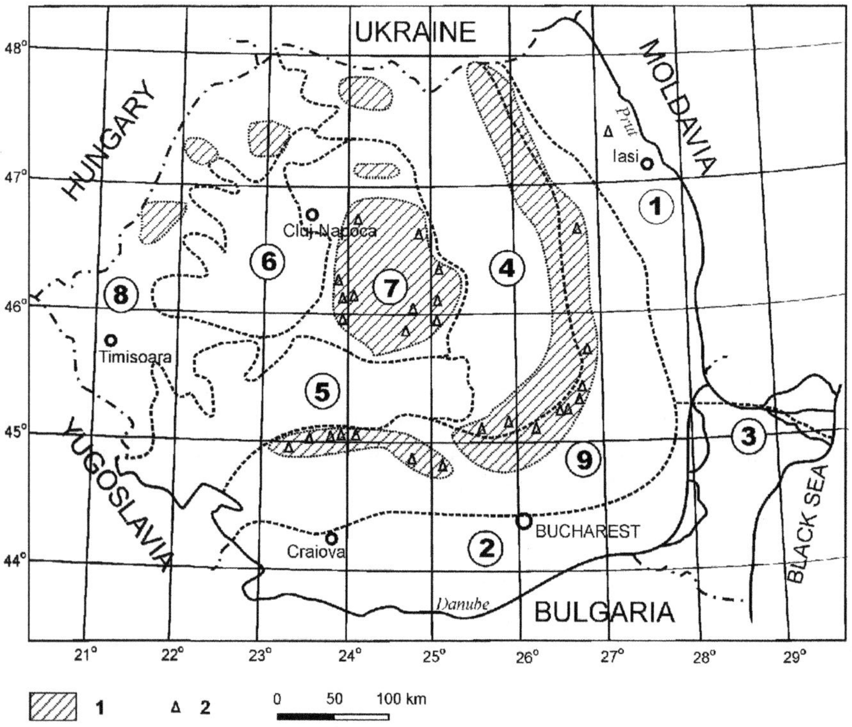

Figure 4. Distribution of mud volcanoes and hydrocarbon seeps in Romania. 1 – hydrocarbon seepage area; 2 – mud volcanoes; structural units identical to fig. 1 (after Paraschiv, 1984).

The more or less fluid mud is liberated through gryphons situated on top of conical constructions, up to 10 meters high, or circular pools filled with muddy water. The activity is quiescent most of the time, with rare explosive episodes. The total surface occupied by recent or ancient mud volcanic products is about 2.5 km^2, and the total methane output was estimated as being at least 1,200 tonnes per year (Etiope et al, 2004). Another interesting gas emission point is Andreiasu, about 50 km north of Berca. On a surface of 400 m^2, strong gas emissions produce everlasting fires and some small pools with bubbling water. A more detailed description of the peculiarities of Berca-Arbanasi and Andreiasu structures is given by Etiope et al. (2003b), Sencu (1985), Paraschiv (1984). Other less important oil and gas seeps are widespread in the whole area of the Carpathian Foredeep.

Lower flux emissions can be found in Transylvania and on the western side of the Carpathian Foredeep (Oltenia). Isolated and weak seeps were reported from the Pannonian Depression (Western Romania) and Moldavian Plateau (EEP). In Transylvania, the mud volcanoes are generally small and less active. In this region, a special mention should be made for Sarmasel everlasting fire, with very high fluxes, comparable to Andreiasu (Etiope et al, 2003a).

5. RELATIONS BETWEEN MUD VOLCANOES AND SEISMICITY

One can assume that a very important role for the occurrence and activity of mud volcanoes is played by seismicity, taking into account the fact that the major gas emissions in Romania are located in the vicinity of Vrancea earthquake-prone area. Seismic shocks can generate or reactivate deep fractures over the hydrocarbon reservoirs, which facilitate the upward migration of pressurised fluids.

Related to the most recent catastrophic seismic event on March 4, 1977, mud volcanoes appeared for the first time near Slobozia (South-Eastern Romania), giving indications as to the presence of natural gas accumulations in that region. At Berca, on the biggest mud volcanoes in Romania, the reactivation of the main longitudinal fault produced the appearance of new vents and extinguishing of other, previously active, ones. In the Beciu area, north of Berca, the earthquake triggered a mud eruption lasting for 6 hours. In the same area, important mudflows were previously observed, which are not directly connected to earthquakes (Sencu, 1985).

By consequence, mud volcano activity variations seem to be sensitive indicators for strain changes in the shallow crust. However, the practical use of this assumption, mainly as a predictive earthquake tool, is impeded by a series of collateral phenomena. Some of them are briefly emphasized in the following lines.

The weather conditions influence the mud volcano activity. The atmospheric pressure variations and the soil moisture as well can produce changes in soil degassing. As an empirical observation, the mud volcano activity becomes more intense during rainy periods. The general accepted explanation relates the increase of activity to the availability of water. In our interpretation, a non-negligible factor should be the diminution of gas-phase permeability of the sediments due to imbibition. Recent research (Etiope et al, 2002) stressed the importance of diffuse soil degassing versus localized emissions (vents and bubbling gases). Microseepage can be a few times more important than the macroseepage for a mud volcano area. Consequently to soil wetting and a gas permeability reduction, more gas will be driven by the reservoir pressure towards the vents, producing an enhanced activity.

In the case of explosive activity, there is a continuous and slow pressure increment, due to the reservoir energy. The eruption takes place when the equilibrium between the cumulated energy and the resistance of the sealing deposits is broken. This event can be triggered by a change in the strain field induced by an earthquake, but also by an external factor such as atmospheric condition variability or earth tides (Guliyev & Feizullayev, 1997).

6. CONCLUSIONS

The spatial proximity of mud volcanoes and earthquake-prone areas in Romania provides good conditions to study the link between the seismic process and the intensity of earth degassing. Due to the intermediate focal depth of Vrancea earthquakes. Forerunners are hard to distinguish by means of the usual instrumental methods. Mud volcanoes have proved to be sensitive strain indicators, and more attention should be paid to the possibility of interpreting their degassing rate variability as a seismic precursor.

External factors strongly influence mud volcanic activity, making the interpretation of the results very difficult. An appropriate monitoring system must be able to differentiate the effects of the strain field changes from the background and diminish the "noise" introduced by external factors. The installation of a complex system on a representative mud volcano field consisting of pressure transducers coupled with other sensors, measuring the physical and chemical fluid parameters, should provide the opportunity to better correlate mud volcano activity to seismic processes.

REFERENCES

1. Aliyev A. A., Guliev I., Panahi B., 2000, *Mud volcanoes hazards*. Nafta Press, 60 p., Baku.
2. Aliyev A. A., Guliyev I. S., Belov I. S., 2002, *Catalogue of recorded eruptions of mud volcanoes of Azerbaijan (for period of years 1810-2001)*. Nafta-Press, 89 p., Baku
3. Balan S., Cristescu V., Cornea I., 1982, *The 1977 March 4, earthquake in Romania*. Academiei Publ. House, 516 p., Bucharest.
4. Constantinescu L., Enescu D., 1985, *The Vrancea earthquakes within a scientific and technologic framework. (in Romanian)* Academiei Publ. House, 230 p., Bucharest.
5. Diaconescu C., Knapp J.H., Keller G.R., Stephenson R., MocanuV., Raileanu V., Matenco L., Bala A., Prodehl C., Hauser F., Dinu C., Wenzel F., Harder S., 2001, Intermediate depth seismicity in the Vrancea zone of Romania: a geodynamic paradox. *Eos.Trans.AGU*, **82** (47), Fall Meet. Suppl., Abstract S51A-0587.
6. Enescu B. D., Enescu D., Constantin A.P., 1999, The use of electromagnetic data for short-term prediction of Vrancea (Romania) earthquakes: preliminary data. *Earth, Planets, Space*, **51**, 10, 1099-1117.
7. Etiope G., Caracausi A., Italiano F., Favara R., Baciu C., 2002, Methane emission from the mud volcanoes of Sicily (Italy). *Geoph. Res. Lett.*, **29(8)**, 14,340-14,343.
8. Etiope G., Caracausi A., Italiano F., Favara R., Baciu C., 2003a, Reply to comment by A.

Kopf on "Methane emission from the mud volcanoes of Sicily (Italy)" and notice on CH4 flux data from European mud volcanoes. *Geoph. Res. Lett.*, **30(2)**, 1094, doi:10.1029/2002GL016287.

9. Etiope G., Baciu C., Caracausi A., Italiano F., Cosma C., 2004, Gas flux from mud volcanoes in eastern Romania. *Terra Nova*, **16**, 4, 179-184.
10. Gold T., Soter S., 1985, Fluid ascent through the solid lithosphere and its relation to earthquakes. *Pageoph.*, **122**, 492-530.
11. Guliyev I.S., Feizullayev A.A., 1997, *All about mud volcanoes*. Baku Publ. House, Nafta Press, 52 p., Baku.
12. Gutenberg B., Richter C. F., 1965, *Seismicity of the Earth and associated phenomena*. Hafner Publ. Co., 310 p.
13. Knapp J. H., Diaconescu C., Hauser F., Prodehl C., Raileanu V., Matenco L., Bala A., Keller G. R., Stephenson R., Mocanu V., Dinu C., 2001, The Vrancea zone, Romania: intermediate depth seismicity in search of a viable subduction zone. *Eos. Trans. AGU*, **82** (47), Fall Meet. Suppl., Abstract S42D-11.
14. Morelli C., Cornea I., Lazarescu V., 1975, A seismotectonic comparison between the Eastern Carpathians and Sicilian-Calabrian arc bends. *Bull. Geof. Teor. Appl.*, **18** (66): 169-178.
15. Paraschiv D., 1984, On the natural degasification of the hydrocarbon-bearing deposits in Romania. *An. Inst. Geol. Geofiz.*, **LXIV**: 215-220.
16. Sandulescu M., 1984, *Geotectonics of Romania (in Romanian)*, Tehnica Publ. House, 336 p., Bucharest.
17. Sandulescu M., 2003, The geodynamic evolution of the Central and South-East European Tethyan Chains. European Geoph. Soc., *Geoph. Res. Abstr.*, **5**, 07020.
18. Sencu, V., 1985, *Mud volcanoes from Berca*. (in Romanian), Sport-Tourism Publ. House, 21 p., Bucharest.

MUD VOLCANISM, GEODYNAMICS AND SEISMICITY OF AZERBAIJAN AND THE CASPIAN SEA REGION

Behrouz M. Panahi
Geology Institute, Azerbaijan National Academy of Sciences
29A, H. Javid Ave, Baku 370143, Azerbaijan, e-mail: bpanahi@gia.ab.az

Abstract: A dramatic activation of geodynamic processes of a natural and technogenic origin has been observed within the Azerbaijan and the Caspian Sea area for the last decade. It is evident this activation observed due to the unique geodynamics of the region which associated with peculiarities of geological structure - in particular, with a very thick and young sedimentary cover. Geological and thermodynamic conditions of sedimentary cover have formed the situation of an extremely non-balanced phase and mechanical instability. Most part of the earthquakes sources are confined to the sedimentary cover and mud volcanism is widely spread there. Space and time correlation of seismicity and mud volcanism testify that relationship mechanism is of a specific character and may be stipulated by re-organization of acting stress fields as a result of internal processes in the sedimentary thick.

Key words: earthquake, eruption, mud volcanism, seismicity

1. INTRODUCTION

There are number publications devoted to the study of geodynamics of the Caucasus, including regions of the Caspian and Black Sea. However, a unique approach to the geodynamic interpretation of the tectonic structure of the region had not been worked out. Apparently, it was stipulated by the heterogeneous structure of the whole region that is reflected in the results of geophysical, geochemical, petrologic, structural and other investigations.

On the other hand, the difference of opinions on the geodynamic conditions of the region (sometimes – alternative) can be explained by the very simple interpretation of initial data. In this respect, it is necessary to operate with correct initial seismological information that is a basic element for modern geodynamic reconstructions and focus on the solution of the most important tasks, such as earthquake sources, deep and epicenter spatial distribution, focal mechanism solution, seismic regime, distribution of mud volcanoes and also the activity and intensity of eruptions, correlation of eruptions of mud volcanoes and earthquakes.

The solution of the tasks mentioned above gives grounds to ask whether

G. Martinelli and B. Panahi (eds.), Mud Volcanoes, Geodynamics and Seismicity, 89–104

the modern tectonic processes are inherited of old geologic ages or they are of an accidental character. In correlation with the data on geomagnetism and the global positioning system it allows us to assess the application of existing geotectonic conceptions to the region and to develop our own point of view. Thus, we can state the main purpose of the present paper that is to reveal the peculiarities of seismicity and time-space correlation of the earthquakes and eruptions of mud volcanoes in the Azerbaijan and Caspian Sea region.

2. MUD VOLCANOES DISTRIBUTION

The modern structure of Azerbaijan territory was formed at the last stage of the Alpine period. The territory of Azerbaijan covers the eastern part of Caucasus and adjacent part of Caspian Sea. In tectonic terms the meganticlinoriums of Lesser Caucasus and Great Caucasus and the Kura intermountain depression were developed within the considered territory. As an eastwards continuation of the meganticlinorium the Great Caucasus may be considered the southeastern subsidence passing through the Absheron-Prebalkhan zone of uplifts in Western Turkmenistan. Gusar-Kelkor trough, representing the main marginal tectonic element of joint zone of epihercynian Skif-Turan platform and Alpine folded area, adjoins to this zone from northern side.

The South-Caspian Depression, representing an abyssal sedimentary basin, is a largest tectonic element of the considered region. Mud volcanism developed within the boundary zone of Eastern Georgia and Western Azerbaijan (interfluve of the Kura and Iori rivers), Eastern Azerbaijan (Shamakha-Gobustan area, Absheron peninsula, Lower Kura trough, Precaspian trough), Absheron-Cheleken trough, Baku archipelago and Western Turkmenistan (Fig. 1). In tectonic terms (Rahmanov, 1987) the mentioned areas locate within the South-Caspian area of earth crust's warping, including Kura, South-Caspian and Western Turkmenistan intermountain depressions, Absheron-Prebalkhan interpericlinal and Shamakha-Gobustan rear troughs.

3. SEISMICITY

The seismicity of Azerbaijan and the Caspian Sea region was analyzed using the database of Geology Institute of Azerbaijan National Academy of Sciences, including the information on more than 18,000 seismic events collected from the results of observations of existing seismological networks including international monitoring systems such as NEIC, IRIS, CTBT, GS RAS etc., detailed seismological data and also from macroseismic information on earthquakes occurring during last 4000 years.

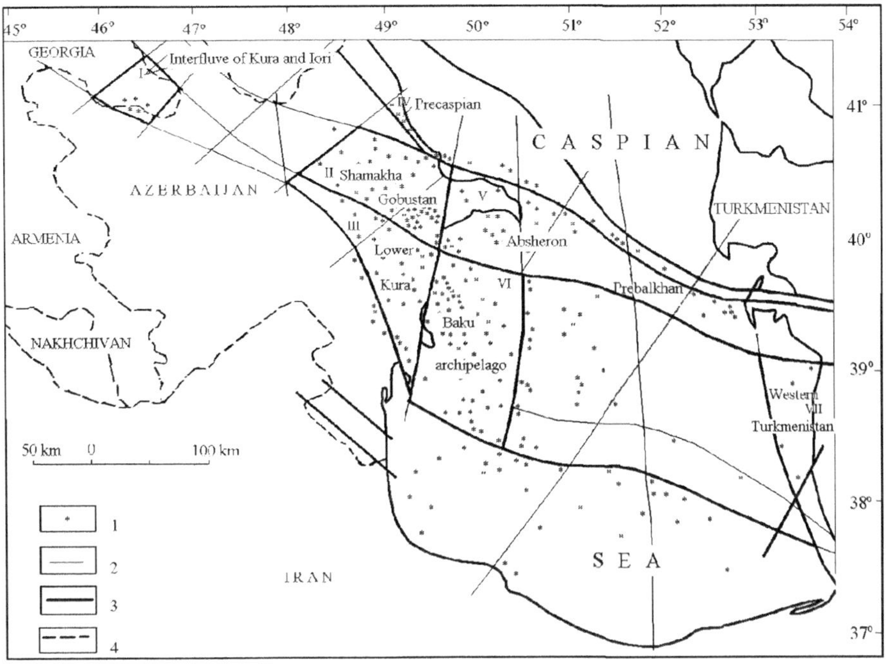

Figure 1. Mud volcanoes space distribution within Azerbaijan and adjacent area of the Caspian Sea
1 - mud volcanoes, 2 – deep faults, 3 – boundaries of mud volcano areas, 4 – political margins
I to VII - mud volcanoe areas: I - Interflow of Kura and Iori rivers, II - Shamakha-Gobustan, III – Lower Kura, IV – Precaspian, V - Absheron-Prebalkhan, VI - Baku archipelago, VII - Western Turkmenia.

Seismic events from local and remote sources of folded structures of the Great Caucasus, the Lesser Caucasus, Kopetdag and Alborz were felt within Azerbaijan and the Caspian Sea region with various magnitudes. In the number of destructive onshore earthquakes we can mention Gandja (427, 1235), Gey-Gel (1139), Shamakha (1667, 1669, 1828, 1859, 1872, 1902), Nakhichevan (1841), Mashtaga (1842), Zurnabad (1867), Ismailli (1981) events that caused serious changes in the earth's surface, destructions of constructions and human losses. Besides, the Caspian Sea area earthquakes of 1911, 1931, 1935, 1961, 1963, 1986, 1989 and 2000 of magnitude M = 6-6.5 affected the coastal part with high intensity. Strong far earthquakes of Western Turkmenistan (1895, 2000) and Iran (1990) of magnitude M=7-7.5 were felt on a significant part of the Caspian Sea region and Caucasus. The basic information on the strong events is given in the Table 1.

Table 1. Strong earthquakes of Azerbaijan and surrounding areas

Name of event	Date	Coordinates φ°N	λE	Depth (km)	Magnitude	Intensity (MSK)
Gandja	427	40.5	46.5	12	6.7±1	IX±I
Gey-Gel	1139	40.3	46.3	15	6.8±0.5	IX±I
Gandja	1235	40.6	46.2	10	5.7	VIII±I
Shamakha	1667	40.6	48.6	12	6.9±0.7	IX-X±I
Shamakha	1669	40.6	48.6	5	5.7±1	IX+
Shamakha	1828	40.7	48.5	10	5.7±0.5	VIII
Nakhchivan	1841	39.4	46.2	15	5.7±0.5	VII-VIII+
Mashtaga	1842	40.5	50	4	4.7±0.5	VIII
Shamakha	1859	40.7	48.5	6	5.8±0.5	IX+
Zurnabad	1867	40.5	46.3	15	5.8±0.5	VII-VIII+
Shamakha	1872	40.6	48.7	16	6.6±0.5	VIII-IX+
Krasnovods	1895	39.5	57.3	60	8.2±0.3	X+
Shamakha	1902	40.6	48.6	13	5.7	VIII-IX+
Caspian	1911	41.0	50.5	30	6.4±1	VI-VII±I
Caspian	1931	42.5	50.8	25	6.2±0.5	VII-VIII±I
Caspian	1935	42.1	48.8	30	6.3±0.5	VI±I
Caspian	1961	41.1	50.2	25	6.6±0.5	VII±I
Caspian	1963	41.1	49.8	26	6.2±0.1	VII-VIII±I
Ismailli	1981	40.8	48.0	10	5.4	VII-VIII
Caspian	1986	40.3	51.6	40	6.1	V-VI
Caspian	1989	41.2	52.1	30	6.2	V-VI
Rudbar	1990	37.3	49.4	20-25	7.4	IX-X
Caspian	2000	40.3	50.1	25	6.2	VII-VIII
Nebit-Dag	2000	39.5	57.3	35	7.3	IX+

The earthquakes list of Azerbaijan and the Caspian Sea region includes more than 1,500 events of various magnitudes and a map drawn on the basis of these earthquakes shows a common character of seismicity in different tectonic areas (Fig. 2).

The magnitude distribution of earthquakes (Table 2) demonstrates that seismic events of M=3-3.5 represent the overwhelming majority (up to 85%) in the total quantity of earthquakes, and the earthquakes of magnitude M>6.5 were not recorded within the areas of the mud volcanoes location in Azerbaijan and the Caspian Sea region. According to the data analysis (Table 1) the high-magnitude earthquakes (M>6.5) had not been marked within the mentioned areas.

4. SEISMIC REGIME

Quantitative seismicity estimation of the areas of mud volcanoes location in Azerbaijan was considered through analysis of earthquakes repetition, seismic activity maps and released seismic energy (Panahi, 1998).

The analysis of the earthquake repetition allows us to state the lower seismicity level of mud-volcanic areas in Azerbaijan and the Caspian Sea region (Table 3). However, earthquakes of moderate magnitudes are possible in these areas, too. The values of long-term parameters of seismic regime had been calculated for the areas of mud volcanoes location of Azerbaijan and the Caspian Sea region and included in Table 3. According to the Table, the magnitude values of the maximum expected earthquake of the areas mentioned above change from 3.7 to 4.7 (for a 100 years period). This fact also confirms the lower seismicity level of these areas. The moderate values of magnitudes 4.6≤M≤5.8 had been obtained for the period of 1,000 years. The prognostic magnitude values of maximum earthquakes in these areas also have moderate values 5.3≤M≤6.5 (Tab.3).

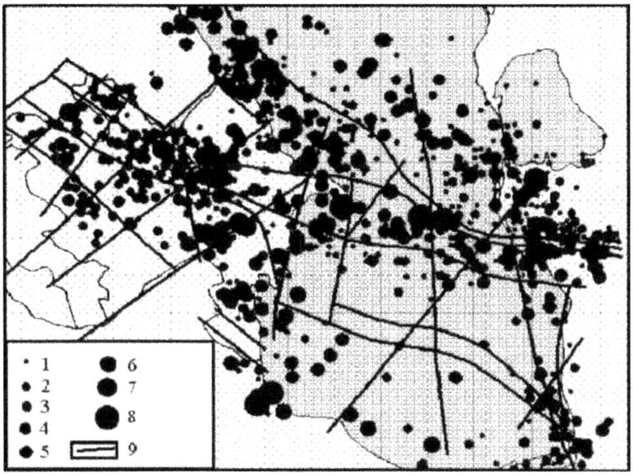

Figure 2. Epicenters space distribution within Azerbaijan and adjacent area of the Caspian Sea
1-8 – Magnitude: 1- 3, 2 –3.5, 3 – 4, 4 – 4.5, 5 – 5, 6 – 5.5, 7 – 6, 8 > 6.5; 9 – deep faults

Table 2. Magnitude distribution of earthquakes

Mud volcano Areas	Area	M	A	G	N	I	T	U	D E
	S (km^2)	3-3.5	3.5-4	4-4.5	4.5-5	5-5.5	5.5-6	6-6.5	ΣN
Shamakha-Gobustan	9825	161	65	19	7	3			225
Precaspian	5800	52	23	3	4	1	1		84
Lower Kura	9200	117	31	11	9	1			169
Kura and Iori Rivers interfluve	18000	185	101	18	9	4			317
Absheron-Prebalkhan	33725	158	94	36	21	6	1	2	318
Baku archipelago	21800	45	10	3	5	2			65
Western Turkmenistan	25200	26	23	3	2				54

Table 3. Calculated parameters of some earthquakes of studied area

Area	γ	α	β	M_{100}	M_{1000}	M_{10000}
Shamakha-Gobustan	-0.69	0.4	-1.24	4.6	5.4	6.2
Precaspian	-0.50	0.18	-0.9	4.7	5.8	6.5
Lower Kura	-0.69	0.24	-1.24	4.4	5.2	6.1
Interfluve of Kura and Iori rivers	-0.59	0.22	-1.06	4.7	5.5	6.4
Absheron-Prebalkhan	-0.51	0.14	-0.92	4.6	5.7	6.5
Baku archipelago	-0.53	0.042	-0.95	4	5.1	6.1
Western Turkmenistan	-064	0.025	-1.15	3.7	4.6	5.3

The maps of seismic activity of the Caspian Sea territory (Fig. 3) had been drawn by the "summing method of constant accuracy" (Riznichennko and Zakharova, 1971).

To assess the seismic activity of the territory separately for weak M≤3.5 and/or higher magnitude M>3.5 earthquakes was done by this method of quantitative seismicity assessment.

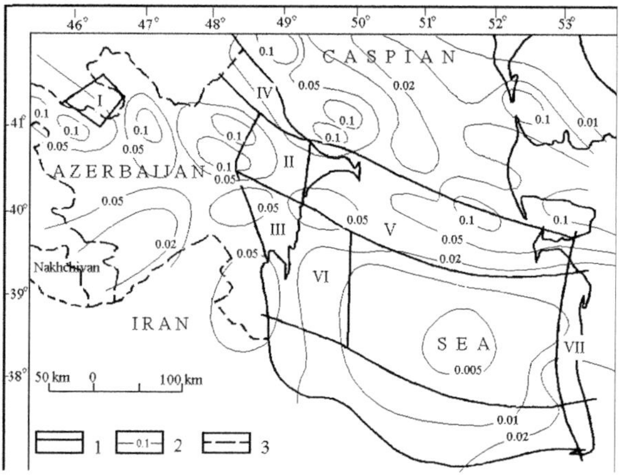

Figure 3. Map of the 0.005, 0.01, 0.02, 0.05 and 0.1 isolines of the parameter A_{10} of seismic activity of Azerbaijan and the Caspian Sea
1 – boundaries of mud volcano areas; 2 – isolines of seismic activity (A_{10}); 3 - geographic and political margins

According to results of seismic quantitative assessment (Panahi and Kasparov, 1986) we can state the location of seismic zones of seismic activity (for lower energy earthquakes) within internal parts of separate blocks of earth crust, and coincidence of active zones (for higher energy earthquakes) with deep fault zones. Referring to that we can suppose the existence of a relationship between higher magnitude earthquakes and most weakened zones of the earth's crust. Contrarywise, the events of lower magnitudes show relations mostly with the internal parts of the blocks of earth's crust that are considered to be responsible for earthquakes preparation processes.

Quantitative seismicity estimations (Panahi, 1998) showed the moderate level of seismicity in Eastern Azerbaijan and adjacent areas of the South Caspian Depression in comparison with high-seismic areas of Anatolia, Albors, Kopetdag and Balkhan.

The different modifications of released seismic energy mapping can be used for the assessment of the seismic regime. Shenkaryova's method Shenkaryova, 1966) was adopted for present study. This method was developed on information about the depths of the known earthquakes sources of the region (Fig. 4).

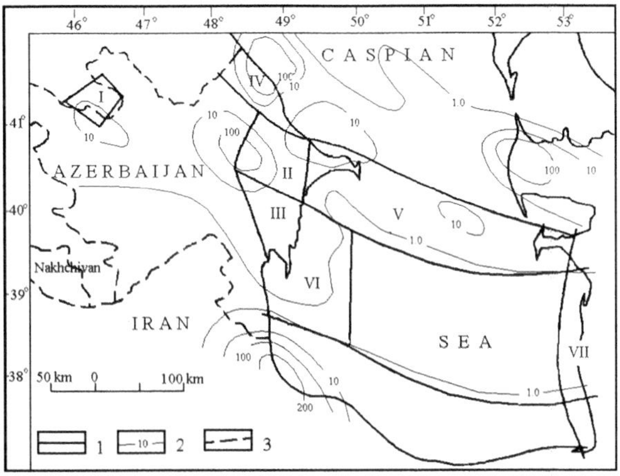

Figure 4. Map showing the 1, 10, 100 and 200 contours of released seismic energy in the eastern Azerbaijan and the Caspian Sea (drawn in units of $N_m=10^6$ J/km^3×yr). 1 – boundaries of mud volcano areas; 2 – isolines of reliased energy (N_m); 3 - geographic and political margins

According to data the central part of the South-Caspian depression is attributed to the areas of minimum values of released seismic energy $N_m \leq 10^6$ j/km^3×y. Maximum values of $N_m \geq 200\times10^6$ j/km^3×y are characteristic for the south-west edge of the Alborz seismogenic zone. The earthquakes of November 4, 1978 and May 4, 1980 of magnitude M≥6 are connected to this zone. The eastern flank of this zone is characterized by average values of released energy $10^6 \leq N_m \leq 100\times10^6$ j/km^3×y. Earthquakes of magnitude M=5.4 and epicentral intensity I_0=VIII degrees occurred within this zone. Absheron-Prebalkhan threshold and Baku archipelago zones are distinguished by lower values ($10^6 \leq N_m \leq 10\times10^6$ j/km^3×y). There is no event known there with magnitude more than M≤5.The part of the Middle Caspian depression is characterized by minimum values of released seismic energy. The Derbend (marine) and Krasnovodsk peninsula areas of higher values

($100\times10^6 \leq N_m \leq 200\times10^6$ j/km$^3\times$y) corresponding to different elements of interaction zone of Skif-Turan platform and Alpine folding are within this zone. The earthquakes of April 9, 1933 and Februrary 14, 1938 of magnitude M=6-6.5 and intensity of I_0=VII-VIII degrees occurred here. The central part of the Mid-Caspian depression also revealed a sub-latitudinal area of lower values of released energy ($10^6 \leq N_m \leq 10\times10^6$ j/km$^3\times$y). Finally the areas of mud volcanoes location of Azerbaijan and the Caspian Sea region are characterized by the minimum and low values of N_m=(1.0÷100) x 10^6 j/km^{3}year.

A detailed study of strong and weak earthquakes' sources of the present segment of the Mediterranean mobile belt showed their location in the earth's crust at the intervals 0-10 and 15-20 km. This relation is more expressive in Eastern Azerbaijan and adjacent area of the South Caspian Depression (Fig. 5), where 87% of the total number of events corresponds to the mentioned depths intervals.

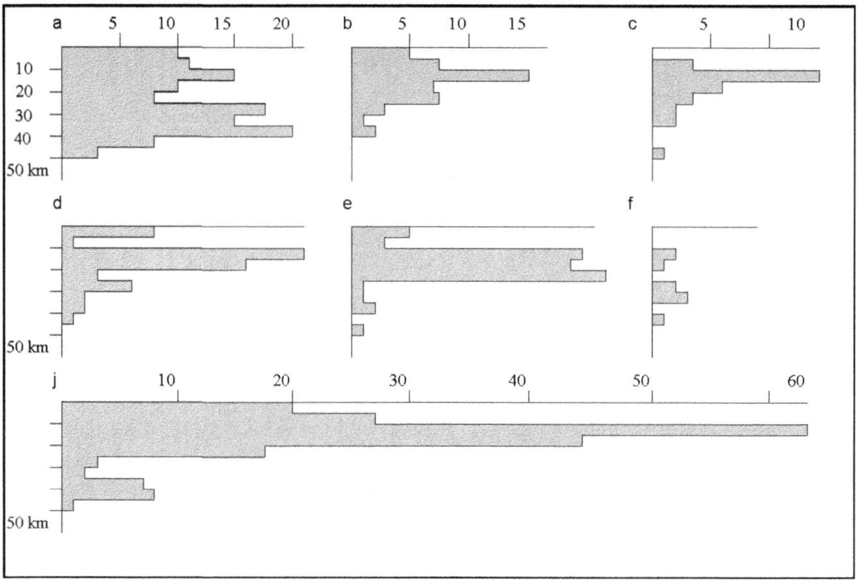

Figure 5. Earthquakes depth distribution for some areas of Azerbaijan and adjacent areas of the Caspian Sea
a-Absheron-Prebalkhan, b-Precaspian, c-Baku archipelago, d-Lower Kura, e-Kura and Iori rivers interflows, f-Western Turkmenia, j-Shamakha-Gobustan

Epicenters space distribution in the region doesn't give us a reason to assert their relation especially with boundaries of lithosphere plates, microplates or great blocks (inter-plate). On the contrary, the connection of seismicity with internal parts of great structural elements (intra-plate or intra-block seismicity) more clearly developed here.

As a whole the territory of Azerbaijan and the Caspian Sea region are characterized by moderate level seismicity. This fact could be considered as its basic difference from other segments of Mediterranean mobile belt, where magnitude values exceed 7 degrees.

5. CORRELATION OF MUD VOLCANO ERUPTIONS AND EARTHQUAKES

For a long time discussions on space and time correlation of seismicity and eruptions of mud volcanoes the eruptions that coincided with earthquakes , due to the time of their origin, were mostly mentioned (Table 4). The cases when eruptions happened before the earthquakes or occurred just after them were not discussed widely.

Table 4. Parameters of earthquakes following by eruptions of mud volcanoes

Date of events	Coordinates φ°N	λ°E	H km	M	I MSK	Location area
26.08.1833						Burma
23.03.1839						Burma
26-29 07.1843						Burma
31.12.1881						Burma
08.07.1895	39.5	53.7	60	8.2	X	Uzun-Ada/ Western Turkmenistan
21.09.1897						Kalimantan/ Indonesia
13.02.1902	40.7	48.6	30	6.5	VIII-IX	Shamakha/ Azerbaijan
11.09.1927	44.3	34.3	17	6.8	IX	Kerch/Crimea
03.02.1931	39.2	176.4		7.9	IX	Raukumara/ New-Zealand
1934						Baratang/India
27.11.1945	24.5	63		8.3	X	Makran/ Iran-Pakistan
19.03.1959						Sakhalin

Mud volcanic eruptions mostly accompany remote earthquakes and coincide with them at the time of their origin. They can accompany only the remote earthquakes of magnitude M>6 and intensity I≥VIII degrees of MSK scale whose source depths do not exceed the earth crust thickness .

Besides the typical mud volcanoes in the various regions of the world the "mudvolcanic manifestation" are distinguished. They outwardly resemble the mud volcanoes but differ from them by sizc, valuc of outpouring , etc. The term "mudvolcanic manifestation" is applied in the traditional sense to the extruding of mud and sand on the ground surface as a result of earthquakes mainly in the areas with no mud volcanoes that forms sandy fountains and characteristic floods of water and sandy mixes on the earth surface. The mentioned appearance was observed within the Mediterranean mobile belt during earthquakes occurred in Italy, Romania etc. Mudvolcanic manifestation in Bulgaria accompanied the earthquakes in Kresna, Yambol and Trakiya. Small mudvolcanic manifestation of 5 m in diameter was formed during the Ismailli earthquake of 29.11.1981 in Azerbaijan.

The mudvolcanic manifestations are also marked during Middle-Baikal (09.08.1959) and Altai (04.12.1957) earthquakes. Same effects accompanying mudvolcanic manifestations were noted at the earthquakes of 18.04.1906 in San Francisco, of 28.03.1964 in Alaska, and also of 22.05.1960 in Chile. The mudvolcanic manifestations are marked within the maximum shaking area (VII-VIII degrees of MSK) of the Ismailli earthquake in Azerbaijan in 1981. It is characteristic that, at intensity less than VII-VIII degrees of MSK scale, these manifestations were not marked.

The depths of the earthquake sources change over a wide range and most of them are of a crustal nature. However, the mudvolcanic manifestations caused by intermediate and deep earthquakes are also known (Harkovsca et al., 1982).Mudvolcanic manifestations mainly accompany the strong local earthquakes of magnitude M>5.4 and intensity I>VII-VIII and source depths located in the earth's crust and upper mantle (Table 5).

Table 5. Parameters of earthquakes followed by mud volcanic manifestations

Date	φ°N	λ°E	H	M	Io, MSK	Place of event	Ref.
18.4.1906	38	123		8.3	X+	San Francisco	[8]
5.10.1948	37.9	58.3	18	7.3	IX+	Ashgabad	[5]
04.12. 1957	45.1	99.4	25	8.6	XII	Gobi-Altai	[4]
09.08. 1959	52.7	106.9	20	6.8	IX	Mid. Baikal	[23]
22.05. 1960				8.4		Chile	[21]
28.03. 1964	61.1	147.7	20	8.4		Alaska	[3]
06.05. 1976	46.7	13.2		6.5	IX+	Friuli	[13]
04.03. 1977	45.8	26.8	95	7.2	VIII	Vrancea	[6]
29.11. 1981	40.8	48	10	5.4	VII+	Azerbaijan	[11]
22.02. 1984				6	VIII+	Turkmenia	[20]
07.12. 1988				7.1	X	Armenia	[7]
14.06. 1990	48.1	84.7	35	7	VIII	Zaysan	[18]
27.5. 1995	52.9	143.3	15	7.2	IX	Sakhalin	[2]

Sometimes mud volcano eruptions occur after the earthquakes. In this case connection between both phenomena are stated through the basic criteria of their correlation as a distance between epicenter and the nearest mud volcano, magnitude value and sizes of source area. The connection of mud volcano with the source area of the event is supposed.

The eruptions of mud volcanoes occurring after earthquakes had been revealed as a result of investigations in the East Azerbaijan and the Caspian Sea region. Also the dependence between the time origin of the seismic events and the eruptions and the magnitude value of earthquakes was reported.

The sporadic and sparse information on earthquakes or fluctuations in the ground surface observed after eruptions are found in the descriptions of eruptions of mud volcanoes (Agabekov et al., 1960; Ridd, 1970; Rahmanov et al., 1992). The eruption of Arakihi Road (New Zealand), Makarova bank (Caspian sea) volcanoes etc., may be counted from them. The local event was marked by all components of mobile seismic station "Cherepakha" (frequencies range of 2-4 h) after the short-lived Keyreki mud volcano eruption of 26.02.1989. So we can consider the influence of mudvolcanic activity on tectonic or seismicity background of mudvolcanic areas as a factor strengthening seismicity (Gorshkov, 1984).

In some cases the mudvolcanic eruptions may enliven the earthquake sources. Earthquakes stimulated by eruptions of mud volcanoes have a macroseismic epicentral intensity lesser than $I_0 \leq V$ degrees of MSK, and M=2.5-3. Two events, of M=2-3.5 occurring in Eastern Azerbaijan at the maximum close distances (Δ=10-15 km) from erupted mud volcanoes Lokbatan and Buzovna are known. One may suppose that this fact predetermines the moderate energy of earthquakes caused.

Besides the carried out calculations, it is shown that the earthquakes caused by eruptions of mud volcanoes are shallow and have depth intervals at 8-12 kms. This is in close relation with the information on "roots" of mud volcanoes located at depths of up to 15 kms.

Thus, the eruptions of mud volcanoes can be accompanied by earthquakes of a low and moderate energy class M=2.5-3 at the distances Δ=25-40 km, and rarely by M=4 at the distances Δ=10-15 km. The depths of the earthquake sources are mainly shallow (Table 6).

The overwhelming majority of mud volcanoes paroxysms are accompanied by the tremors and large rock bursts, visual on the surface, though the energy of ground fluctuations caused by them is insignificant. These ground motions are very local and are not recorded by remote and distant seismic stations. Such tremors are subdivided into the special type of mudvolcanic earthquakes, spatially and genetically connected with eruptions or other signs of mudvolcanic activation, due to fact that they cannot be identified with normal earthquakes. According to macroseismic data the maximum intensity of tremors accompanying the eruptions does not exceed VI degrees. The value of the maximum magnitude of the tremors does not exceed 3-3.5 (Table 7).

Table 6. Earthquakes occurred after eruptions of mud volcanoes

Mud volcano	Date of eruption	Date of earthquake	M	I, MSK	H, km	Δ, dys	Δ, km
Delaniz	07.04.1912	27.04.1912		III-IV		20	
Lokbatan	28.02.1935	31.03.1935		III-IV			30
Lokbatan	18.01.1938	23.02.1938		III-IV			30
Makarova	15.10.1958	28.11.1958		III-V	12	43	15
Koturdag	15.10.1966	28.11.1966		III	11		
Melikcoban	03.10.1967	22.10.1967		IV	8	19	30
Demirchi	01.11.1971	21.12.1971					
Lokbatan	01.10.1972	17.10.1972					
Buzovna	10.09.1973	14.09.1973		III-IV	10	4	10
Davalidag	02.04.1975	03.05.1975				31	30
Dashgil	26.09.1976	15.11.1976				49	35
Lokbatan	31.03.1980	04.04.1980				4	40
Shikhzarli	02.11.1980	28.12.1980				46	30
Hamamdag	25.04.1984	02.05.1984	3.6			7	40
Makarova	20.05.1984	04.07.1984					35
Fersman	23.03.1987	16.05.1987	4				
Ayrantekan	20.03.1988	14.04.1988	3.4				40
Karabetova	02.08.1981	02.09.1981					30

Table 7. General information on earthquakes and eruptions of mud volcanoes occurring after them

Mud volcano	Date of eruption	Date of earthquake	M	I MSK	ΔT, days	Δ, km
Glinyani	23.06.1859	11.06.1859	5.9	VIII-IX	12	115
Bulla	24.06.1859				13	115
Garasu	17.02.1876	03.02.1876	4	V-VI	14	
Ignati Stone	07.10.1920	03.10.1920	4.5	V	4	25
Jautepe	13.09.1927	11.09.1927	6.8	IX	2	
Otmanbozd	01.10.1965	27.08.1965	3.5	IV-V	32	15
Kelani	12.12.1969	06.10.1969	3.4		36	18
Cheildag	04.06.1970	19.04.1970	3.3		45	40
Agnour	01.05.1976	19.04.1976	3.9		34	40
Garasu	28.03.1977	01.03.1977	3.3		27	
Hamamdag	25.04.1984	13.04.1984	3.8	IX	2	

The analysis of the records of ground motions caused by eruptions of mud volcanoes in Azerbaijan allows us to remark that they can arise just before the eruptions and accompany them. It is possible also to reveal the group of fluctuations observed just after eruption. Due to the character of the record these fluctuations are similar to the records of local shallow tectonic earthquakes. The maximum amplitude and the period of fluctuations are related to the beginning of the record (Panahi, 2000). By analogy with volcanic trembling whose basis probably compose the surface waves , the explosive tremors of relatively continuous eruptions can be accepted as mud-volcanic trembling.

There is a reason to suppose that mudvolcanic earthquakes are the result of the process of preparation and eruption of mud volcanoes. Due to that some stages of mudvolcanic eruptions had been revealed: pre-eruptive, eruptive and post-eruptive (Panahi and Rahmanov,1993). The first stage includes imperceptible ground fluctuations and corresponds to the rise of elastic stress in the source of mudvolcanic earthquakes and can be perceived as a weak local shock.

Fluctuations in second stage are usually more powerful and long , their amplitudes and periods are almost twice times larger in comparison with fluctuations of the first phase. In some cases fluctuations are repeated, but every subsequent repetition is weaker than the previous one. The part of mud-volcanic earthquake is identified with the period of the moderate mudvolcanic activity or volcanic trembling and is included in the third stage. The number of shocks falls very fast here, the intensity of fluctuations quickly attenuate and the stage is finished by weak local paroxysmal events.

Figure 6 shows the characteristic of time distribution of seismicity in structural zones of the Caspian Sea region. According to the graph during the mentioned period within the considered area events of magnitude greater than 7 were not recorded and also the total number of events characterized by magnitude M=6-7 is significantly lower in comparison to the ones recorded in the neighboring areas.

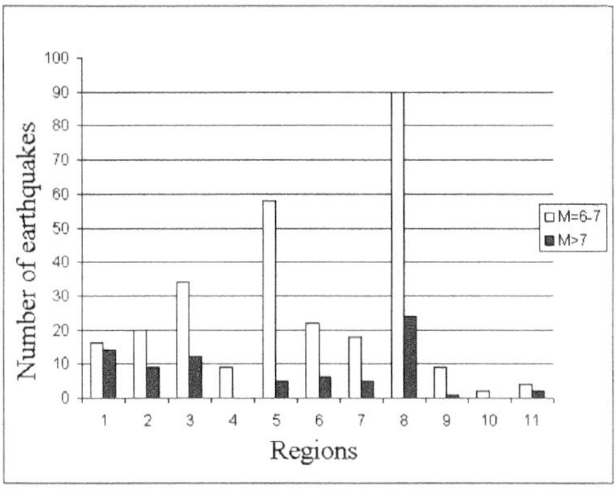

Figure 7. Distribution of earthquakes with magnitude greater than 6 within the structural zones of the Caspian Sea region (total number of events is shown in comparison with the number of earthquakes characterized by M>7) for the period from 2000 B.C. to 2000 A.D.
1-11: Regions; 1-NE Iran, 2-Kopetdag, 3-Alborz, 4-Eastern Azerbaijan and the Caspian Sea, 5-Caucasus, 6-Central Azerbaijan Massif, 7-Zagros, 8-Anatolia, 9-Black Sea, 10-Kerch, 11-Gazli.

6. CONCLUSIONS

The received result shows an important role of mud volcanism in the structure of Azerbaijan and the Caspian Sea region. Active tectonic processes together with the geomaterials plastic conditions in the top part of earth's crust and mud volcanism provide fast relaxation of accumulated stresses that is a cause of moderate level seismicity. However, mud volcanism should not be considered as a phenomenon that increases seismic activity. On the contrary, it is a process promoting the lessening of elastic tensions. In this sense, mud volcanism can be considered as a natural system controlling the moderate level of seismicity of Eastern Azerbaijan and South-Caspian Depression.

Within this areas, due to the composition of their high thickness plastic masses , counted traditionally as aseismogenic, the problem of seismic hazard assessment was considered in relation with remote and distant earthquakes. The seismicity investigation of the areas populated with mud volcanoes (Panahi, 2000) provided the chance to change the existed point of view on their seismicity, mechanism of seismogenesis and conditions of seismic sources formation.

The relation of the main part of the hypocenters to depths intervals 0-10 and 15-20 km allow us to connect the mechanism and conditions of sources formation to sedimentary thickness. Data analysis of the different structural zones of the mentioned regions allows us to consider this connection more clearly with the increase in the sedimentary deposit thickness. Thus seismic activity maps which show the space coincidence of active areas with zones of higher thickness of sedimentary deposits of the Caspian Sea area cab be utilized.

This behaviour characterizes the plastic deposits (thickness about 25-30 km) as a main structural element that is responsible for seismogenesis within the considered area. Deformations acting during the final stage of evolution of depression provide evidence on the main role of modern tectonic activation in the development of depression. As most of the Earth crust structure is represented by plastic masses there are no conditions for long-term stresses accumulation in the considered area.

ACKNOWLEDGMENTS

We are grateful to Adil Aliyev and Rauf Israfilov constructive discussions and to Solmaz Agayeva for some data used in this paper and also to Malahat Efendiyeva for help in preparation of the manuscript.

REFERENCES

1. Agabekov, M., Sultanov, F., and Islamov K., 1960. The geological reasons of earthquake felt in Baku on November 28, 1958, *Letters of Azerbaijan State University, series of geological sciences*, 67-70 (in Russian)
2. Arefiev, S., Rogodzin, E., Tatevossian, R., Rivera, L. and Cisternas, A., 2000. The Neftegorsk (Sakhalin Island) 1995 earthquake: a rare interplate event, *Geophys. Journ. Int.*, **143**, 595-607.
3. Coulter, H., and Migliaccio, R.R., 1967. *Effects of the earthquake of March 27, 1964, at Valdes, Alaska*, U.S. Govt. Print Off., Washington
4. Florensov, N.A., and Solonenko, V.P., (Eds.), 1963. The Gobi-Altai earthquake, Acad. Nauk USSR, Moscow (in Russian), (English translation, U.S. Dept. Commerce, Washington, 1965).
5. Gorshkov, G. P., 1984. Regional seismotectonics of the USSR southern part, Nauka, Moscow. (in Russian).
6. Harkovska, A., Karaguleva, I., and Iliev I., 1982. Sand volcanoes caused by Vrancea earthquakes (March 4, 1977) in North Bulgaria, Sofia. Mar. 87-98 (in English).
7. Kondorskaya, N., and Gorbunova I., 1993. Catastrophic earthquake of December 7, 1988 in Armenia: the analysis of seismic situation in the seismic source zone, *Izvestia Physics of the Solid Earth*, **7**, 17-23 (in Russian).
8. Lukyanov, A.V.,1963. Horizontal movements along faults during modern catastrophic earthquakes, in *Faults and horizontal movements of earth crust*. Proceedings of Geologic Institute of USSR's Academy of Sciences **80**, 123-137 (in Russian).
9. Panahi, B., and Kasparov V., 1986. The problems of the Caspian Sea area seismic regime, *Academy of Sciences Azerbaijan SSR, Earth Sciences Series*, **1**, 91-98 (in Russian).
10. Panahi, B., and Rahmanov, R., 1993. Mudvolcanic earthquakes. M., Science, *Volcanology and seismology*, **2**, 98-103 (in Russian).
11. Panahi, B. , 1998. *Seismicity of the areas of mud volcanoes location*. Thesis of doctor dissertation, IPE, Moscow.
12. Panahi, B., 2000. On spatial and time correlation of earthquakes and mud volcano eruptions and seismic regime of Azerbaijan-Caspian sea region. *Geophysics news in Azerbaijan*, Baku, **1**, 26-29.
13. Pichard, P., Ambraseys, N., and Ziogas G., 1976. The Gemona di Friuli Earthquake of 6 May, 1976, in *Promotion of the study of natural hazards of geophysical origin*. Paris, 111.
14. Rahmanov, R., Safarova, O., Panahi, B., Khalilov, A., Efendiyev, A., Muradova, F., and Shikhlinski, S., 1992. To eruption of mud volcano Keyreki, *Azerbaijan Oil Economy*, **7**, 8-13 (in Russian).
15. Rahmanov, R. , 1987. Mud volcanoes and their importance in forecasting of subsurface petroleum potential, Nedra, Moscow, (in Russian).
16. Ridd, M. , 1970. Mud volcanoes in New Zealand, *Bulletin of AAPG*, **54**, 4, 601-616 .
17. Riznichenko, Y., and Zakharova , 1971. Generalized law of earthquakes repetition. *Izvestia Physics of the Solid Earth*, **3**: 29-38 (in Russian).
18. Rogodzin, E., and Leontyev A., 1992. Zaysan earthquake of 1990. Surface deformations and tectonic position of the source, *Izvestia Physics of the Solid Earth*, **9**, 3-14 (in Russian).
19. Rogodzin, E., 1995. Neftegorsk earthquake of 27(28). 05. 1995: geologic appearances and tectonic position of the source, *Informational and analytic bulletin of Neftegorsk earthquake*. FSSNPZ, Moscow, special issue: 80-94 (in Russian).
20. Rogodzin, E., and Borisov B., 1986. Seismic dislocations in the epicenter zones of West Turkmenian earthquakes. Detailed engineering seismologic investigations. Problems of

engineering seismology. M., Science, Issue 2, 116-135 (in Russian).

21. Segerstrom, R. , 1960. Eruption of water, sand and clay resulting from the earthquake of May 21, 1960 near Conception, Chile. Bulletin of the Geological Society of America, **71**, 12 .
22. Shenkaryova, G., 1966. On the method of seismic data analysis (Caucasus and Turkmenistan areas), *Soviet geology*, **8**, 160-170 (in Russian).
23. Solonenko, V., and Treskov A., 1960. Middle Baikal earthquake of August 29, 1959, Nauka, Moscow, 237 (in Russian).

Chapter 3
SEISMIC HAZARD AND MUD VOLCANISM

MUD VOLCANIC MANIFESTATIONS IN THE MAXIMUM SHAKING AREAS OF STRONG EARTHQUAKES

Eugeny A. Rogozhin
Schmidt United Institute of Physics of the Earth, Russian Academy of Sciences, Bol'shaya Gruzinskaya st. 10, Moscow, 123995 Russia, e-mail: eurog@uipe-ras.scgis.ru

Abstract: New evidence of previously unknown catastrophic Holocene earthquakes in the Elbrus region, the northern Caucasus, is reported. The remains of ancient neptune dikes, other signs of liquefaction, seismic faults, rock avalanches, landslides and dammed pond sediments have been discovered. The ages of four seismic events over the last 5,500 years have been determined by the radiocarbon dating of paleoseismic dislocations. Their recurrence interval was also estimated. The chronological comparison of ancient seismic shocks and volcanic eruptions showed that various manifestations of endogenic activity lack a direct correlation.

Key words: palaeoseismicity, radiocarbon dating, mud volcanic eruptions, Caucasus

1. INTRODUCTION

Several of the strongest earthquakes in the world have demonstrated a special kind of effects along the seismic rupture: e.g. liquefaction, gryphons, mud volcanoes. These effects were well-studied in the epicentral areas of New-Madrid, USA 1811, 1812 (M=8.2-8.3), Ashkhabad, Turkmenistan 1948 (M=7.6), Burun, West Turkmenistan 1984 (M=6.0), Zaisan, East Kazakhstan 1990 (M=7.0), Neftegorsk, Russia, North Sakhalin 1995 (M=7.6) and the other earthquakes (Nikonov et al., 1991; Laverov, 2002; Bogatikov et al., 2003). Data on the ancient liquefaction together with other seismic ruptures could be used for paleoseismological investigations, providing evidence of strong earthquake occurrence in the past and the dating of such events. For the Elbrus Volcano region, situated in the central sector of the northern Caucasus, the task of reconstructing relations of paleoearthquakes and volcanic eruptions

G. Martinelli and B. Panahi (eds.), Mud Volcanoes, Geodynamics and Seismicity, 105–110

in Holocene era was formulated. Dating the ancient eruptions was achieved earlier (Kondorskaya and Shebalina, 1977), and for the paleoseismic analysis the traditional geomorphologic and trenching techniques were used.

Fieldwork performed in 2002 was aimed at geological and geomorphological investigations of the Elbrus region, with particular emphasis on seeking evidence of strong ancient earthquakes (i.e. paleoseismic dislocations (Solonenko, 1973) and sampling for their dating. The radiocarbon (14C) dating of the samples was carried out at the Institute of Geography, Russian Academy of Sciences. The age estimates of strong earthquakes thus obtained were compared with the known periods of volcanic eruptions.

2. METHODOLOGY

The Elbrus Volcano – at the top of the Great Caucasus mounting system, erupted repeatedly in the late Pleistocene and Holocene. Based on dating by various methods, the eruptions took place approximately 39 ± 5, 23 ± 2, 21, 9.2-9.3, 7.8-8.0, 7.2, 6.0, 4.9, and 4.6 ky ago and in the 1st and 2nd centuries B.C. (Kondorskaya and Shebalina, 1977). Effusion of lava flows was accompanied by the eruption of gas and the consequent formation of glowing clouds, ash clouds, and lahar avalanches (mud-stone flows).

The estimation of the seismic hazard in the Caucasus region based on the cluster analysis of the geological, geophysical and seismological data showed that the Elbrus region has a potential seismic source (PSS) with an anticipated maximum magnitude of 7.0 (Rogozhin, 2002). The PPS was very poorly manifested during instrumental and historical periods of observation. As a matter of fact, a vast zone of seismic quiescence without weak or moderate shocks, let alone strong seismic events, is outlined in the Elbrus Volcano area. At the same time, the information on paleoseismic dislocations collected by many authors indicates that very strong earthquakes had probably taken place here in the past (Nikonov et al., 1991; Laverov, 2002; Bogatikov et al., 2003). During the fieldwork, every occurrence related to the seismogenic origin of the geomorphologic structures of relief and geological forms in Quaternary thickness was studied. After being studied they were classified by primary (seismotectonic) and secondary (seismogravitation and vibration) nature. Primary ruptures - seismic fault zones - were excavated by trenches or outcrops were cleaned. Secondary ruptures - rock falls, rock avalanches, neptune dikes, signs of liquefaction, dammed ponds sediment sequences – were detailed, documented and mapped. Collecting of samples from the buried paleosoil, charcoal and pieces of wood allowed radiocarbon dating to be performed. From the time periods of the mass manifestation of different kinds of dislocation, the author took a decision concerning the history of the paleoseismic events. Therefore, the study of interrelations between seismic events and volcanic eruptions is a pressing issue - do these processes accompany each other or are they separated in time?

3. DATA

A seismotectonic field study of volcanic areas in the basins of the Baksan River and the upper part of the Malka River valley revealed paleoseismic dislocations related to both primary (seismotectonic) and secondary (seismogravitation and vibration) processes. The primary dislocations are exhibited by an echelon system of paleoseismic faults oriented in the Transcaucasus southwards direction as well as in the west-north-westwards direction, along main fold axis of the Caucasus fold system. The southern faults complicate the walls and surface of a small, flat graben-like basin (a German airdrome site during World War II) on the north-eastern slope of the volcanic cone (Fig. 1, point 1). The total length of the seismic fault system exceeds 5 km, while separate fractures reach 800-1000 m in length. The faults are characterized by normal and sinistral strike-slip kinematics. They displaced the Paleozoic rocks of the Pshekish-Tyrnyauz fault zone and the moraine-covered surface of the late-Pleistocene (40-45 ky) Kyzylkol dacitic lava tongue. Thus, they make up terraces in the mountain topography and the framework of the airdrome basin. The faults on the airdrome surface are accompanied along their entire length by narrow linear pockets of anomalously thick paleosoil (~1 m) against the background of normal modern soil (5-15 cm). Angular dacite fragments from the uplifted western wall are contained at the paleosoil matrix. This material looks like the specific colluvial wage. The amplitude of vertical seismogenic displacement along this fault system may be more than 1.0-1.5 m. The radiocarbon age of the lower paleosoil unit in the fault zone is 2280 ± 90 yr b.p. (Sample IGAN 2592), suggesting that the tectonic displacement could have happened somewhat earlier.

The west-north-west oriented problematic seismic fault was observed on a top of northern slope of the Baksan river and looks like a typical seismic trench. This structure is about 10 km long. The chain of block landslides accompanies this trench in the upper and middle parts of the Baksan and Terskol river valley slopes.

The secondary paleoseismic dislocations are indicated by numerous landslides of friable deluvium material and bedrock blocks on the slopes of the volcano and in valleys of both the left and right tributaries of the Baksan River, as well as at the sources of the Malka and Kuban' rivers. Additionally, materials of ancient stone avalanches can be seen in the Irik, Yusen'ga, and Adyrsu river valleys. The volume of separate seismogravitational proluvium reaches 15-17 mln. m3. The largest bodies of this type once completely dammed up the Baksan Valley and its tributaries (Adyrsu, Adylsu, Kyrtyk, Irik, and Yusen'ga rivers), resulting in the formation of a cascade of temporary lakes. Sediments of these lakes can still be observed at the surface of Pleistocene alluvial terraces (Fig. 2). Cascades of block landslides on steep (30-40°) slopes of the Baksan Valley were formed as a result of one seismic pulse and did not undergo subsequent displacements. This also testifies to the seismogenic nature of gravitational dislocations.

There are distributed neptune dikes in friable (alluvial, fluvial-glacial and lacustrine) sandy sequences and lahar deposits. Neptunic dikes in Quaternary friable sediments indicate the injection of water-saturated sand into the overlying rock sequence under the impact of compression during seismic wave propagation. Such phenomena are widespread in epicentral zones of strong recent earthquakes within an isoseismal intensity of 8 or more.

4. DISCUSSION

Various relief dislocations (active faults, collapses and stone avalanches, dammed lakes, and neptunian dikes) were repeatedly formed in the Holocene in different parts of the mountainous region. The almost simultaneous and short-lived tectonic events were separated by long-term quiet periods (Fig. 4). The short-term episodes of the deformation of young sediments and topography can be identified, with a great degree of probability, as moments of strong earthquakes.

According to the paleoseismological data, these seismic reactivation periods occurred 5,500, 3,800, 2,300, and 300-400 yrs. ago. The first seismic event in this series induced the formation of an extensive deep lake that flooded the paleosoil at high (40-50 m) terraces of the Baksan River, near the mouth of the Kyldybashsu tributary (Sample IGAN 2610). This event is also probably responsible for the injection of numerous neptunian dikes in the bedded sequence of fluvio-glacial volcanomictic sands on the northern slope of Elbrus Volcano at the end of Ulluchiran glacier and in sandy lenses on alluvial terraces at the mouth of the Kyrtyk River. The second earthquake was accompanied by the filling of two dammed lakes in the Baksan River valley near the present-day Elbrus Settlement area and upstream, near the town of Tyrnyauz 3870±90 yr b.p., Sample IGAN 2588). The third seismic event produced the southern seismic fault at the flat surface of the former German airdrome site on the north-eastern slope of the volcanic cone (2280 ± 90 yr b.p.; Sample IGAN 2592). This event initiated avalanches in the lower reaches of the Adyrsu River, the Tyubele rampart zone, and the sources of the Beitik-Tebe River (2320 ± 70 yr b.p.; Sample IGAN 2589). Finally, the fourth seismic shock induced large collapses at the sources of the Adyrsu and

Yusen'ga rivers, as well as in the Baksan River valley east of the settlement of Elbrus (490 ± 30 yr b.p., Sample IGAN 2611; 430 ± 60 yr b.p., Sample IGAN 2619; 340 ± 150 yr b.p., Sample IGAN 2585; 530 ± 30 yr b.p., Sample 2618). These events generated several dammed lakes near the settlement of Elbrus, at the source of the Adyrsu River, and upstream of the town of Tyrnyauz.

The available historical information confirms a strong earthquake on the northern slope of the Greater Caucasus in 1688 ± 1, i.e., more than 300 yr ago (Kondorskaya and Shebalina, 1977). This event, known as the Terskoe earthquake (coordinates 43.7±1° N and 44.7±1°E; magnitude 5.3 ± 0.7; intensity, 7 ± 1), destroyed several buildings in the Terskii region. It is likely that this precise earthquake gave rise to the above-mentioned young seismic dislocations in the Baksan and Adyrsu river valleys. Their coordinates are very close to the coordinates reported in the catalogue. The earthquake source could be localized in the Elbrus region, and its magnitude could have exceeded 6.0-6.5.

On the whole, polychronous and polygenous seismic dislocations were found over a spacious oval-shaped area of the Elbrus region. Two cones of the Elbrus Volcano are situated in the western part of this area, and its long axis is oriented in the west-northwest (Caucasus) direction, roughly coinciding with the Syltran fault zone, which is considered a magma-controlling structure. The major volcanic edifices of Elbrus are related to this fault zone. The northern boundary of the paleoseismic area extends along the northern wall of the Pshekish-Tyrnyauz Fault. The southern boundary is traced in the middle reaches of the Adylsu River (the Shkhel'da River mouth) and Adyrsu River (the Novyi Dzhailyk alpinist camp area). The western boundary of the oval area extends into the upper part of the mountainous segment of the Kuban River valley. The eastern boundary is not traced beyond the meridian of Tyrnyauz. Thus, the outlined region of paleoseismic dislocations reaches 50-60 km in length and 20-25 km in width. This region coincides approximately with the PSS position revealed by the seismotectonic method. The dimensions of the region affected by paleoseismic dislocations generally correspond to the dimensions of the pleistoseist region of an earthquake with a magnitude of 7.2 and a crustal source (Wells and Coppersmith, 1994). The primary seismic dislocations and gravitational (both seismic and aseismic) slope structures are unknown over a distance of tens of kilometres beyond the above-stated oval region.

5. CONCLUSIONS

The comparison of seismic reactivation periods established from the paleoseismological data discussed above and periods of Holocene activity of Elbrus Volcano shows that strong earthquakes and catastrophic eruptions had occurred there at different times over the last 6 ky. Moreover, a periodical alternation of two forms of endogenic activity has been established. The recurrence period is 1,500-1,900 yrs. for the strongest earthquakes and 1,000-2,000 yrs. for volcanic eruptions. An in-phase relationship has not been observed in the manifestations of these natural catastrophes.

ACKNOWLEDGEMENTS

This work was supported by the Russian Foundation for Basic Research (project no. 02-05-64946).

REFERENCES

1. Bogatikov, O.A., Rogozhin, E.A., Gurbanov, A.G., et al.,2003. About correlation of Quaternary Volcanism and earthquakes on the North Caucasus, in *Materials of the 36th Tectonic Conference*, Moscow: GEOS, vol. 1, 48-54 (in Russian).
2. Catastrophic Processes and Their Impact on the Environment, 2002 in *Volcanism*, Laverov, N.P., (ed.), Moscow: *Region. Obshchest. Organizatsiya Uchenykh Probl. Prikl. Geofiz.*, (in Russian).
3. Nikonov, A.A., 1991. Paleoseismic Ruptures in axial part of Main Caucasian Ridge (Elbrus region*), Dokl. Akad. Nauk SSSR* (Reports of USSR Academy of Sciences), **319**, no. 5, 1183-1186.
4. *New Catalog of Strong Earthquakes on the USSR Territory*,1977. Kondorskaya, N.V. and Shebalina, N.V., (eds.), Moscow: Nauka, (in Russian).
5. Rogozhin, E.A., 2002. Modern Geodynamics and Potential Earthquake Sources of Caucasian region. In *Modern Mathematical and Geological Models of the Environment*, Moscow: United Institute of Physics of the Earth., 244-254 (in Russian).
6. Solonenko, V.P., 1973. Paleoseismology. in *Izvestia: Physics of the Solid Earth*, no. 9, 3-16 (in Russian).
7. Wells, D.L. and Coppersmith, K.J., 1994. New empirical relationships among magnitude, rupture length, rupture width, rupture area, and surface displacement, *Bull. Seismol. Soc. Am.*, **84**, no. 4, 974-1002.

THE AREAS OF MUD VOLCANISM IN THE SOUTH CASPIAN AND BLACK SEA: SEISMICITY AND NEW TECHNOLOGY FOR SEISMIC RISK ESTIMATION

Leonid E. Levin[1], Leonid N. Solodilov[1],
Behrouz M. Panahi[2], Nadejda V. Kondorskaya[3]

1 Fedynsky Centre of Regional Geophysical and Geoecological Researches (Centre GEON), 4, Chisty per., Moscow 119034, Russia

2 Geology Institute, Azerbaijan National Academy of Sciences, 29-A H. Javid Ave., Baku 370143, Azerbaijan

3 Schmidt United Institute of Physics of the Earth, Russian Academy of Sciences, 10 Bolshaya Gruzinskaya str., Moscow 123810, Russia

Abstract: Technology for a seismicity study includes the following sections: a study of the newest geodynamics for the central part of the Alpine belt of Eurasia for a long time interval; differentiation of the lithosphere into elastic-brittle and plastic-viscous layers; determination of the peculiar features of seismic wave energy distribution for the instrumental period (1900 - 2000); differentiation of regional lineaments and faults by seismic activity; analysis of direction of earthquake foci migration in the plastic-viscous layer of the lithosphere; determination of space position of seismic hazard zones; calculation of predicted coordinates of high seismic potential sites and expected earthquake magnitudes in these sites.

Key words: Geodynamics, mud volcanism, seismic hazard

1. GEODYNAMIC SETTING AND MUD VOLCANISM

The Caspian – Black Sea region is characterized by the presence of many mud volcanoes and by a complex geodynamic setting. Volcanoes and mud diapirs are developed, predominantly in the zones where the sedimentary cover is of great (from 10-15 to 20-30 km) thickness. The geodynamic setting is distinguished by a reverse correlation between the belts of pseudosubduction, i.e. subduction without spreading (Khain, 2003) and regions of mud volcanism.

For the considered region, two main belts of pseudosubduction are envisaged: the northern - from the Absheron - Balkhan sill and Eastern Caucasus to the Mountain Crimea and Dobrudja inclusively, with a probable

G. Martinelli and B. Panahi (eds.), Mud Volcanoes, Geodynamics and Seismicity, 111–121

continuation in the Vranch zone of the Eastern Carpathians, the southern - from the Alborz and Talesh to the orogens of Pontides and further towards the area of the northern Rhodopian massif. In the south, outside of the Caspian - Black Sea region, another belt of pseudosubduction along the orogens of Zagros and Taurides is developed. In these belts, the depths of earthquake hypocentres range from 50-60 to more than 100 km (Fig. 1).

Upwelling of the partially melted part of the asthenosphere above zones of pseudosubduction causes a high concentration of earthquake hypocentres in the elastic-brittle layer of the lithosphere with maximum value of released seismic wave energy of up to 10^{21} - 10^{23} erg • km^{-2} • $year^{-1}$. The regions of mud volcanism are found predominantly in the inner areas of the South Caspian and Azov - Black Sea sedimentary basins, where subduction and upwelling of the asthenosphere occurred in geological past, more likely, up to Early Cenozoic.

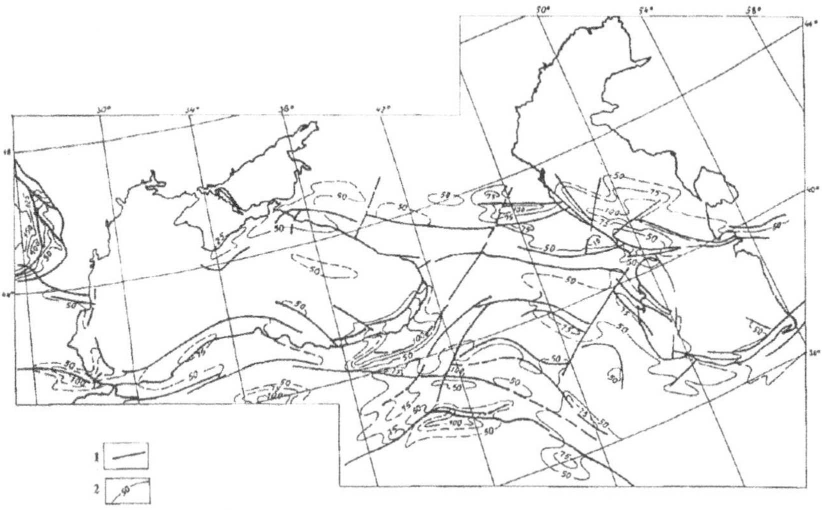

Figure 1. The scheme of pseudosubduction zones structure in the Caspian - Black Sea region (Levin, 2003). 1 - Regional lineaments, 2 - Isolines of pseudosubduction zones structure, km

In the directions from the mentioned belts to the regions of mud volcanism, values of released energy of seismic waves decrease to 10^{17} - 10^{18} erg • km^{-2} • year $^{-1}$, but the hazard of ecological disaster remains (Fig. 2).

In the Caspian region, mud volcanoes are concentrated in the south-westernmost part of the North-Absheron trough, along the structures of the Absheron-Balkhan sill, the western part of the South Caspian deep-sea basin. Single mud volcanoes are found in the Pre-Alborz deep and to the south of the Godin uplift. All of them are located in the areas with thickness of the Pliocene-Quaternary sediments of more than 7-8 km and characterize a higher thermal regime.

The Azov - Black Sea basin is distinguished by the presence of mud volcanoes predominantly in the deep-sea basins and partly on the north-western and north-eastern shelves. All these volcanoes are confined to those regions where the thickness of the Miocene - Quaternary sediments exceeds 5 km.

In these regions, the peculiarities of the seismicity of the mud volcanism areas have been the subject of specialized analysis (Panahi, 1988; Panahi, 1998; Panahi and Kasparov, 1988; Panahi and Rakhmanov, 2000).

In the mentioned works, the main conclusions were achieved from the comparison of earthquake parameters for a long period with a sequence of mud volcano eruptions. These are due to the following factors:

a) hypocentres of earthquakes are located at depths between 10 and 20 km in the sedimentary cover and between 40 and 50 km in the upper horizons of the earth crust;

b) eruptions of volcanoes are fixed either till the moment of earthquakes or after this moment not revealing an empirical relationship;

c) the intensity of the seismic effect is VI-VII and does not exceed VIII degree at the MSK scale, but can reach VIII ± 1 degree depending on the structure of bottom sediments;

d) the energetic class of events does not exceed K≤15, values of magnitudes of most earthquakes are between 4.0 and 5.9 and sometimes ≤6.4;

therefore, the mud volcano regions are characterized by a lower and moderate level of seismic activity. Such an activity is in a reverse relationship with the thermal regime of the lithosphere reflected in a decrease of the elastic-brittle layer thickness.

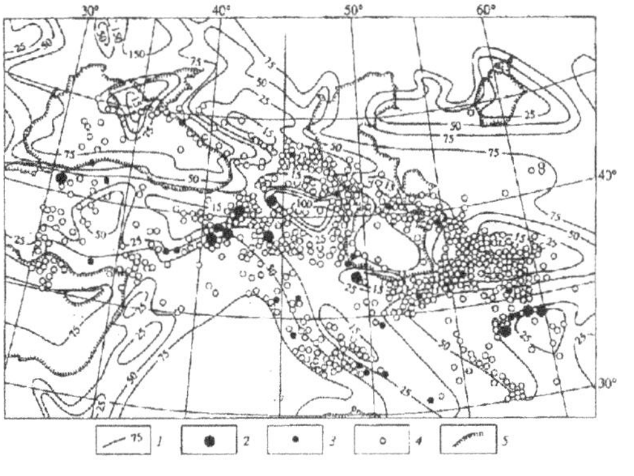

Figure 2. The Caspian - Black Sea region: seismicity in the brittle layer of the lithosphere during 1964 - 1993.
1 - 600°C depth contour lines, km; 2 - 4 - earthquakes with magnitudes ≥6.5; 6.4 - 5.0 and 5.0 respectively; 5 - Caspian, Black and Mediterranean Seas shorelines.

2. CHARACTERISTICS OF REGIONAL SEISMICITY

Pseudo-subduction is associated with the intrusion of the cold mantle part of the continental lithosphere into the relatively heated plastic-viscous layer of the mantle and sometimes directly into the asthenosphere. Compared with early known types of subduction, this process has been distinguished earlier as so-called C-subduction (Khain and Lobkovsky, 1994). Destruction of the cold sheets of the above-asthenosphere mantle is responsible for the origination of earthquakes with large magnitudes at depths ranging from 50 to more than 100 km. All of this also is confirmed by the physical parameters of earthquake foci (Tuliani, 1999).

In the Caspian – Black Sea region, pseudosubduction has different directions. For the belt of the Absheron – Balkhan sill – Southern Crimea, these are the northern and north-eastern directions controlled by a movement of the Arabian plate to the north. Only in the Vranch zone is pseudo-subduction directed to the west. The belt of the Alborz – northern part of the Rhodopian massif is uneven by this indicator in different part. Along the Alborz and Pontides pseudo-subduction is directed to the south, along the Talesh – to the south-west and along the limitation of the Rhodopian massif – to the north. For the Alborz and Talesh these directions had been established earlier (Jackson et al., 2002). These different directions of pseudosubduction are most likely to have been caused by two processes: movements of the lithosphere plates in the region with s displacement of the collision zones (i.e. a combination of pseudo-subductions with thrusts); the influence of the asthenosphere upwelling (mantle diapirism) on a direction of pseudosubduction in the central parts of the deep-sea basins of the South Caspian and Black Sea which is most likely to have occurred here , until the Early Cenozoic. An upwelling presence is confirmed by an estimation of depths of the asthenosphere top of 40 km for the South Caspian (Ashirov et al., 1976) and 75-125 km in the deep-sea basins of the Black Sea (Belousov et al., 1988).

The shallow-focus earthquakes in the elastic-brittle layer of the lithosphere with hypocentre depths up to 30 km are developed above zones of pseudo-subduction. As has previously been established for other seismoactive regions of the Earth, up to 90% of earthquake foci and the released energy of seismic waves is contained in this layer alone (Kondorskaya and Kireev, 1985; Wang and Molnar, 1983). In the Caspian – Black Sea region, five types of seismoactive zones are established in this layer (Levin et al., 2000): zones along the collision boundaries of plates; those along the strike of the boundary between the Eastern Black Sea and South Caspian microplates with horizontal displacements; those of intraplate seismicity on the frame of recent upwelling zones (axial zones of orogens) with a sharp thickness gradient of the elastic-brittle layer; those located directly above the mentioned upwelling zones; those of intraplate seismicity along the regional lineaments with horizontal displacement. For the South and Middle Caspian basins, the greatest seismic hazards are Absheron – Cheleken – Kum-Dagh zone and the region of the south of the North Absheron graben with high-magnitude earthquakes in the subduction plate. To the west of these tectonic elements, the zone of the Transcaucasus transverse uplift and to the south of them, the Alborz and its surroundings, belong to the analogous category. For the Azov – Black Sea sedimentary basin, the highest seismic hazards are the zones of its surroundings: the Mountain Crimes, the Pontides with the North Anatolian lineament, the Eastern Carpathians with the Vranch zone.

3. ESTIMATION OF THE EARTHQUAKE RISK IN THE CASPIAN REGION

Detailed estimation of the earthquake risk by the new technology has so far only been made for the region of the Caspian Sea and its frame. This problem requires, first of all, an analysis of the thermal regime of the lithosphere on data of heat flow and generalization of data on the earthquakes for the instrumental period (i.e. from 1900 to 2000).

Data on the earthquakes (date, epicentre coordinates, focus depth, magnitude) were taken from a number of catalogs: SCETAC; the specialized catalog of Northern Eurasia; operative reports of the Geophysical Service of Russia (Kondorskaya and Ulomov, 1993; Kondorskaya et al., 1994).

A calculation of the thermal regime (depths of isotherms 600 and 1200^{0}C) was made by 50x50 km squares with the use of a standard equation of relationship between the heat flow and temperature at any depth.

A calculation of the released energy of the seismic waves was carried out by the quadratic equation of the relationship between the logarithm of seismic energy and magnitude (Tuliani, 1999).

A distribution of energy is controlled by an alternation of belts with high (up to 10^{21}) and low (10^{17} erg • km^{-2} • year^{-1}) values which have a south-eastern orientation and are traced from the Scythian to the Turanian plate through the Middle Caspian (Fig. 3). These belts also embrace the orogen of the Caucasus.

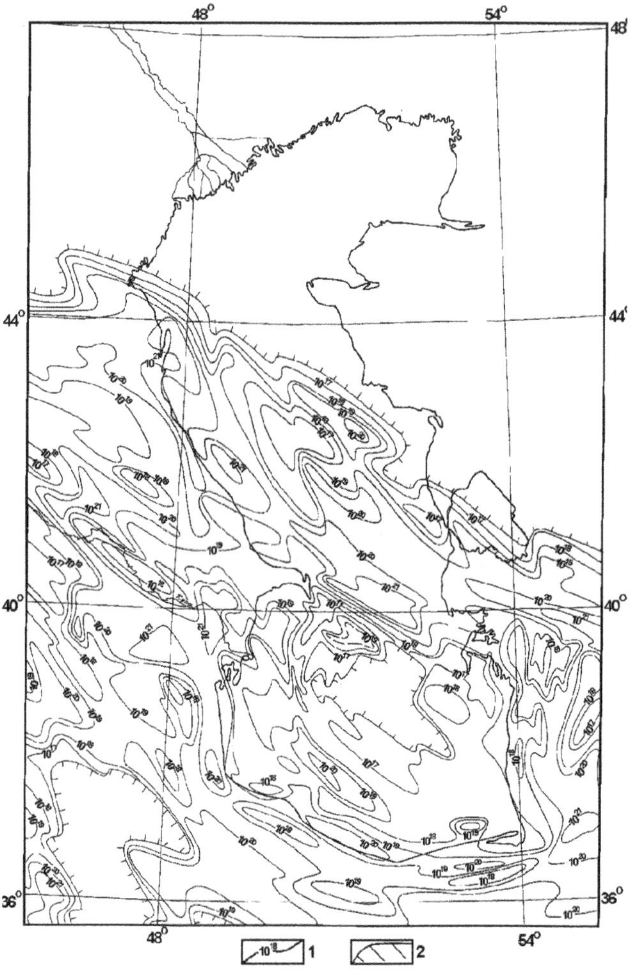

Figure 3. The Caspian region: the map of released energy of seismic waves for the period of 1900-2000 years:
1 - isolines of total energy of seismic waves for 1900-2000 years, erg •.km^{-2}• year^{-1};
2 - contour of area of earthquakes with M≥2 absence

The South Caspian is characterized by a ring system of belts including the Talesh and Alborz orogens with values of 10^{19} to 10^{20} erg • km^{-2} • year^{-1}. The belts of increased energy correspond to the zones 75 km thick of the elastic-brittle layer and the belts of decreased energy to the zones with a of 20-40 km thickness of this layer (Fig. 3).

Regional faults are traced along the narrow zones of thickness gradient of the elastic-brittle layer and values of released energy of seismic waves. Faults are subdivided into four groups: extremely high activity, high activity, low or without seismic activity (Fig. 4).

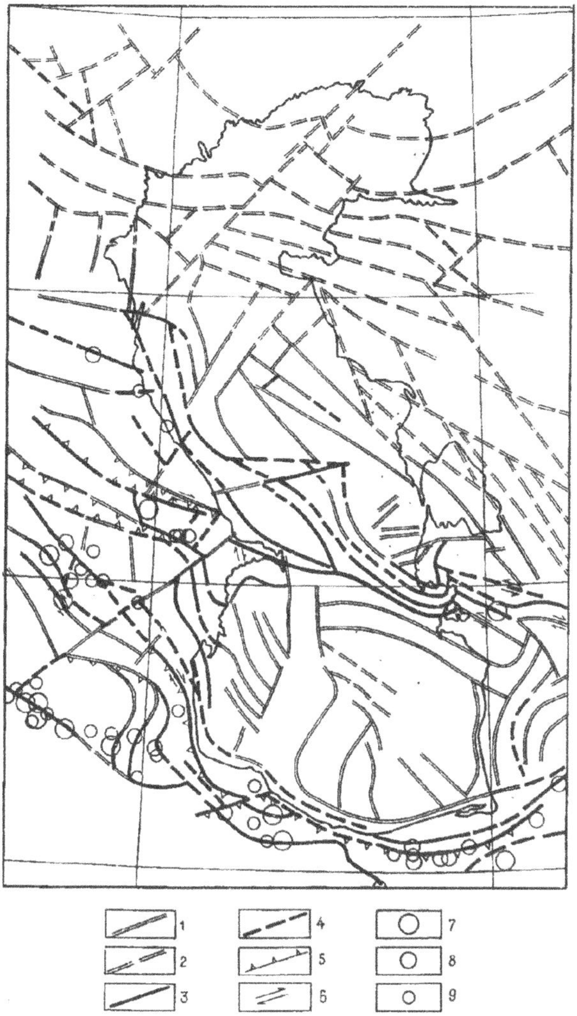

Figure 4. Differentiation of regional faults by the value of magnitude and released energy: 1-4 activity of regional faults: 1 – extremely high, 2 – high, 3 – middle, 4 – low and no activity; 5 – regional thrusts; 6 – direction of relative or apparent lateral displacement along fault; 7-9 magnitude of historical earthquakes: 7 – M > 7, 8 – M = 6-7, 9 – V = 5.5 - 6.0

For the west of the Caspian region, migration of earthquake hypocentres is envisaged in two directions: the north-eastern and south-eastern. In the south-east and south of the region, the analogous migration has three directions: submeridional, north-western and north-eastern. Some directions have south-western orientation being connected with uneven distribution of stresses in the plastic-viscous layer.

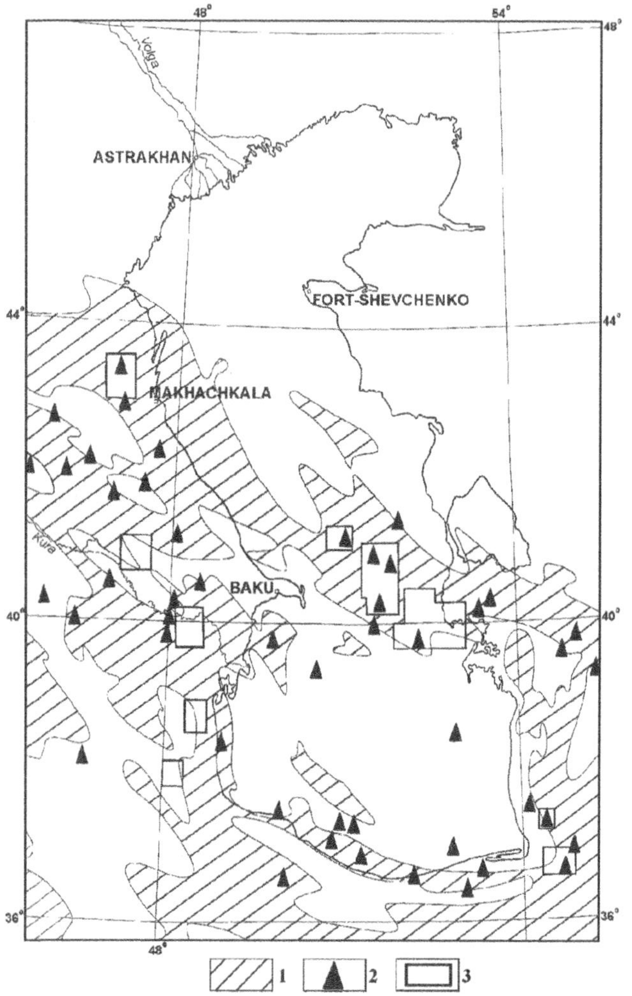

Figure 5. Caspian region: the map of highest seismicity belts.
1 – belts of highest seismicity; 2 – high seismic potential sites with M > 5.5 up to 2005 year; 3 – areas with inferred earthquakes up to the year 2005

Directions of the migration coincide with the seismic hazard zones which form intersecting systems. The intersections are high seismic potential sites, as had been established earlier by L.I. Tuliani (1999). These sites are distributed over a wide area of the frame of the South and Middle Caspian. About 90% of the high seismic potential sites are located in zones of block conjugation with different thicknesses of the elastic-brittle layer of the lithosphere where the gradient reaches 20 km. Zones of sharp gradients of released seismic wave energy can also be traced here (Figure 3).

In the period 1998-2000 about 43 high seismic potential sites with predicted magnitude of more than 5.5 were distinguished in the interfaulted blocks. Since the second half of 1998 until December 2000, 5 destructive earthquakes were recorded in the region. These earthquakes confirmed the prediction of the coordinates of high seismic potential sites and the intensity of events. A difference between the observed and predicted coordinates ranges from 5 to 25 km. This value does not exceed a volume of earthquake foci (Levin and Solodilov, 2003; Levin et al., 2002).

In the Caspian region, on the basis of the geological-geophysical parameters of the structure and distribution of the hydrocarbon fields, 9 high density belts of total initial potential hydrocarbon resources have been identified.

A comparison of these belts with the belts and areas of increased seismicity provides grounds for a differentiation of the region, including the Caspian Sea proper, by a degree of seismic hazard during petroleum industry development.

Five belts with an increased density of initial potential resources are the most seismic hazard (Fig. 5):

1. the belt of the Absheron-Balkhan sill together with North Absheron trough, including the western continuation of the first of them into the Kura graben ,where high seismic potential sites with a probable magnitude ≥ 5.5 are envisaged;

2. the belt of the eastern Tersko-Caspian graben including the Tersko-Sulak and the Dagestan grabens where high seismic potential sites with probable magnitude ≥ 5.5 are inferred;

3. two belts along the western and eastern flanges of the South Caspian sedimentary basin together with sublatitudinal Pre-Alborz deep with high seismic potential sites similar by intensity;

4. the interesting north-western belt within the Middle Caspian sedimentary basin including the Middle Caspian deep-sea basin, where high seismic potential sites with a magnitude 5.5 are also inferred.

The debatable issue is the estimate of seismicity for the belt of increased resource density on the marine continuation of the Southern Mangyshlak together with the Peschanomysskoe uplift and Kazakh bay trough. The disputable issue is defined by the presence of a number of seismic hazard zones with a north-eastern orientation combined with data on historical earthquakes.

4. CONCLUSIONS

Seismicity of the Caspian-Black Sea region is controlled by a combination of interrelated processes: pseudo-subduction, interaction of three main layers of the tectonosphere – asthenosphere, plastic-viscous and elastic-brittle.

Directions of pseudo-subduction in separate sectors appear to be different - the northern, southern and western.

Concentration up to 90% of earthquakes hypocentres and released energy of seismic waves occurs at depths varying from 10 to 30 km in the elastic-brittle layer of the lithosphere.

Regional faults by magnitudes of earthquakes and energy of seismic waves are subdivided into four groups: extremely high activity, high activity, middle activity, low or no seismic activity.

At the frame of the Caspian Sea, the following tectonic elements are of high seismicity: the Alborz and Talesh orogens, the zones of thrusts on the northern and southern flanges of the Greater Caucasus.

The South and Middle Caspian basins are characterized by a presence of five belts of increased resource density and high seismicity.

The South Caspian and Black Sea deep-sea basins are characterized by a low seismic activity probably associated with a great thickness of the sedimentary cover, the presence of mud volcanoes, the low thickness of the elastic-brittle layer of the lithosphere.

The region of the Caspian Sea and its frame appear to be sharply differentiated by seismic activity and its correlation with belts of increased density of initial potential hydrocarbon resources.

The estimation of the seismic hazard with a prediction of coordinates of high seismic potential sites and intensity of probable earthquakes on the basis of complex technology tested in the region of the Caspian Sea, can be used in any other seismoactive region of the Earth.

REFERENCES

1. Ashirov, T., Dubrovsky, V.G., and Smirnov, Ya.B., 1976. Geothermic and geoelectric investigations in the South Caspian deep-sea basin and nature of the layer of the increased conductivity. *Reports of the Ac. Sci. USSR*, **226**, N 2, 401-404.
2. Belousov, V.V., Volvovsky, B.S., Arhipov , I.V. et al., 1988. Structure and evolution of the Earth's crust and upper mantle of the Black Sea, in *Black Sea*, Finetti, I.R., (ed.), *Black Sea,* Bollettino di. Geofisica Teorica e Applicata, Vol. XXX, 109-196.
3. Khain, V.E., 2003. Pseudosubduction and its role in development of orogenic belts, In *Tectonics and geodynamics of the continental lithosphere*" M., GEOS, 270-271.
4. Khain ,V.E., and Lobkovsky, L.I., 1994. Conditions of Existence of the Residual Mantle Seismicity of the Alpine Belt in Eurasia, *Geotectonics*, N 3, 12-20.
5. Jackson, J., Priestley, K., Allen, M., and Berberian , M., 2002. Active tectonics of the South Caspian Basin., *Geoph. Int.*, N 148, 214-245.
6. Kondorskaya, N.V., Kireev, I.A., 1985. On the estimation of the earthquake maximum magnitude based on a joint analysis of seismological and geothermal parameters., *Tectonophysics*, **121**, N 1, 79-86.

7. Kondorskaya, N.V., and Ulomov, V.I. , 1993. Specialized Catalog of Earthquakes in North Eurasia since Ancient Times to 1990: A System of Geoscience Data Bases, Fed. Reg. N 0229 70 33 56. http//www.scgis.ru//system of data bases.
8. Kondorskaya, N.V., Shebalin , N.V. et al., 1994. Special Catalog for GSHAP. Test Area "Caucasus" from 2000 BC through 1993. http.//www.scgis.ru//system of data bases.
9. Levin, L.E., and Solodilov, L.N., 2003. Seismicity and estimation of earthquakes risk in the oil/gas bearing basins. *Exploration and Earth Depths Protection Journal*. N 2, 21-23.
10. Levin, L.E., Solodilov, L.N., Kondorskaya , N.V. , 2000. The Central Part of Alpine Belt of Eurasia: Geodynamics, Heat Flow Seismicity, *Volcanology and Seismology*, **22**, 179-193
11. Levin, L.E., Solodilov, L.N., Kondorskaya, N.K., Gasanov , A.G. and Panahi , B.M., 2002. Thermal Regime of the Lithosphere and Prediction of Seismic Hazard in the Caspian Region, in *Abs. Inter. Conf. Petroleum Geology and Hydrocarbon Potential of the Caspian and Black Sea Regions*", Soc. Petr. Geol , Baku, Azerbaijan.
12. Tuliani , L.I., 1999. Seismicity and seismic hazard on the basis of thermodynamic and rheological parameters of tectonosphere, Naucnyj mir , Moscow, 210 p. (in Russian).
13. Panahi , B.M. , 1998. Seismotectonics of the territory of the Caspian Sea. Ph. D. Thesis, Baku.
14. Panahi, B.M., 1988. Seismicity of areas of mud volcanoes development. B. Sc. Thesis, Moscow.
15. Panahi , B.M., and Kasparov, V.A., 1988. Problems of seismic regime on the territory of the Caspian Sea. *News Ac. Sci. Azerbaijan SSR, series of Sciences of the Earth*, N 1, 91-98.
16. Panahi, B.M., and Rakhmanov, R.R., 2000. Seismicity of areas of mud volcanoes development, In *Prediction and control of geodynamic and ecological environments in the region of the Caspian Sea in connection of development of oil and gas industry*, Scientific World, 111-126.
17. Wang Ping Chen and Molnar, P., 1983. Focal depths of intracontinental and interplate earthquakes and their implication for the thermal and mechanical properties of the lithosphere, *Jour. Geoph. Res*., 88, N 5, 4183-4214.

RECENT SEISMIC ACTIVITY IN ALBANIA AND ITS FEATURES

Siasi Kociu
Department of Geology and Geophysics University of Missouri- Rolla 125 Mc-Nutt Hall Rolla-Missouri USA, e-mail: kocius@umr.edu

Abstract: Data on historical earthquakes in Albania are very scarce. The total number of historical earthquakes (with I_0>VIII degrees) prior to1800 is only 21 events, and 77 events (with I_0>VII degrees), and for the 19th century. 1037 events with M>4.1, were recorded during period 1900-1995.Albania is characterized by a continuous and sporadic seismic activity. During past and recent earthquakes a lot of liquefaction phenomena including "pseudo" volcanoes of sand and water have been observed in the Preadriatic area of Albania.

Although there is very little or even a lack of evidence concerning mud-volcanoes in Albania, traces can be found in places crossed by the most important seismogenic belts . For this purpose the calibration of past seismic activity with recent activity has to be calibrated. For a continuous seismic activity, the calibration of past seismic activity with recent activity can be performed through the third extreme value distribution, while for sporadic seismic activity the "quiescence" zones are more important for such a calibration. Besides the geological and tectonic settings, some additional evidence such as fault rupturing on the free surface, induced ground failure phenomena, archaeological ruins in some ancient historical cities, etc. can be used to assess the size of the missing strong events.

The seismic activity maps showed large linear foci in Albania.

Key words: Past and recent seismic activity, "quiescence" zones, "pseudo-volcanoes"

1. INTRODUCTION

Due to the lack of historical data and the incompleteness of many earthquake catalogues, many scholars have been using different approaches in order to have a full set of data for the missing events. Some are using priory given values for missing events (Tezcan, 1996). One of the methods for the seismic hazard assessment through M_{max} is Gumbel's extreme value theory (Gumbel, 1958), which requires yearly maxims.

2. THE CALIBRATION OF SEISMIC ACTIVITY

2.1 The earthquakes catalogues

For the seismic activity assessment of Albania, the following were used:

G. Martinelli and B. Panahi (eds.), Mud Volcanoes, Geodynamics and Seismicity, 123–133

The catalogue of earthquakes in Albania including the events with Io≥7 degrees for the period prior up to 1900 and events with Io≥6 degrees (or M≥4.1±1/3) for the period 1900-1970 (Sulstarova and Kociu, 1975),

The catalogue of earthquakes with M>4.0±1/3 for the Balkan Region for the period 1971-1990 (Kociu et al., 1991), updated for the period 1991-1995.

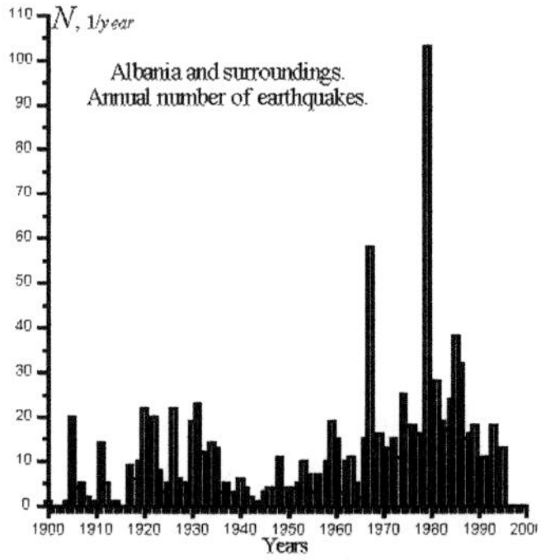

Figure 1. Yearly number of earthquakes for Albania and surrounding areas (Zavyalov et al., 2002)

2.2 Check-up of raw catalogue homogeneity

The completeness of earthquake catalogues is affected by the uncertainties concerning the coordinates, time of occurrence, earthquake magnitudes , determining the **resolution**, and the cutoff magnitude (**M_{cut}**), defined as the lowest magnitude of earthquakes determining the **sensitivity** of an earthquake catalogue.

The evolution of **M_{cut}** in space and time was studied for the earthquake catalogue of Albania for 1900-1995 (Fig.2)

It was established that an **M_{cut}** decrease from 4.6 at the beginning of last century up to 3.6-3.8 at the beginning of 1990's. For the recent period 1971-1995**M_{cut}** has varied from M_{cut}=3.6 (Central Albania) to M_{cut}=3.9 (for northeastern Albania)

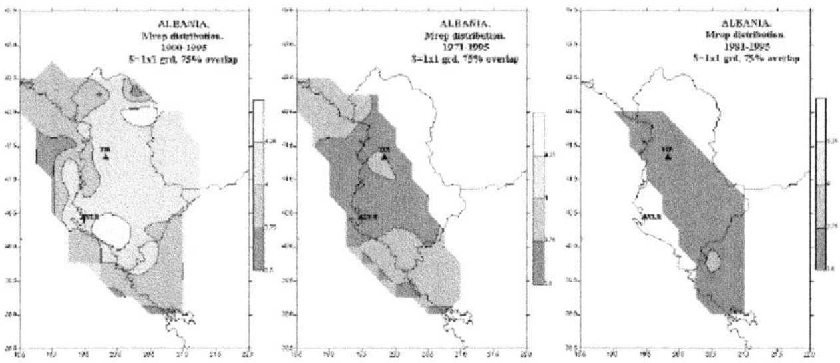

Figure 2. Cut-off magnitudes for different periods of observation for Albania (Zavyalov et al., 2002)

2.3 Normalized seismic activity data (A_k)

For the regions with a continuous seismic activity, it is difficult to correlate the space distribution of recent seismic activity with particular faults, due to a great number of epicenters and lower values of $\mathbf{M_{cut}}$.

For the regions with sporadic seismic activity, where there is a lack of strong events, the complementary information for strong missing events can be achieved, mainly based on normalized recent seismic activity data. This information was used to fill the gap of strong missing data at every grid point, where normalized seismic activity data were computed. The distribution of magnitudes for truncated exponential distribution corresponding to 3-d extreme value distribution (Kociu and Varfi, 1984) is:

$$n(M \geq M_{\min}) = \frac{n_o \mathrm{EXP}[-\beta(M - M_{\min})]}{\{1 - \mathrm{EXP}[-\beta(M_{\max} - M_{\min})]\}} \qquad (2.3.1)$$

For: $M_{\min} \leq M_j \leq M_{\max}$

where: $no = 10^a$ (2.3.1.a)

and log $N = a - bM$ (2.3.1.b)

The corresponding normalized seismic activity (A_k) (for M_{min}=3.5, S=1000 km^2 and T= 1 year) (Kociu, 1994) is:

$$A_k = \frac{A_o}{\{1 - \mathrm{EXP}[-\beta(M_{\max} - (M_{\min} - 1/3))]\}} \qquad (2.3.2)$$

where:

$$Ao = \frac{n(M \geq M_{\min})1000[1 - \mathrm{EXP}(-2\beta/3)]}{\{\pi R^2 T \mathrm{EXP}[-\beta(M - M_{\min})]\}}$$

is the normalized seismic activity for simple exponential distribution of magnitudes. For the computation of a map Ak=0.05,0.1,0.3, 0.5, 1.0, 3.0 were accepted according to (2.3.2) at every grid point.

2.4 Assessment of b-values

To compute the normalized seismic activity we have to know the coefficients a and b of Richter or Gumbel's type of distribution of magnitudes (1.3.1.b) (see Table 1).

Table 1. Event number and b-values for different periods

Periods of observation	*Total number of events*	M_{min}	*Events with* $M>M_{min}$	*b-value (See 1.3.1.b)*
Prior to 1900	98	5.5	99	0.82
1900-1970	577	4.8	286	0.76
1900-1995	1037	4.1	617	0.79
1971-1995	460	4.1	331	0.66

The b-values for different periods: prior to 1990 (b=0.82), 1900-1970 (b=0.76) and 1900-1995 (b=0.79) are close to each other. Lower b=0.66 (period 1971-1995), shows an increase of seismic activity, in northwestern Albania (where the earthquake of April 15, 1979 occurred) and in south-western Albania.

2.5 Equal accuracy mapping

Normalized seismic activity maps were computed taking the same accuracy/number of epicenters (P=55% or n ($M \geq M_{min}$)=5).

For the period prior to 1900 (Fig.3), normalized seismic activity is concentrated in some very well-known active knots along the Adriatic-Ionic coast (such as Shkodra, Durres, Vlora, Himara, Konispol-Sagiadha) and less inland (Elbasani and Ohrid lake).

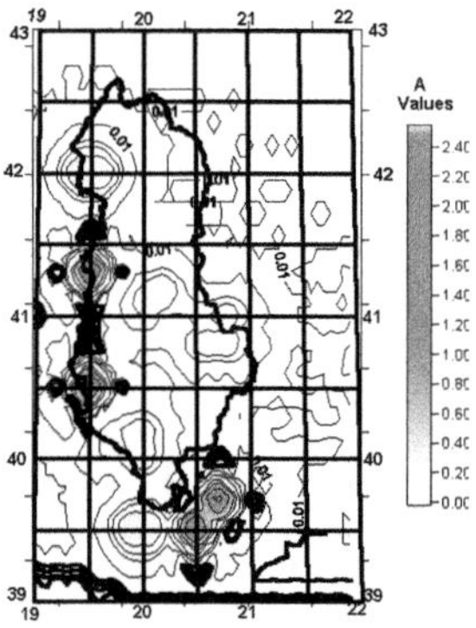

Figure 3. The map of normalized seismic activity for the period prior to 1900 ($I_{omin} \geq$VII degrees or $M_{min} \geq 5.2$)

The seismic activity maps for the longest instrumental observations: 1900-1970 (Fig.4a) and 1900-1905 (Fig.4b), indicate on the well-known seismogenic belts:

a) the Adriatic and Ionian sea coast
b) Korca -Ohrid-Peshkopia zone, both intersected by
c) Vlora-Elbasan-Diber transversal of NE-SW strike

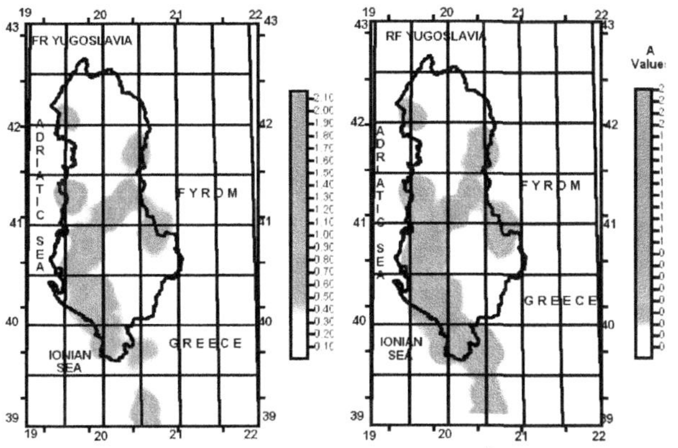

Figure 4. The maps of normalized seismic activity ($M_{min} \geq 5.1 \pm 1/3$, N=5, P=55%) for different periods: (a) 1900-1970, b) 1900-1995

2.6 The non-equal accuracy mapping

The seismic activity maps were computed on the basis of different threshold magnitudes (M_{min}) and number of n ($M >= M_{min}$), i.e. with different accuracies (see Table 2)

Table 2. Parameters of seismic activity maps with different accuracies

Period	M_{min}	*b*	$n(M>=M_{min})$	*Accuracy P%*	*Figures*
Prior to 1900	5.5	0.82	5	55	1.3
1900-1970	4.8	0.76	5	55	1.4a
	4.8	0.76	25	80	1.5a
1900-1995	4.8	0.79	5	55	1.4b
	4.8	0.79	25	80	1.5b
1971-1995	4.1	0.66	25	80	1.6a
	4.8	0.66	25	80	1.6b

The accuracy increase from 55% to 80% shows more on the continuous seismic activity zones, and less on the "quiescence" zones (the areas of the April 15, 1979 earthquake are missing).

The recent period 1971-1995 (Fig.6a, 6b) shows a concentrated activity in north-western Albania (April 15, 1979 earthquake, with considerable liquefaction phenomena observed) and in north-western Greece, representing more recent seismogenic zones.

Therefore, the representative seismic activity maps valid for the seismic hazard assessment have to be computed, mainly using $M_{min}=5.1 \pm 1/3$ independent events for the longest period (Fig.4b)

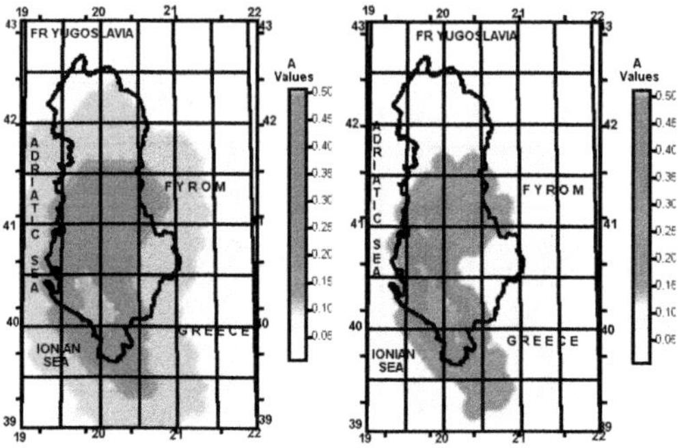

a b

*Figure 5.*The maps of normalized seismic activity (M_{min}≥5.1±1/3, N=25, P=80%) for different periods: (a) 1900-1970, b) 1900-1995

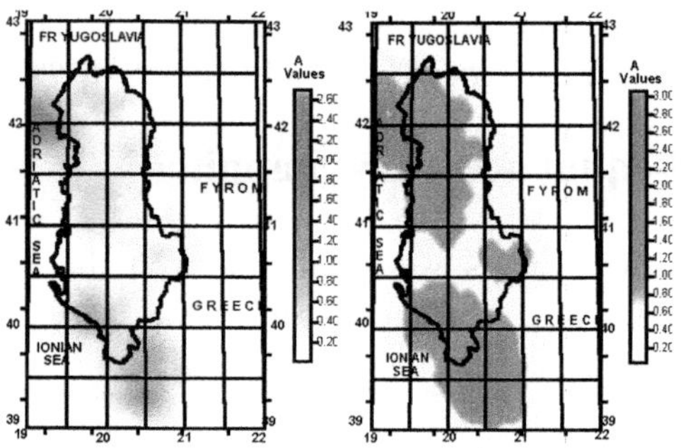

a b

*Figure 6.*The map of recent normalized seismic activity (N=25,P=80%) (The period: a) 1971-1995(M_{min}≥4.1±1/3,b) 1900-1995(M_{min}≥5.1±1/3)

2.7 The calibration of sporadic seismic activity

Some cases show the importance of "quiescence" zones in Albania (Fig.7)

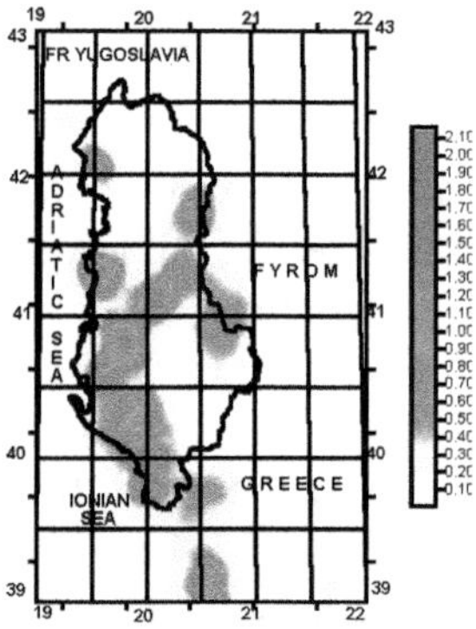

Figure 7. The map of normalized seismic activity for the period 1900-1970 (Fig. 4.a) and the "quiescence" zones

1. The earthquake of November 30 created a 10km fault on the free surface, in a "quiescence" zone, which was previously considered as locked at Elbasan (Sulstarova and Kociaj, 1969,1980), confirming a big linear Vlore-Diber transversal fault (Sulstarova and Kociaj, 1980)

2.The focus of the main shock of the April 15, 1979 earthquake was located in a seismic "quiescence" between active zones (M_{max}=7.1) (Kociaj, 1983b).

3.The Vlora zone, situated in the knot of Vlore-Diber transverse fault and Adriatic-Ionic longitudinal seismogenic zone, characterized by past strong earthquakes in 1833 (Io=IX+) and in 1851(Io=IX)), according to the seismic activity maps (Fig.2.4-2.6), represents a "quiescence" zone, which could be reactivated and where copious liquefaction phenomena could be observed. On the other hand, this zone may represent a zone for future investigations concerning mud-volcanoes.

3. M_{MAX} AND NORMALIZED SEISMIC ACTIVITY

For a continuous seismic activity, the higher the seismic activity, the higher should be the M_{max} M_{max} to be found by Gumbel's extreme value theory, which requires yearly maxims. Due to the lack of yearly maxims, normalized seismic activity data can be used (Fig.4b). For A_k=0.3, 0.5, 1.0 values, the missing annual magnitude values (M_j) were substituted by: M_j=3.5±1/3 .

3.1 M_{max} according the first Gumbel distribution

The first Gumbel distribution assumes M_{min}=0 and $M_{max}=\infty$.
Based on the first distribution of extreme values:

$$\Phi^{I}(u) = \mathrm{EXP}[-\mathrm{EXP}(-\beta(M-u))] \tag{3.1.1}$$

There are two parameters that can be found by the least-square method (Kociu, 1994):

u - Annual modal maximum magnitude, the most frequently occurring magnitude within a return period of T=1 year, and

$$H\beta = b\ln(10) \tag{3.1.2}$$

M_{max} can be found as suggested by Tezcan (1996):

$$M_{max} = \frac{a + \log T_r}{b} \tag{3.1.3}$$

where: $\alpha = n(0) = n_0$ and T_r is the mean return period.

3.1.1. Space distribution of b values according to $\Phi^{I}(u)$

The areas of the lowest **b** values coincide with the highest M_{max} (Fig.8), as follows:

- south-western and north-western Albania,
- seismogenic belt Vlorë-Elbasan-Dibër and Ohrid-Prespa lakes.

The axes of these areas (Fig.4b) represent the great linear seismic foci of Albania (Kociaj, 1983a).

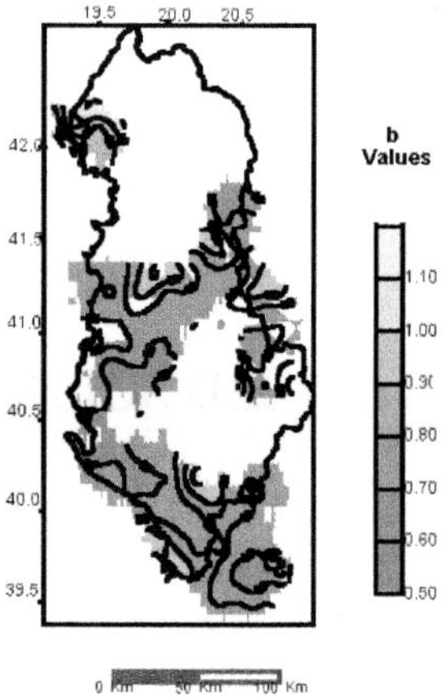

Figure 8. The map of b-values according to

3.2 M_{max} according to the third Gumbel distribution

The sensitivity, and Mmax limit Mmin by an upper boundary of the strength of the rocks to be fractured. These limitations are considered by the third distribution of extreme values (Drakopoulos and Makropoulos, 1982):

$$\Phi^{III}(u) = \text{EXP}\left[\frac{-(M_{\max} - M_j)}{(M_{\max} - u)}\right]^k = G(j) = \frac{(j - 0.44)}{N + 0.12} \qquad (3.2.1)$$

where:

M_j - set of observed annual maxim magnitudes,

$M_{max}=M_{III}$, u and coefficient k, were found at any grid point, taking into account the normalized seismic activity for the longest period of observations: 1800-1988 (Kociu, 1994) (see Table 3). From this Table it can be seen that:

- there is a large difference between M_{max} values found by (3.1.1) and (3.2.1), respectively M_I and M_{III}

- the difference $(M_{max}=M_{III})-M_t=0.3$ is constant and equal to the magnitude determination error: ±1/3. It implies the use either of $M_{max}=M_{III}$ or M_t

Table 3. Comparison of M_{max} found by seismological data with M_G= M_{max} found from neotectonic and seismotectonic data (Aliaj, 1988)

Neotectonic setting	M_G	M_{max}	$M_t^* \pm 0.2$	M_{max}-M_G	M_{max}-M_t
		Zones under compression			
Northward from Shkodra-Peja transform fault	7.0-7.5 (7.2±0.2)	7.0-7.5 (7.2±0.2)	7.0±0.2	0	+0.3
Adriatic-Ionic Coastal zone:					
Vlora-Butrinti zone	6.0-7.0 (6.5±0.5)	7.0-7.5 (7.2±0.2)	7.0±0.2	+0.7	+0.3
Lushnje-Elbasan-Diber zone	6.0-7.0 (6.5±0.5)	6.0-7.0 (6.5±0.5)	6.5±0.2	+0.2	+0.3
		Zones under horizontal extension			
Peshkopi Pogradec-Korca zone	6.0-7.0 (6.5±0.5)	7.0-7.5 (7.2±0.2)	7.0±0.2	+0.7	+0.3
Shkoder-Mat-Librazhd-Devoll zone	5.0-5.5 (5.2±0.2)	5.1-5.7 (5.4±0.2	5.3±0.2	+0.2	-0.3

Note:

$$Mt = \frac{\{\log[\Sigma Ej(Joule)] - 4.8\}}{1.5} \quad (3.2.2) \text{ (Kociu, 1994)}$$

4. EVIDENCE FROM PAST EARTHQUAKE SIZES

4.1 The free surface rupturing

The evidence of free surface rupturing only started recently. There are some observations of rupturing traces of about 10km, on the free surface during the strong earthquakes of June, 1, 1905 (M=6.6) (Kociaj and Sulstarova, 1980) and November 30, 1967 (M=6.7) (Sulstrova and Kociaj, 1980). Based on observed free surface rupturing the magnitudes of strong earthquakes can be evaluated.

4.2 Ground failure and liquefaction phenomena

A lot of ground failure and liquefaction phenomena were observed, especially in Pre-Adriatic area of Albania during the earthquake of June 1, 1905 (M=6.6,Io=IX) (Kociaj and Sulstarova, 1980), explained 75 years later on as liquefaction phenomena when another very strong earthquake in April 15,1979 (Ms=6.9,Io≥IX) hit this area. Liquefaction phenomena were observed even during the less strong earthquakes of September 11959 (M=6.4) and of March 181962 (M=6.0). For many inhabited centres in the

Pre-Adriatic area in Albania, where similar conditions exist, ground failure and liquefaction phenomena are possible in the future (such as Vlora City). This zone represents a site where investigations into possible mud-volcanoes need to be concentrated in the future

4.3 Thickness of archaeological ruins in ancient cities

According to geophysical investigations (Kociaj and Kapllani, 1983), the thickness of the archaeological ruins in Duress City was determined from 5-20m. As this City suffered greatly in the past from many strong destructive earthquakes, such thicknesses may be used as one of the parameters in determining the size of M_{max} for other ancient cities as well.

5. CONCLUSIONS

Normalized seismic activity data can successfully be used for the determination of maximum expected magnitudes (M_{max}) in the regions of continuous seismic activity.

The great linear foci, poorly represented in normalized seismic activity maps, may be reactivated in the "quiescence" zones.

For Albania observed differences M_{max}-M_G=0-1.0. When M_{max}-M_G =0 the seismological cycle may coincide with the neotectonic cycle.

The size of strong earthquakes of the past may be evaluated by additional evidence such as rupturing on free surface, ground failure phenomena and the thickness of archaeological ruins in ancient cities.

Future investigations into possible mud-volcanoes have to be concentrated in the crossing of seismogenic belts. One of the most promising sites is the Vlora zone.

The whole of the pre-Adriatic area of Albania characterized with poor Quaternary sediments and shallow underground water, the liquefaction potential is expected to be very high during future strong earthquakes.

REFERENCES

1. Aliaj, Sh., 1988. *Neotektonika dhe Sizmotektonika e Shqiperise,* Funds of Seismological Institute, Tirana, 1988 (In Albanian).
2. Drakopoulos J., and Makropoulos,K.,1982. Seismicity and seismic hazard evaluation, in: *Final report of WGA UNDP-UNESCO projects RER/79/014*, Unesco, Paris.
3. Gumbel, E.J., 1958. *Statistics of Extremes*, Columbia University Press, New York.
4. Kociaj, S. and Sulstarova, E., 1980. The earthquake of June 1, 1905, Shkodra, Albania, *Tectonophysics,* **67**, 319-332.

5. Kociaj, S.,1983. Some features of anomalous seismic activity before strong earthquakes in Albania, In: *Proceedings of UNDP-UNESCO seminar on earthquake prediction and instrumentation,* Thessalonica, Greece, 1-48.
6. Kociaj, S.,1983. On the main characteristics of the focus of April 15, 1979 earthquake derived from instrumental data, In: *Proceedings of the XVII Gen. Ass. of ESC,* Budapest, 109-114
7. Kociaj, S. and Varfi, P., 1984. Aspekte teorike te vleresimit te aktivitetit sizmik, *Buletini Shkencave te Natyres, Universiteti i Tiranes*, **3**, Tirane.
8. Kociu, S. et al., 1991. Catalogue *of earthquakes with M>=4.0, for the period 1971-1990 in Balkan Region"* (PC file) Seismological Institute, Tirana.
9. Kociaj, S., and Kapllani, L., 1993. Geophysical and engineering seismological methods for buried archaeological ruins: prospecting in Albania, In: *Geophysical Exploration of Archaeological Sites*, Vieweg Publishers Monograph, Wiesbaden, 161-169.
10. Kociu, S., 1994. Maximum expected magnitudes mapping using seismic activity data". In: *Proceedings of 2nd Workshop "Statistical models and methods in Seismology: Applications on prevention and forecasting of earthquakes*, Cephalonia Greece, 64-72.
11. Sulstarova, E., and Kociaj S., 1975. *Katallogu i termeteve te Shqiperise* Botim i Qendres Sizmologjike te Akademise se Shkencave, Tirane.
12. Sulstarova E., and Kociaj, S., 1980.The Dibra (Albania) earthquake of November 30, 1967, *Tectonophysics*, **67**, 333-343.
13. Tezcan. S.S.,1996. *Probability analysis of earthquake magnitudes*, TDV/TR 96-001.TDV, Technical University of Istanbul, Faculty of Civil Engineering, Ayazaga, Istanbul Turkey.
14. *The earthquake of April 15,1979*, Publ. House"8 Nentori" Tirana, 1983.
15. Zavyalov, A.D., Kociu, S., and Smirnov. V.B. Space-time magnitude cutoff evolution of Albania earthquakes catalog for 1900-1995, in *Proceedings XXVIIIth General Assembly of ESC, Genoa, Italy, September 2002* (in press).

ASSESSMENT OF SEISMIC HAZARD IN AREAS OF MUD VOLCANO LOCATION ON THE BASIS OF GEOPHYSICAL DATA

Tofik A. Ismailzadeh[1], Mayya .I. Isayeva[2]

1 United Institute of Physics of the Earth, Russian Academy of Sciences, Bolshaya Gruzinskaya 10 , 123995 Moscow, Russia tel (095) 2549105, fax (095) 2556040

2 Geology Institute, Azerbaijan National Academy of Sciences, H. Javid 29-A, 370143 Baku, Azerbaijan

Abstract: The paleomagnetic peculiarities of bottom sediments of the South Caspian Sea and their geochronological sections have been studied in present paper. Correlations between separate oil-gas bearing structures near mud volcanoes as well as quantitative determinations of age according to paleomagnetic data are also given in present paper. The association of local changes of paleomagnetic trends with tectonic movements has been shown.

Keywords: mud volcanoes, paleomagnetic peculiarities, residual magnetization, sedimentary basin, saturation

1. INTRODUCTION

The South-Caspian and Azov-Black sea sedimentary basins are characterized by a high potential of oil and gas content in combination with significant areas of mud volcanism and high seismicity (Levin et al., this volume). Earthquakes more than M=5.0 can result in the destruction of different marine oil industry constructions (drilling platforms, oil and gas pipelines) and can cause ecological disasters. The estimated intensity of bottom shakes near offshore oil fields is I=7.7 up to 9.0 in the Absheron-Balkhan sill. Hazard mitigation can be achieved by seismic zonation of bottom sea and by improvements in constructions. Seismic zonation can be improved by paleomagnetic prospections. In a zone of Assheron-Cheleken Sill and adjacent areas of South Caspian Depression (Fig.1) earthquakes epicenters and paleomagnetic sampling points are shown.

2. METHODS

The adopted methods include the following research: the definition of natural residual magnetization, the demagnetization of natural residual

G. Martinelli and B. Panahi (eds.), Mud Volcanoes, Geodynamics and Seismicity, 135–140

magnetization in alternating and constant magnetic fields, the temperature demagnetization of residual magnetization of saturation Irs (T^0) and magnetization of saturation Ir (T^0) with definition of Curie points and temperature of phase transformations Tn. Relative error of measures was no more than 3%. Research also includes the definition of density of bottom sediments.

Calibration of methods has been performed on core samples of the upper Pliocene and Pleistocene deposits gathered from the core of nine boreholes drilled on uplifts of the Garasu and Bulla mud volcano islands.

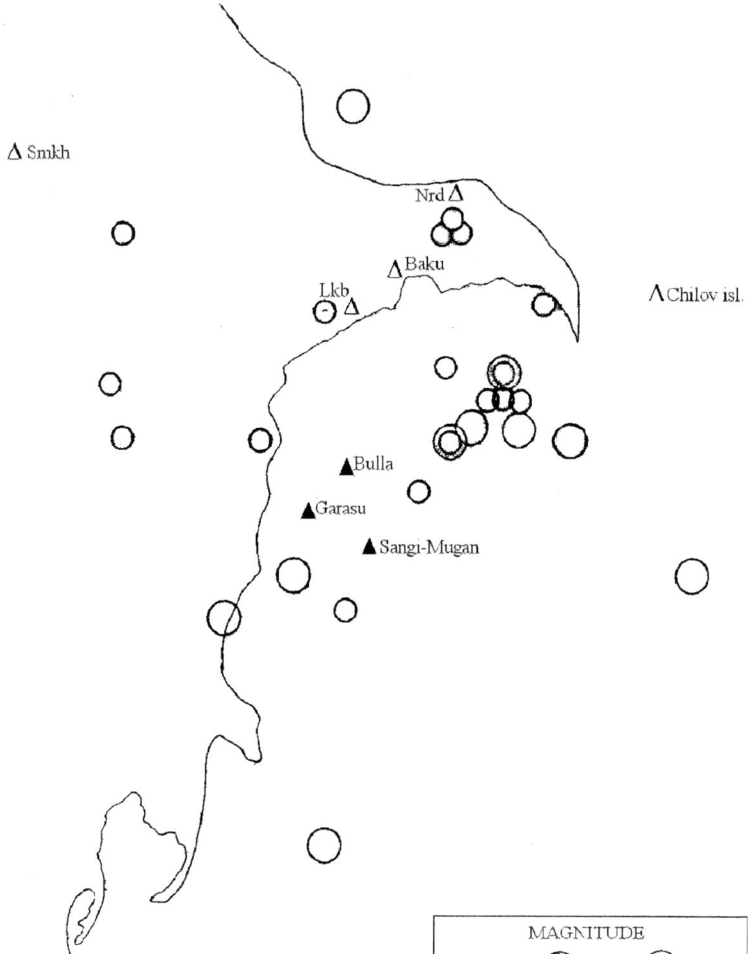

Figure 1. Location of Eastern Azerbaijan earthquakes of M>4 (black triangles show the location of sample collection for paleomagnetic investigations).

The uplift of the Garasu island is characterized by the non-coincidence of the structural plan on horizon in the upper and middle Pliocene and is complicated by faults of different directions. In the north-west and north-east the rocks of Pliocene productive series are at sea-bottom in direct contact with the Pleistocene deposits.

The Bulla island is represented by a large fold, almost symmetrical in the longitudinal section. The uplift is complicated by one longitudinal and several cross faults with vertical shift up to 240 m. Discordances in stratigraphic horizons of Pleistocene deposits can be clarified by paleomagnetic researches (Kok, 1969; Ismail-Zadeh et al., 2001).

Curves of normal magnetization of Pleistocene sediments (wells 473-478) show that saturation of residual magnetization occurs in the interval 1,000-1,500 e. In alternating fields 75-100e half of the residual magnetization In is removed, and in field-H = 350 e. 10-15 % In remains (Fig. 2). Here magnetic soft ferromagnetic minerals prevail in the composition of the magnetic fraction of Pleistocene sediments (Fig. 3).

Thermomagnetic analysis was conducted for the definition of the Curie point using two heatings for Irs and Is. The curve of the first heating is of gentle form and a weak magnetic phase can be noted at 200 C°. Curve Irs (T_0) of repeated heating does not essentially differ from the first heating. The Curie point on the two curves Is (T^0) is 580 C°. Transition from ferrum hydroxides to superparamagnetic hemalite and hydrohelite into hemalite is probably connected with the thermal treatment of samples (Fig. 3).

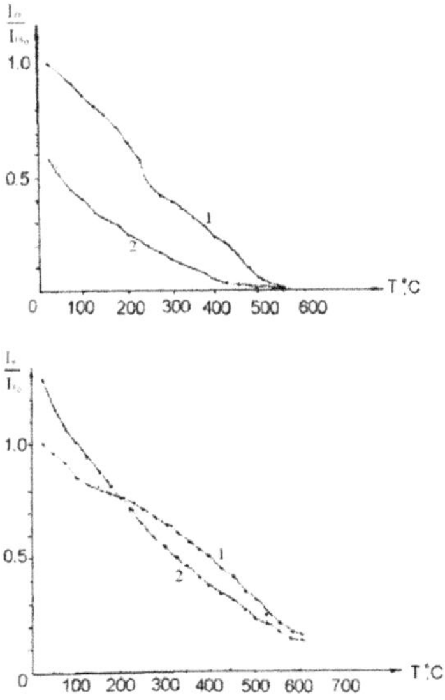

Figure 2. Typical demagnetization curves (In) of samples of the Caspian Sea Pleistocene bottom sediments in the variable magnetic field

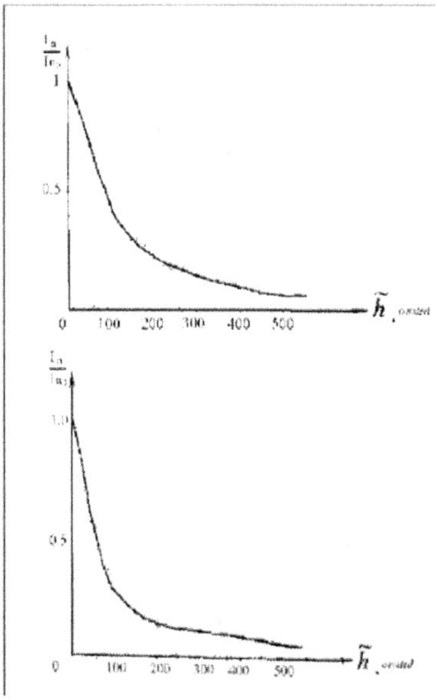

Figure 3. Typical curves of thermo-magnetic analysis Irs (T°) and Is (T°)

The magnetic fraction of Pleistocene sediment samples is stable to heating and does not undergo essential phase changes by high temperature heating. Magnetic and density characteristics of Pleistocene and Upper Pliocene sediments are identical. The examples are represented by changes of these characteristics on well section 478, 473, 477 and 476.

A reserve zone magnetic polarity has been recognized in the bottom deposits of the South Caspian and can be correlated with the Etrussia event and is of stratigraphic and correlative importance for the upper Pleistocene and Holocene deposits.

The Heterbourg event also is well-registered in the sections and were recognized early in the bottom sediments of the Black Sea, the Indian Ocean and the Caspian Sea. Two zones of reverse polarity correlated with Blake and Dnepr Chagan have been defined.

3. RESULTS OF THE STUDIES

Detailed paleomagnetic studies of sediments of the South Caspian Sea have enabled us to correlate oil-gas bearing structures with corresponding age of separate horizons. Three paleomagnetic zones have been identified in sections. The first one corresponds to the event of reversed polarity of Etrussian (2,7-3,0 thousands years), the second one to Heterbourg event (10-12 thousands years), the third one is known as Bleik event (108-114 thousand years).

These events are synchronous to New Caspian, Upper Khvalyn and Upper Khazar horizons of stratigraphic scale of Pleistocene. Changes in magnetic susceptibility and natural residual magnetization appear to be similar and reflect the concentration of magnetic minerals in the sediments. Different levels of K susceptibility decrease in sections and are linked with vertical movements on different sides of the faults. Synchronous levels of magnetic susceptibility decrease reflecting a single setting of sedimentation.

Curve variations of residual magnetization D declination and inclination J form linear areas, frequently parallel to axis of ordinates. At the same time, there are linear intervals, where paleomagnetic directions do not differ from the recent magnetic dipole. On the other hand, there are cases when paleomagnetic directions of synchronous intervals of the section differ from directions of dipole field and simultaneously from each other as well. These differences in paleomagnetic directions at a distance of 200 m between wells cannot be associated with geomagnetic phenomena and are linked with tectonic movements of small blocks with possible turns around their vertical axis.

Paleomagnetic inclinations vary within $\pm 15^0$ relative trend line. In sections of wells 474, 475 (Garasu uplift) average values of inclinations virtually do not differ from recent geomagnetic dipole ($j=64^0$). The section of wells 473, 474, 476, 477 and 478 is characterized by low values of inclination. Possible changes of inclination are probably tectonically originated.

4. CONCLUSIONS

Detailed paleomagnetic studies of bottom sediments of South Caspian Sea have enabled us to attribute their geochronological sections. Correlations between separate oil-gas bearing structures near mud volcanoes and determination of age was performed by paleomagnetic data.

An association of local changes of paleomagnetic trends with tectonic movements was revealed.

The methodology can be used during further regional and detailed seismic zonation of bottom sediments for determination of tectonic movements in recent stages of geological history.

REFERENCES

1. Kok, A., 1969. Geomagnetic reversals-science, v.163, N° 3864, 237-245.
2. Ismail-Zadeh, T.A, Petrova, G.N, and Sharonova, Z.V, 2001.The Binding of borehole samples and sedimentary strata sections to the time-scale using characteristics of sedimentary rocks, in *Geophysics XXI Century*, New World, 354-359.
3. Levin, L., Solodilov, L., Panahi, B., Kondorskaya, N., in Martinelli, G. and Panahi, B., (Eds.), *Mud Volcanoes, Geodynamics and Seismicity. Proceedings of NATO Advanced Research Workshop, Baku, May 20-22, 2003.*

Chapter 4
GREENHOUSE EFFECTS OF MUD VOLCANISM

METHANE EMISSION FROM MUD VOLCANOES
Towards a global estimate

Giuseppe Etiope
Istituto Nazionale di Geofisica e Vulcanologia, Via Vigna Murata, 605, 00143 Roma, Italy, e-mail: Etiope@ingv.it

Abstract: Natural emission of methane from geologic sources has been recently recognised as an important component of the atmospheric methane budget. While some authors have focused their attention only on submarine gas seepage and gas hydrates, recent studies have suggested that also mud volcanoes (MVs) on land and microseepage in hydrocarbon-prone areas are significant methane sources. A new global estimate of methane flux from MVs has recently been made on the basis of new experimental flux data (including diffuse microseepage around craters and vents), and a new data-set on MV sizes. The emission results conservatively to be between 6 and 9 Mt/y, that is, the same level as the ocean and hydrates sources. The global geological methane flux, including MVs, submarine seepage, microseepage in petroliferous basins and geothermal flux, would conservatively amount to 35-45 Mt/y, that is, comparable to other sources or sinks considered in the IPCC tables.

Key words: mud volcanoes, methane sources, gas flux, greenhouse gas

1. INTRODUCTION

The International Panel of Climate Change in its official tables includes exclusively gas-hydrates as a geological source of methane, emitting around 5-10 Mt/y (IPCC, 2001). However, during the last decade, several studies have acknowledged the importance of other geological sources on a global scale. Firstly, submarine gas seepages, being the best estimates in the order of 14 or 20 Mt/y (Judd, 2000; Kvenvolden et al., 2001). More recently also mud volcanoes (MVs), microseepage in hydrocarbon-prone basins and geothermal flux have been altogether considered a non-negligible methane source, in the order of at least 12-23 Mt/y (Etiope and Klusman, 2002). The sum of both offshore and onshore sources would be conservatively to the order of 30-50 Mt/y. Another estimate, not including microseepage, suggests 13.1 to 35.6

G. Martinelli and B. Panahi (eds.), Mud Volcanoes, Geodynamics and Seismicity, 141–146

Mt/y (Judd et al., 2002). These numbers are, indeed, of the same level as or higher than other sources or sinks considered in the IPCC tables (IPCC, 2001), such as biomass burning (40 Mt/y), termites (20 Mt/y), oceans (10 Mt/y), soil uptake (30 Mt/y).

Mud volcanoes and microseepage are two poorly investigated methane sources, although early estimates suggest a great potential. The emission from onshore mud volcanoes, including flux from vents and soil degassing, was directly measured in detail only from 2001, in Italy and Romania, thanks to a NATO Project (Etiope et al., 2002; 2003; 2004). These studies have allowed us to understand that the pervasive microseepage flux from the soil surrounding the MV gas manifestations (craters, gryphons, bubbling pools) is a fundamental component of MV gas output, and that, in total, MVs typically emit a specific flux in the range 10^2 to 10^3 t km^{-2} y^{-1}. This finding resulted to be a key element towards improving global flux estimates (Etiope and Milkov, 2004) as discussed in this work.

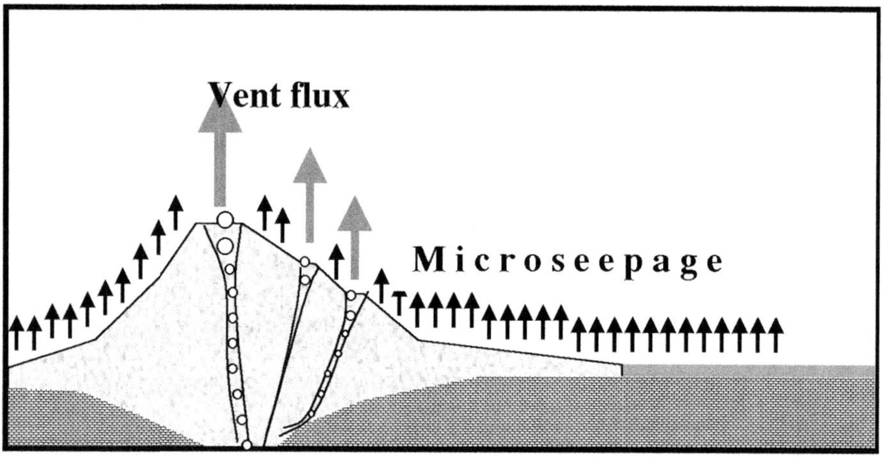

Figure 1. Sketch showing gas emission from mud volcanoes. Gas is released from vents (gryphons, bubbling pools) and from the soil throughout vast areas, also out of the muddy cover. Microseepage may represent from 50% to 90% of the total gas output.

2. FLUX FROM VENTS AND MICROSEEPAGE

Methane flux to the atmosphere was measured in 10 MV fields in Italy and Romania (Etiope et al., 2002; 2003; 2004). The flux was evaluated in more than 200 gas vents (gryphons, gases bubbling in water-pools and dry emission of everlasting fires) and from diffuse microseepage along the MV flanks and in the surrounding soil, for more than 4 km^2 altogether. The output in pools was measured by means of a stainless-steel funnel connected to a bottle having a preset volume. As the volume is known, the time the bubbling gas took to displace the water from the bottle provided the flow rate. For vents inaccessible to direct measurements, a theoretical model was derived and tested to calculate the gas flux from bubble trains as a function of bubble diameter and bursting frequency (Etiope et al., 2004).

The gas output from dry vents and diffuse soil degassing was measured by the closed-chamber method, similar to the "Crill system" (Norman et al., 1997; Etiope, 1999). Gas accumulated in the chambers was collected two to three times by syringes following time intervals from a few seconds to 10 minutes from the box emplacement, and analyzed by gas chromatography.

The measured output from single vents is on average 0.4 t y^{-1} in western Sicily and 4 t y^{-1} in eastern Romania, where the maximum vent-flux recorded was 28 t y^{-1}. Vent-flux is highly variable in relation to gryphon sizes, the number of bubble trains, bubble dimension and frequency. The pervasive flux from soil ranges generally to the order of 10^2-10^3 mg m^{-2} d^{-1} but restricted zones of high degassing exist with mean values of 10^4-10^5 mg m^{-2} d^{-1}. Basically, the flux of a mean vent of 1 t y^{-1} (typically a train of one 5-cm-bubble per second) is equivalent to the microseepage from about 4000 m^2 of soil. Microseepage is clearly a major way of degassing being from 50% to 90% of the total MV output. The results are summarised in Table 1.

Table 1. Summary flux data from the main mud volcanoes in Italy (western Sicily: Maccalube, Bissana, Bivona fire) and Romania (Paclele Mari, Paclele Mici, Fierbatori, Andreiasu fire)

Site	Area measured	N. MV fields	N. vents	Output from vents (t/y)	Output from soil (t/y)	TOTAL (t/y)
Western Sicily	1.4 km^2	3	≥ 50	20 (0.4 x vent)	380 (270 x km^2)	400
Eastern Romania	2.55 km^2	4	≥ 150	588 (4 x vent)	601 (236 x km^2)	1200

The soil CH_4 positive flux is significantly high even at large distances (more than 1 km) from the MV domes, suggesting that microseepage exists over wide areas. Many local or regional flux values reported for MVs in Azerbaijan (Dadashev, 1963; Jakubov et al., 1971; Holland et al., 1997) did not consider such a microseepage and therefore are likely to be strongly underestimated.

3. ESTIMATING GLOBAL MV OUTPUT

Hitherto, four independent estimates of global methane flux from MVs have been reported (Dimitrov, 2002; Etiope and Klusman, 2002; Kopf, 2002; Milkov et al., 2003).

The estimates of Dimitrov and Kopf are based on undefined data-sets and the calculation procedures are not completely described. The results are quite different being 10.3-12.6 Mt y^{-1} and 0.08-1.4 Mt y^{-1}, respectively. The problem is that both are based on local estimates, very limited data-sets, and unreferenced estimations of eruptive flux. Kopf's estimate considers mainly submarine MVs and does not indicate the global output to the atmosphere.

The estimates of Etiope and Klusman (2002) and Milkov et al. (2003) are based on wider data-sets, including the first experimental flux data of Sicilian MVs (Etiope et al., 2002), and statistical elaboration. The results are coherent, being 2-10 Mt/y and 6 Mt/y, respectively.

More recently Etiope and Milkov (2004) have converged these estimates on the basis of new flux data (Etiope et al., 2003; 2004), partially described in the previous section, and elaborating a classification of MV sizes, in terms of area (km^2), based on a compilation of 120 data on onshore and offshore MVs (from Azerbaijan, Sakhalin, Panama, Adriatic Sea, Italy, Mexico, Romania, Irian-Java Papua N.G., Ukraine, Pakistan, Turkmenistan, Venezuela, Taiwan, Alaska, India, Trinidad).

A comparative analysis suggests that most of the flux values of Azeri MVs (Dadashev, 1963) are strongly underestimated or unrepresentative of whole MVs, since they are too low in relation to the area covered, in comparison to the data from the Italian and Romanian MVs (Etiope et al., 2002; 2003; 2004). It is evident that the flux values do not refer to the total number of vents and definitely do not include microseepage.

The procedure followed for the new global flux estimate has been based on:

- the definition of MV number and classes (small, medium, large) in relation to the size (area);
- calculation, for each MV class, of global onshore flux based on experimental data on specific flux (gas amount/time/area), including emission from vents and diffuse microseepage;
- calculation, for each MV class, of global offshore flux based on estimates of emission from vents and bubble-dissolution models.

The global methane flux from onshore and shallow submarine MVs was, therefore, estimated conservatively between 6 and 9 Mt y^{-1}, that is the same level of ocean and hydrate sources. The uncertainties are related to

1. limited quiescent flux data-set,
2. the lack of measurements during eruptive events,
3. poorly constrained numbers of MVs and their area,
4. the short-distance spatial variability and short-term temporal variability in flux rates.

The total geological source, including MVs, submarine seepage (Kvenvolden et al. 2001), microseepage in hydrocarbon-prone areas and geothermal sources (Etiope and Klusman, 2002), would amount to 35-45 Mt y^{-1}. This value is likely to be still underestimated but it is comparable or higher than several sources and sinks considered by the IPCC (e.g., termites, soil uptake, oceans, hydrates). We believe the time has come to add this parameter to the IPCC official tables. The actual role of geological sources in the atmospheric methane budget can be accurately assessed after further flux measurements from other MVs in the world and from soil microseepage in petroliferous areas. Such a target could be reached through co-operation programmes and projects including terrestrial and marine investigations, and by using technological tools recently developed for gas detection and monitoring.

Table 2. Global MV flux estimates

	flux to the atmosphere	
	quiescent degassing	quiescence+eruptions
Dimitrov (2002)	0.33-2.64	10.3-12.6
Etiope and Klusman (2002)[a]	>1-2[b]	5-10[c]
Kopf (2002)	-	-
Milkov et al., (2003)	2.9	6
Etiope and Milkov (2004)	-	>6-9[c]

[a] onshore MV only.
[b] flux from vents.
[c] flux from vents + flank microseepage.

REFERENCES

1. Dimitrov, L.I., 2002. Mud volcanoes - the most important pathway for degassing deeply buried sediments., *Earth-Science Rev.*, **59**, 49-76.
2. Etiope, G., 1999. Subsoil CO_2 and CH_4 and their advective transfer from faulted grassland to the atmosphere, *J. Geophys. Res.*, **104D**, 16889-16894.
3. Etiope, G., Caracausi, A., Favara, R., Italiano, F., and Baciu, C. 2002. Methane emission from the mud volcanoes of Sicily (Italy), *Geoph. Res. Lett.*, **29**, 8, 10.1029/2001GL014340.
4. Etiope, G., Caracausi, A., Favara, R., Italiano, F., and Baciu , C. 2003. Reply to comment by A. Kopf on "Methane emission from the mud volcanoes of Sicily (Italy)", and notice on CH_4 flux data from European mud volcanoes, *Geoph. Res. Lett.*, **30** 2, 1094, doi: 10.1029/2002GL016287.
5. Etiope, G., Baciu, C., Caracausi, A., Italiano, F., and Cosma C., 2004. Gas flux to the atmosphere from mud volcanoes in Eastern Romania, *Terra Nova*, **16**, 4, 179-184.
6. Etiope, G. and Klusman, R.W., 2002. Geologic emissions of methane into the atmosphere., *Chemosphere*, **49**, 777-789.
7. Etiope, G. and Milkov, A.V. , 2004. A new estimate of global methane flux from onshore and shallow submarine mud volcanoes to the atmosphere, *Environm. Geology*, in press.
8. Intergovernmental Panel on Climate Change , 2001. In *Climate Change 2001: The scientific basis*, Houghton, J.T., Ding, Y., Griggs, D.J., Noguer, M., van der Linden, P.J., Dai , X., Maskell , K., Johnson, C.A. (Eds.) Cambridge Univ. Press., Cambridge
9. Jakubov, A.A., Alizade, A.A., and Zeinalov, M.M., 1971. *Mud volcanoes of the Azerbaijan SSR, Atlas, Elm*, Publishing House of the Academy of Sciences of Azerbaijan SSR, Baku (in Russian).

10. Judd , A. G., 2000. Geological sources of methane. In: *Atmospheric Methane: its role in the global environment*, Khalil, M. (Ed.), Springer-Verlag, New York, NY, 280-303.
11. Judd, A.G., Hovland, M., Dimitrov, L.I., Garcia Gil , S., Jukes, V. , 2002. The geological methane budget at Continental Margins and its influence on climate change, *Geofluids*, **2**, 109-126.
12. Kopf, A.J., 2002. Significance of mud volcanism, *Rev. Geophysics*, **40**, 2, 1005, doi: 10.1029/2000RG000093.
13. Kvenvolden, K.A., Lorenson, T.D. and Reeburgh, W. , 2001. Attention turns to naturally occurring methane seepage, *EOS*, **82**, 457.
14. Milkov, A.V., Sassen, R., Apanasovich , T.V., Dadashev, F.G. , 2003. Global gas flux from mud volcanoes: a significant source of fossil methane in the atmosphere and the ocean, *Geoph. Res. Lett.*, **30** , 2, 1037, doi:10.1029/2002GL016358.
15. Norman, J.M., Kucharik, C.J., Gower, S.T., Baldocchi , D.D., Crill, P.M., Rayment, M., Savage, K., and Striegl, R.G., 1997. A comparison of six methods for measuring soil-surface carbon dioxide fluxes, *J. Geophys. Res.*, **102D**, 28771-28777.

GAS EMISSIONS FROM MUD VOLCANOES
Significance to Global Climate Change

Alan Judd
School of Marine Science & Technology, Ridley Building, University of Newcastle upon Tyne, Newcastle upon Tyne NE1 7RU, U.K, e-mail: alanjudd@ncl.ac.uk

Abstract: There are about 1,000 mud volcanoes on land and 5,500 offshore - mostly in deep water. Activity varies between gentle emissions and violent eruptions accompanied by the release of enormous volumes of gas - mainly (85%+) methane and carbon dioxide. Global gas emissions are provisionally estimated to exceed 27 billion cubic metres per year, of which more than 23 billion (15.8 Tg) is methane. More than 70% of this is from short-lived eruptions, about 30% of which ignite to produce flames tens or hundreds of metres high. The majority of the methane is emitted by submarine mud volcanoes, most in deep water. About 11.4 Tg per year is lost to the hydrosphere, but a tentatively estimated 3.6 Tg per year escapes to the atmosphere. So, mud volcanoes are significant sources of atmospheric methane, much of it 'fossil'. Contributions are thought to increase when sea level is low (in glaciations), providing negative feedback to global warming and working to limit climatic extremes.

Keywords: mud volcanoes, atmospheric methane, global climate change, natural gas emissions

1. INTRODUCTION

Gas emissions from mud volcanoes on land make direction contributions to the atmosphere. This is significant because the most common gases emitted by mud volcanoes are methane and carbon dioxide, two of the most important 'Greenhouse gases'. However, relatively little has been done to quantify mud volcano gas emissions and their role as contributors to atmospheric gas budgets. Notable contributions have been made by a few authors, most recently Dimitrov (2002a and b) and Milkov (2000, Milkov et al., 2003); these and other papers provide the raw data used in this review.

The purpose of this paper is to estimate the amount and fate of the emitted gases, and to consider whether or not mud volcanoes are significant compared to other natural and anthropogenic sources of Greenhouse gases.

2. MUD VOLCANO DISTRIBUTION

Mud volcanoes provide a mechanism for defluidising thick sequences of sediments in the compressive zones associated with convergent plate boundaries, and where sediment accumulation rates exceed normal compaction rates, for example in deep sea fans. Onshore examples have been

G. Martinelli and B. Panahi (eds.), Mud Volcanoes, Geodynamics and Seismicity, 147–157

recognised for centuries, as have those in shallow waters. However, relatively recent advances in seabed mapping technology have made it possible to identify examples in the deep oceans. Their true distribution offshore is still a matter of speculation as vast areas of the deep ocean have yet to be mapped in detail. Dimitrov (2002a) presented a list of occurrences which suggested that the total number of mud volcanoes on land is more than 926, and that offshore there are more than 268 confirmed, and probably at least another 572. However, continued compilations of data led Dimitrov (2002b) to increase these estimates to about 1,100 onshore and in shallow water, and between 150 and 700 in deeper waters. Milkov (2000) based an alternative estimate on the density of mud volcanoes in well mapped areas of the deep sea: the Barbados Accretionary Prism (450 to 150,000 km^{-2}), the Black Sea (about 30 individuals), the Caspian Sea (60), and the Eastern Mediterranean Sea (up to 150). By extrapolation, he estimated that there are between 1,000 and 100,000 individual mud volcanoes in the deep sea.

For the purposes of the estimations made later in this paper the numbers presented in Table 1 will be considered representative of the true figure.

Table 1. Estimated global number of mud volcanoes

On land	*Continental shelves*	*Continental slopes and deep oceans*	**Total**
1,000	500	5,000	**6,500**

3. MUD VOLCANO ACTIVITY

There is a wide variety of shapes and sizes of mud volcano. The smallest are less than a metre high, whereas the larger ones are more than 500 m tall, and three or four kilometres wide at the base. The style of activity is equally varied. Kalinko (1964) identified three main types of mud volcano on the basis of the activity, each type being named after a particular individual:

- Lokbatan type: characterised by infrequent explosive activity, and the ignition of the emitted gases. Lokbatan itself has erupted 23 times since 1829, most recently in October 2001. Each event has been short-lived, generally lasting no more than a few hours. Some have been violent, shooting flames up to 500 m into the air (Aliyev et al., 2002).
- Chikishlyar-type: characterized by gentle and continuous activity during which gas bubbles are produced at a more or less uniform rate.
- Schugin-type: a transitional type characterised by continuous weak activity interrupted by brief eruptive phases.

The frequency of Lokbatan-type eruptions varies considerably from one individual to another. Some seem to be created by a single event and never erupt again, whilst others erupt as frequently as once per year. 76 (about 25%) of the 300 or so known mud volcanoes in Azerbaijan (including those offshore) experience eruptions (Dimitrov, 2002b). Bagirov et al. (1996) calculated that there are, on average, 9.7 eruptions per year in Azerbaijan, of which 3.4 are major events.

For the purposes of estimation it will be assumed here that about 75% of mud volcanoes are continuously active, that 3% erupt every year, and one third of these eruptions are 'strong'; see Table 2. As Dimitrov (2002b) found evidence of 11 strong eruptions in the year 2001 (8 in Azerbaijan, 2 in Barbados and 1 in New Zealand), it seems that there is some justification in accepting these estimates.

Table 2. Estimated average global number of mud volcano eruptions per year

	On land	*Continental shelves*	*Continental slopes and deep oceans*	***Total***
Continuous emissions	750	125	1,250	**2,125**
'Normal' eruptions	20	10	100	**130**
'Strong' eruptions	10	5	50	**65**

4. GAS EMISSIONS

The gas flux from gentle Chikishlyar-type emissions can be estimated from available data, however the data set is very small and the flux from individual mud volcanoes may vary considerably over time. Guliyev and Feizullayev (1997) reported daily emission rates from the Dashgil mud volcano in Azerbaijan that varied from 200 to 40,000 m^3 day^{-1}, and Hovland et al. (1997) reported an independent estimate of 800 m^3 yr^{-1} (i.e. only 2.2 m^3 day^{-1}) from the same mud volcano. Milkov et al. (2003) identified flux estimates from 36 mud volcanoes. These varied considerably from 100 (10^2) to 10^7 m^3 yr^{-1}. Statistical analyses suggested that these fitted a lognormal distribution, as is to be expected of natural data populations. Based on this analysis they suggested a mean annual gas flux of 3.6 x 10^6 m^3 gas per quiescent mud volcano.

Estimating the flux from eruptive emissions is significantly more difficult, and has probably never been achieved with any degree of certainty. The following estimates, reported by Sokolov et al. (1969) and Guliyev and Feizullayev (1997), refer to mud volcano eruptions in Azerbaijan:

- Bolshoi Maraza, 1902: 120 x 10^6 m^3
- Touragai, 1946: 500 x 10^6 m^3
- Great Kjanizdag, 1950: 100 x 10^6 m^3
- Makarov Bank (Bakhardeniz), 1958: 300 x 10^6 m^3
- Duvannyi Island, 1961: 65 x 10^6 m^3

These were individual eruption events lasting no more than a few hours.

Dimitrov (2002a) suggested that the average volume emitted during a single strong eruption is about 340 x 10^6 m^3, but Dadashev (1963; cited by Milkov et al., 2003) suggested a lower figure: 250 x 10^6 m^3. Bearing in mind the difficulty in estimating these volumes, and the tendency to quote only those eruptions that ignite (i.e. volume estimates are based on the size of the flame), it is probable that these estimates are valid only for the stronger eruptions.More normal eruptions may produce significantly less gas, perhaps one tenth of this figure. The flux values presented in Table 3 will be used for estimations.

Table 3. Estimated gas flux from mud volcanoes

Continuous emissions [m^3 yr^{-1}]	*'Normal' eruptions* [m^3]	*'Strong' eruptions* [m^3]
3.6 x 10^6 *per mud volcano*	25 x 10^6 *per eruption*	250 x 10^6 *per eruption*

5. GAS COMPOSITION

The compositions of the gases emitted by mud volcanoes vary from region to region, and between individuals within a particular region. However most are dominated by methane, which generally accounts for more than 85% of the total gas composition. Dimitrov (2002b) noted that in some petroleum-rich provinces, such as the Southern Caspian Basin and the Gulf of Mexico, higher hydrocarbon gases (and in some cases liquid petroleum) are present, but the most common gas, other than methane, is usually carbon dioxide. Nitrogen and hydrogen sulphide may be present in significant concentrations (<54% N_2 in the South Alaska mud volcanoes), but most often they are 'trace' gases. The available data set is very small, but in the absence of a proper inventory, it is necessary to speculate. For the purpose of the estimations made later in this paper an 'average' composition, presented in Table 4, has been derived from published data: Aliyev et al. (2002), Dimitrov (2002b), Etiope et al. (2002), Milkov et al. (2003), Etiope et al. (2004), Martinelli and Judd (2004).

Table 4. Indicative composition of mud volcano gases

Methane	*CO_2*	*Nitrogen*	*Higher Hydrocarbons*
85.5%	9.5%	4.5%	0.5%

6. ESTIMATING GLOBAL GAS EMISSIONS

The estimations made in the preceding sections make it possible to summarize the speculations about the global gas flux from mud volcanoes. Table 5 indicates the total gas flux; Table 6 summarizes the likely flux of each gas.

Table 5. Global annual gas flux from mud volcanoes

Flux $m^3 \times 10^6$	*Continuous emissions*	*Normal eruptions*	*Strong eruptions*	***Total***
Land	2700	500	2500	5700
Continental shelves	450	250	1250	1950
Deep oceans	4500	2500	12500	19500
Total	**7650**	**3250**	**16250**	**27150**

Table 6. Global annual fluxes of the principal gases

Flux $m^3 \times 10^6$	*Methane*	*Carbon dioxide*	*Nitrogen*	*Higher hydrocarbons*
Land	4875	540	255	30
Continental shelves	1670	185	85	10
Deep oceans	16670	1850	880	100
Total	**23215**	**2575**	**1220**	**140**

Clearly methane is the most important gas. The volume of methane produced is large enough to demand closer attention, particularly because methane is an important 'Greenhouse gas'. However, it is not sufficient to accept the total methane flux without considering its the fate. This depends upon both the rate of emission and where it is emitted (particularly the water depth). A full analysis of the estimates for methane is presented in Table 7. Emissions of carbon dioxide, another important 'Greenhouse' gas, are substantial but not significant in the context of global budgets.

Table 7. Global fluxes of the methane

Methane Flux $m^3 \times 10^6$	*Continuous emissions*	*Normal eruptions*	*Strong eruptions*	***Total***
Land	*2310*	*430*	*2135*	*4875*
Continental shelves	*385*	*215*	*1070*	*1670*
Deep oceans	*3845*	*2135*	*10690*	*16670*
Total	**6540**	**2780**	**13895**	**23215**

7. THE FATE OF EMITTED METHANE

Consider methane emissions from mud volcanoes on land. As the gas is vented directly into the atmosphere these emissions should clearly be included in global methane source inventories. However, it has been noted above that the gases emitted during some (about one third, according to Dimitrov, 2002b) strong eruptions ignite. So, about one third of this contribution should be deducted. As there have been reports of flames rising from the sea surface above considerable water depths, a similar reduction should also be made in the estimated contribution of strong eruptions on the continental shelf. In fact, on ignition this methane is oxidised, producing carbon dioxide. Consequently a contribution should be transferred to the carbon dioxide budget.

Methane bubbling from quiescent submarine mud volcanoes may dissolve in the water column. The rate of dissolution is dependent on a number of factors, most important of which are bubble size on release from the seabed and water depth (Leifer and Patro, 2002). It is probably realistic to expect about 50% of the methane from continuous emissions on the continental shelf to be lost to the hydrosphere. The probability of methane from deep-water mud volcanoes reaching the atmosphere is significantly lower. As Milkov (2000) pointed out, gas hydrates have been discovered at a number of deep-sea mud volcanoes; it is likely that most of the methane initially emitted by Chikishlyar-type mud volcanoes is sequestered by hydrates, the remainder probably being lost to solution in the water column. However, a continuous flux into the gas hydrate stability zone (GHSZ) will, over time, exhaust the supply of water needed to make new hydrate, so the methane flux to the water will gradually increase to match the flux from beneath the GHSZ. Methane bubbles escaping from the seabed within the GHSZ will be protected from solution by an 'armour' of gas hydrate. However, this armour will dissociate before the surface is reached, and it is probable that the methane will be lost to the water anyway.

Gas from eruptions is much more likely to pass through the water column as a consequence not only of the increased flux rate, but also because the rising gas would likely result in upwelling flows of water, which may significantly reduce loss by solution (Leifer and Patro, 2002). It is anticipated that 80% of methane from 'normal' eruptions on the continental shelf would escape to the atmosphere. However, most, if not all the methane from 'strong' eruptions is likely to burst through to the sea surface, although some of this (perhaps a third) will be lost on ignition. In deep water, eruptive gases probably pass rapidly through the GHSZ zone to escape through the seabed, but evidence of the oxidation in the water column of methane-rich 'megaplumes' originating from the Hydrate Ridge gas hydrate field (Oregon Subduction Zone - see Suess et al., 1999) suggests that little eruptive methane would reach the sea surface.

The figures in Table 8 indicate the possible fate of the methane emitted by mud volcanoes. The estimated annual contribution of methane to the atmosphere is 3.6 Tg, about 23% of the total surface (land surface and seabed) flux. In addition, mud volcanoes make a significant contribution to the methane content of the oceans (about 11.4 Tg).

Table 8. The fate of methane emitted by mud volcanoes

Methane Flux $m^3 \times 10^6$	*Continuous emissions*	*Normal eruptions*	*Strong eruptions*	***Total*** $m^3 \times 10^6$ *[Tg]**
Land				
To atmosphere	2310	430	1425	**4165** **[2.83]**
Loss on ignition	0	0	710	**710** **[0.48]**
Continental shelves				
To atmosphere	195	170	715	**1080** **[0.73]**
Loss on ignition	0	0	355	**355** **[0.24]**
Loss to hydrosphere	190	45	0	**235** **[0.16]**
Deep oceans				
To atmosphere	0	0	100	**100** **[0.07]**
Loss to hydrosphere	3845	2135	10590	**16570** **[11.24]**
Total				
To atmosphere	2505	600	2240	**5345** **[3.6]**
Loss on ignition	0	0	1065	**1065** **[0.7]**
Loss to hydrosphere	4035	2180	10590	**16805** **[11.4]**
Total	**6540** **[4.4]**	**2780** **[1.9]**	**13895** **[9.4]**	**23215** **[15.8]**

* assumes methane density of 0.6785 kg m^{-3} at 1 atmosphere, 15°C

8. THE SIGNIFICANCE OF METHANE EMISSIONS

The estimate of the annual contribution of methane to the atmosphere derived in this paper (3.6 Tg) is comparable to the 5.06 Tg estimated by Dimitrov (2002b). Previous estimates by the present author (9.7 to 12.9 Tg by Judd, 2000; 7.8 to 10.1 Tg by Judd et al., 2002) extend this range. The 30.5 Tg suggested by Milkov et al. (2003) does not account for losses to the hydrosphere or on ignition. Assuming losses of 77% (as above), this should be revised to 7 Tg emitted to the atmosphere. Etiope and Klusman (2002) suggested 2 Tg from quiescent mud volcanoes, and speculated that eruptive events could make this figure 2 to 10 times higher.

These estimates all fall within a remarkably small range (3.8 to 12.9 Tg per year) comparable to those of other natural and anthropogenic sources of atmospheric methane (see Table 9). They suggest that mud volcanoes provide 0.8 to 2.6% of the total budget (503 Tg yr^{-1}). If the higher end of the range is accepted, mud volcanoes are more significant than some of the acknowledged sources. Consequently, they deserve attention. Specifically, it is important to acquire new data in order to resolve the differences between these preliminary estimates, and to undertake a more robust audit. The need for a more robust audit is even more apparent when a different approach to flux estimation is taken. Kopf (2002) estimated the volume of mud extruded by mud volcanoes, the porosity of this mud, and therefore the volume of gas emitted. His suggestion of ~1 to 20 x 10^8 m^3 methane yr^{-1} (about 0.07 to 1.4 Tg) is substantially lower than the estimates quoted above once loss to the hydrosphere is taken into account.

Table 9. Estimates of global sources of atmospheric methane (after Khalil, 2000). Geological sources a contribution from mud volcanoes (Judd, 2000)

Source	*Contribution* *Tg yr^{-1}*	*Percentage of total*
Wetlands	100	19.9%
Termites	20	4.0%
Oceans	4	0.8%
Marine sediments	5	1.0%
Geological	14	2.8%
Wild fire	2	0.4%
Total Natural Sources	***145***	***28.8%***
Rice	60	11.9%
Animals	81	16.1%
Manure	14	2.8%
Landfills	22	4.4%
Wastewater treatment	25	5.0%
Biomass burning	50	9.9%
Coal mining	46	9.1%
Natural gas	30	6.0%
Other anthropogenic	13	2.6%
Low temperature fuels	17	3.4%
Total Anthropogenic	***358***	***71.2%***
Total	***503***	

Because the methane emitted by mud volcanoes is mainly derived from deep within the underlying sediments, it will be 'fossil' (^{14}C-depleted) methane. Fossil methane accounts for approximately 20% of the total source (Houghton et al., 1996), yet a considerable proportion (30 to 50 Tg) is "*unaccounted for*" (Crutzen, 1991). The authorities on atmospheric methane budgets, such as the IPCC (Inter-governmental Panel for Climate Change) tend to assume that all the fossil methane in the atmosphere is released by the fossil fuels industries. Clearly this is not the case as mud volcanoes and other geological sources such as natural gas seeps contribute (see Judd, 2000; Kvenvolden et al., 2001; and Judd et al., 2002 for further discussions).

The contribution of 11.4 Tg of methane to the hydrosphere is also significant. It is anticipated that areas with active mud volcanoes, especially in the relatively shallow waters of the continental shelves, will be characterized by an elevated methane concentration in the water, with localized peaks in concentration that may not be recognized by the regional-scale surveys. Consequently, the poorly-constrained estimates of the flux of methane from the oceans to the atmosphere may be underestimated. However, the majority of methane from deep water mud volcanoes, and quiescent mud volcanoes on the continental shelves will be oxidized in the water column. This may have two implications. Firstly, methane oxidation is probably microbial mediated, so the availability of methane will contribute to local biomass and productivity. Secondly, methane oxidation may affect ocean carbon dioxide budgets, and ocean - atmosphere fluxes.

Judd et al. (2002) pointed out that, because the majority of mud volcanoes happen to lie outside the areas affected by the Quaternary glaciations, there is no reason to suppose that their contributions to atmospheric methane will have changed significantly, other than as a result in changes in eustatic sea level. However, lower sea levels during glacial periods will have reduced the number of mud volcanoes below sea level, and enabled a larger proportion of methane from submarine mud volcanoes to avoid solution in the water column. Also, this will have adjusted the proportion of methane sequestered by gas hydrates. As with natural gas seeps, contributions to the atmosphere will have increased during glacial periods and reduced during inter-glacials. This negative feedback will have worked to minimize extremes of climate, contributing to the 'geological thermostat' proposed by Judd et al. (2002).

9. CONCLUSIONS

Mud volcanoes contribute various gases to the atmosphere; most significantly methane, but also carbon dioxide and nitrogen, plus minor amounts of higher hydrocarbon gases. The methane contribution is estimated here as 3.6 Tg, yr^{-1}, but elsewhere at up to 12.9 Tg yr^{-1}. The majority of this is 'fossil' methane. This is mainly achieved by direct venting, but the large contribution (about 11.4 Tg yr^{-1}) to the hydrosphere may also affect air:sea methane fluxes. This contribution makes up between 0.8 and 2.6% of the global methane budget - more than some more widely recognized sources of methane. Fluctuations in mud volcano emissions between glacial and interglacial periods provide some negative feedback to the 'Greenhouse Effect' and therefore work (along with other geological sources of methane, such as natural seeps) to minimize extremes of climate.

These contributions are sufficiently significant to justify a concerted international effort to improve the inventory of gas emissions from mud volcanoes both on land and offshore.

REFERENCES

1. Aliyev, A., Guliyev, I.S. and Belov, I.S., 2002. *Catalogue of recorded eruptions of mud volcanoes of Azerbaijan (for period of years 1810 to 2001).* Nafta Press, Baku, Azerbaijan.
2. Bagirov, E., Nadirov, R., and Lerche, I. ,1996. Flaming eruptions and ejections from mud volcanoes in Azerbaijan: statistical risk assessment from the historical records, *Energy Exploration and Exploitation* **14**, 535-583.
3. Crutzen, P.J., 1991. Methane's sinks and sources. *Nature*, **350**, 380-381.
4. Dadashev, F.G., 1963. *Hydrocarbon gases of mud volcanoes of Azerbaijan.* Azerneshr, Baku (in Russian).
5. Dimitrov, L.I. , 2002a. Mud volcanoes - the most important pathway for degassing deeply buried sediments, *Earth-Science Reviews*, **59**, 49-76.
6. Dimitrov, L.I., 2002b. Mud volcanoes - a sizeable source of atmospheric methane. *VII International Conference on Gas in Marine Sediments, Baku, Azerbaijan, October 2002, Abstracts volume 33.* (also submitted to *Geo-Marine Letters*).
7. Etiope, G. and Klusman, R.W., 2002. Geologic emissions of methane to the atmosphere. *Chemosphere*, **49**, 777-789.
8. Etiope, G., Caracausi, A., Favara, R., Italiano, F., and Baciu, C., 2002. Methane emission from the mud volcanoes of Sicily (Italy), *Geophysical Research Letters*, **29**, 14340-14343, doi:10.1029/2001GL014340.
9. Etiope, G., Baciu, C., Caracausi, A., Italiano, F., and Cosma, C. , 2004. Gas flux to the atmosphere from mud volcanoes in Eastern Romania, *Terra Nova*, **16**, 4, 179-184.
10. Guliyev, I.S. and Feizullayev, A.A., 1997. *All about mud volcanoes.* Bafta Press, Baku, Azerbaijan.
11. Houghton, J.T., Meiro Filho, L.G., Callander, B.A., Harris, N., Kattenberg, A., and Maskell, K., 1996. *Climate Change 1995: The Science of Climate Change*, Cambridge University Press for the Inter-Governmental Panel for Climate Change, Cambridge.
12. Hovland, M., Hill, A., and Stokes, D. ,1997. The structure and geomorphology of the Dashgil mud volcano, Azerbaijan, *Geomorphology*, **21**, 1 - 15.
13. Judd, A.G., 2000. Geological sources of methane. In *Atmospheric methane: its role in the global environment.*, Khalil, M.A.K., (ed.) Springer, Berlin, 280-303.
14. Judd, A.G., Hovland, M., Dimitrov, L.I., García-Gil, S., and Jukes, V., 2002. The geological methane budget at continental margins and its influence on climate change, *Geofluids*, **2**, 109-126.
15. Kalinko, N., 1964. Gryazevie vulkani, prichini ih vozzniknovenia, razvitia I zatuhania. *VNIGRI*, **40**, 30-54 (in Russian).
16. Khalil, M.A.K. , 2000. Atmospheric methane: an introduction. In *Atmospheric methane: its role in the global environment* , Khalil, M.A.K. (ed.) Springer, Berlin, 1-8.
17. Kopf, A.J., 2002. Significance of mud volcanism, *Reviews of Geophysics*, **40** (2), doi: 10.1029/2000RG000093.
18. Kvenvolden, K.A., Lorenson, T.D., and Reeburgh, W. , 2001. Attention turns to naturally occurring methane seepage. *EOS*, **82**, 457.

19. Leifer, I. and Patro, R.K., 2002. The bubble mechanism for methane transport from the shallow sea bed to the surface: a review and sensitivity study, *Continental Shelf Research*, **22**, 2409-2428.
20. Martinelli, G. and Judd, A.G., 2004. Mud volcanoes of Italy. *Geological Journal*, **39**, 49-61.
21. Milkov, A.V., 2000. Worldwide distribution of submarine mud volcanoes and associated gas hydrates. *Marine Geology*, **167**, 29-42.
22. Milkov, A.V., Sassen, R., Apanasovich, T.V., and Dadashev, F.G., 2003. Global gas flux from mud volcanoes: a significant source of fossil methane in the atmosphere and ocean. *Geophysical Research Letters*, **30**: doi:10,1029/2002GL016358.
23. Sokolov, V.A., Buniat-Zade, Z.A., Goedekian, A.A., and Dadashev, F.G. , 1969. The origin of gases of mud volcanoes and the regularities of their powerful eruptions, In *Advances in Organic Chemistry*, Schenk, P. , Havemar, I., (eds). Pergamon Press: Oxford , 473-484.24.
24. Suess, E., Torres, M.E., Bohrmann, G., Collier, R.W., Greinert, J., Linke, P., Rehder. G., Trehu, A., Wallmann, K., Winckler,G., and Zuelger, E. , 1999. Gas hydrate destabilization: enhanced dewatering, benthic material turnover and large methane plumes at the Cascadia convergent margin, *Earth and Planetary Science Letters*, **170**, 1-15.

Chapter 5
MONITORING TECHNIQUES OF MUD VOLCANISM

MUD VOLCANOES OF PAKISTAN - AN OVERVIEW
A report on three centuries of historic and recent investigations in Pakistan

George Delisle
Bundesanstalt für Geowissenschaften und Rohstoffe (BGR), Stilleweg 2, 30655 Hannover, Germany

Abstract: The earliest reports on mud volcanoes in Pakistan stem from the year 1840. Despite recurrent visits ever since, progress in studying these features in detail has been slow. Increased interest in mud volcanoes occurred in the aftermath of an earthquake in 1945, which initiated the sudden emergence of islands offshore from the Makran Desert. These islands were built from highly viscous mud with a high gas content. The mud was driven up by high buoyancy forces. The study of mud volcanoes in Pakistan intensified concurrently with marine cruises offshore from the Pakistani coast, particularly during the last 15 years, by which mud volcanoes on the abyssal plain and widespread gas seeps in the shallow waters of the upper continental slope were identified. The discharged gases on and offshore are predominantly of bacterial origin and appear to rise from the upper km of the subsurface. Little is known about the periodicity of gas discharge. A new instrumentation to monitor gas flow periodicities is briefly described.

Keywords: Mud volcano, Makran, Pakistan, Chandragup, mud flow, gas seep, gas flux periodicity

1. THE GEOLOGICAL SETTING

The western part of Pakistan is made up of one of the largest accretionary complexes of this world. Accretion probably started during Cretaceous times (DeJong, 1982), when the oceanic crust of the Arabian Sea began to subduct under the Eurasian land mass. The accretionary complex reaches today about 400 km inland as measured from the Pakistani south coast (Taponnier et al., 1982). The youngest part of the complex extends about 100 km seawards, which shows very gentle topography along the shallow upper slope and a system of upthrusted ridges along the lower slope, until

G. Martinelli and B. Panahi (eds.), Mud Volcanoes, Geodynamics and Seismicity, 159–169

it terminates at the accretionary front at a water depth of typically 3000 m (White and Louden, 1982; Roeser et al., 1997; Gaedicke et al., 2002). The thickness of the sedimentary cover of the approaching oceanic plate today exceeds 6 km. Judging from the enormous amount of accreted land mass in the Makran desert, conditions probably did not differ significantly in the past. The subduction angle of the oceanic plate and its sedimentary cover with the Eurasian land mass is with 2° very shallow (Minshull and White, 1989). The subduction rate is estimated to be about 4 cm yr-1 (DeMets et al., 1990). Seismic imaging of the downgoing slab suggests that about the upper 3 km of the sedimentary cover are currently being scraped off at the accretionary front and a new upthrusted ridge has been built over time (Gaedicke et al., 2002). The sediments along the coastal zone are of Plio- to Pleistocene age (Hunting Report, 1960). The coastal zone of the Makran desert is the site of several large mud volcanoes on land as well as of periodically emerging mud volcanoes in shallow water (Fig. 1), which, exposed to vigorous wave action during the monsoon periods, disappear within months. This paper attempts to describe the long history of investigation of the Pakistani mud volcanoes and outlines the open questions associated with the mode of emplacement of these features, despite the long observation period.

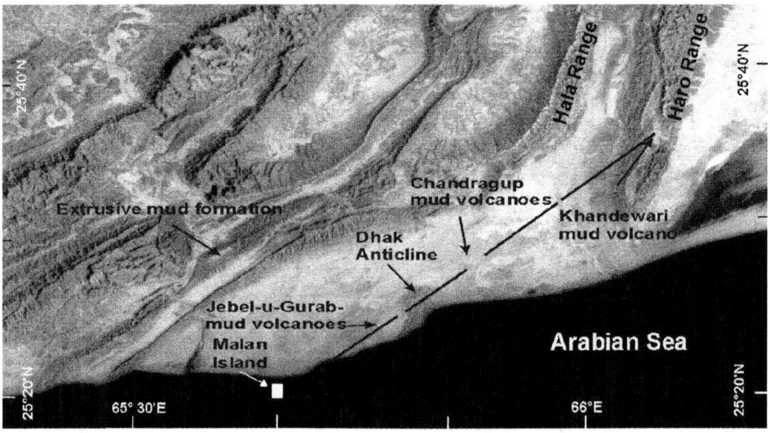

Figure 1. Satellite picture (LANDSAT-MSS scene no. 165-042) of the coastal region with a location map of mud volcanoes and main geological features.

2. THE EARLY DISCOVERIES

Captain Hart from the 2nd Gernadiers (Bombay Army) on his journey from "Kurrachee" (present-day Karachi) across the Makran Desert to Hinglaj (a small religious location with low-level mud volcano activity on the coast near the mouth of the Hingol River) early in the year 1840 passed the site of the Chandragup mud volcanoes, as they are known today (Hart, 1840). His extensive description of the site allows us to draw some conclusions about the long-term behavior of large mud volcanoes, which he did not recognize as such, but noted the existence of three "hills of extremely light-colored earth". The largest mud volcano (Chandragup I) possessed (as it does now) a mud lake occupying the crater and was said to show low level gas exhalation ("a few small bubbles… in a quarter of a minute… near the southern edge"). Today, the most active part of the approximately 16.5 m wide mud lake is near the northern edge. Local people had reported some periodicity in gas exhalation to Hart ("I was told, that every Monday the jets rose with greater rapidity"). In connection with other unvisited mud volcanoes Hart was also told that "the Mahomedans state that they are affected by the tide". Today, visual observation suggests a periodicity in bubbling activity in the mud lake of Chandragup I to the order of about several hours (which actually suggests that the connection with the tide of the near ocean may be viable), but no rigorous measurements are so far available. A smaller mud volcano, resting on the south-western foot of Chandragup I and probably built by a side fissure was described as possessing a shallow mud lake within the crater with minimal gas exhalation activity. Infrequent periods of overflow were reported by the local people. Today, the crater floor has dried up and no sign of recent overflows were detected during two visits by the author in 1998 and 1999. The third mud volcano in the area (Chandragup II) in 1840 possessed a different shape as compared with today. The structure about 50 m in height was found to be occupied by a large circular crater with "two distinctive pools of uneven size, divided by a mound of earth, one containing the liquid mud and the other clear water". This observation is of great interest as it suggests that these two pools, which were located close together were then connected to two different source areas producing two different types of fluid.

A report of a second visit by Stiffe (1874) contains the first drawing of Chandragup I (Fig. 2), which clearly suggests an unchanged shape ever since (Fig. 3).

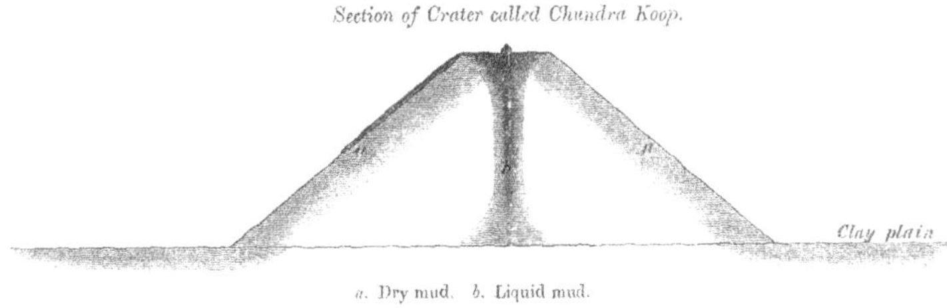

Figure 2. Reproduction of drawing by Stiffe (1874) of Chandragup I

Figure 3. Chandragup I as seen in 1998 from the northeast. Dark streaks on the left side represent recent mud flows.

Stiffe visited the mud lake on the crater (Fig. 4), estimated its diameter at 100 ft (its actual size today is 16.5 m) and measured a lake depth of 50 to 60 feet (in reasonable agreement with a bathymetric survey in 1998 (Delisle et al., 2002), which gave a maximum depth of 29 m).

Figure 4. View on the mud lake occupying the crater of Chandragup I. The rim of the crater is formed by soft and slowly hardening mud (foreground).

New light on the mechanisms involved with the natural rise of mud was shed by the large Quetta earthquake (magnitude 7.5) near Kalat in central Pakistan in the early morning of May 31st 1935. Skrine (1936) reports the Thok mud volcano near Kalat in central Pakistan to burst with a loud roar, clouds of smoke (probably steam) and a vigorous outpouring of reportedly hot mud for hours, forming a 300 yard long mud tongue.

3. MODERN OBSERVATIONS

The first aerial photographs of the Chandragup complex were published by Harrison in 1944. The surface features of Chandragup are strikingly similar to today's configuration . The mud lake of Chandragup I was comparatively active in comparison to later observations and documented to have spilled over in the northern and south-western direction. The lakes in Chandragup II, previously separated by a barrier had coalesced into one mud lake.

Another strong earthquake on 28 November 1945 with magnitude 8.2 along the Makran coast initiated the first professional survey of the mud volcano phenomenon (Sondhi, 1947). The earthquake caused islands to emerge offshore from the Makran coast. At Hinglaj, a fireball "caused by the ignition of a large volume of gas which erupted with such great force that the flames leapt thousands of feet high into the sky". The area was then surveyed from the airplane and attempts by three warships were made to study the nature of the arisen islands. The coordinates of the islands were given by Sondhi as:

Hingol Island: 25° 18'N; 65° 41' E
Ormara Islands: 25°10'N; 64°12'5"E and 25° 9' 5"N; 64° 14'5"E
Gwadar Island: 25°10'45"N; 62°15'50"E.

Figure 5. Western Ormara Island after their emergence in 1945.

Figure 5 shows the western Ormara Island after emergence in 1945 as seen from the airplane. Its structure, interpreted by Sondhi as a result of "earth waves created by the earthquake" are most likely to be created by the outflow of highly viscous mud from a fracture, whereby the mud wall, after reaching a certain height collapsed to the side and was moved sideways by the following extruding mud mass. Visitors to the island noted the presence of gas, whose analysis yielded a predominantly methane gas composition. The volume of the extruded masses were each to the order of several 100 000 m^3 of mud. The islands were eroded away by wave action within about three months. A picture of Chandragup I and the secondary crater on its southwestern foot was also taken during a flight survey (Fig. 6). Clearly seen from the air (but much less obvious on the ground) is a structural step surrounding the volcano complex, which is interpreted as evidence of land subsidence as a consequence of the extrusion of the large mud volume of the volcanoes.

Based on ground observations, the vertical displacement is about 1 to 1.5 m.

Figure 6. Aerial photograph of Chandragup I taken in 1945 (taken by Sondhi, 1947). Notice the rim circling the mud volcano complex.

Visits by Snead in 1959 and 1960 for the first time documented the mud extrusive features of the Kandewari mud volcano in the Haro Range, whose mud flow tongues were reported to extend to two miles from the vent (Snead, 1964). Ellouz reported (personal communication, 2001) the presence of a fresh mud tongue that had apparently developed early in the year 2001. The outflow of mud from this volcano seems to be more frequent and vigorous over time in comparison to the coastal mud volcano complexes (Chandragup and Jebel u Gurab - see below). The location of numerous mud volcanoes along major faults in the Hala Range (Fig. 1) were also described by Snead. He mentions the presence of a group of mud volcanoes eleven miles north of Ormara and a series of 16 m high mud cones near the Pakistani-Iranian border and near the delta of the Dasht River. During his visit of the Chandragup volcanoes, a low level of gas exhalation activity was observed.

The author visited the Chandragup volcanoes twice in 1998 and 1999. Gas composition of fluids from Chandragup I and II was measured. Methane of bacterial origin with traces of higher hydrocarbons was found (for data see Delisle et al., 2002). The analysis of the fluid for $\delta^{11}B$ gave a value of +24 ±1 per mil (Gaebler, personal communication) and points to an only moderately deep source area (see e.g. Lavrushin et al., 1996). In addition, the bathymetry of the mud lake in the crater of Chandragup I was determined. The crater lake terminates at a water depth of 29 m in a fissure. Two measurements in 1998 and 1999 showed nearly isothermal conditions in the mud lake at near-ambient temperatures. In 1998, the crater lake poured over the eastern side of the crater rim and had developed a mud streak along the flank that reached to the crater base. The overflowing mud, however, dries up within hours, crumbles and is eventually blown away by the wind. (see for details Delisle et al., 2002). The main mass of all Chandragup mud volcanoes was built by mud that was apparently very viscous during emplacement, since the all edifices are built by ropy-shaped and meter-thick flows. The current episodic mud flows from Chandragup are of low viscosity and produce only centimeter-thick layers.

From the photographic evidence, the episodic overflow events have clearly not added mass to the volcano structure over the last 70 years, and most likely not during the last 160 years.

A distinct change in gas exhalation activity was noted between May 1998 and November 1999 at Chandragup II. While the activity in 1998 was low, vigorous gas venting occurred in 1999 (see also Delisle et al., 2002).

Finally, north-east of Chandragup I, we find a structure resembling the apparent deeply eroded remnants of a former mud volcano. Visible today are two about 25m high mud walls crossing themselves at near 90° angles. It appears that these walls were formed from solidified former feeder channels. Were this true it would again suggest that mud rises up along fractures and, in this case, apparently along the junction of two deep fractures.

A minor mud volcano field exists at Jebel u Gurab about 7 km to the west-southwest of the Chandragup mud volcano complex. Six mud volcanoes with an elevation of no more than 8 m group in an area of about 500 m by 300 m. The effluents only consist of muddy water. This location might even be the site of a future emplacement of a larger mud volcano. At this time the conduits from below only allow the discharge of fluids from the subsurface. Gas or fluid pressures might be insufficient to widen the conduits to allow for the rise of viscous mud.

All the above-mentioned mud volcanoes line up along a line, which is parallel to the Dhak Anticline (see Fig. 1), suggesting that the location of mud volcanoes is strongly controlled by tectonic lineaments.

Further inland, one formation, whose material composition is well comparable to the above-described mud volcanoes, had previously been termed the "Extrusive Mud Formation" by the Hunting Survey (1960). The extent of this formation and the location of mud volcanoes have been mapped by Bannert et al. (1992). This formation occurs over distances of tens of km along fault structures and would, if it were extruded, represent by far the largest, but also in terms of shape, a very atypical extrusional mud volcano feature. Delisle et al. (2002) have presented arguments in favour of this mass being considered a sedimentary unit of the Parkini Formation. The "Extrusive Mud Formation" probably represents the exposed equivalent of formation at depth, which today acts as the formation source for the now active mud volcanoes.

4. THE RISE OF MALAN ISLAND IN 1999

On March 15th 1999 a new island, built by an extrusive, highly viscous mud flow, emerged and was called Malan Island. Its position (25°20.1'N; 65°40.5'E) is nearly identical to the one of the previously mentioned Hingol Island from the year 1945. The 1999 eruption was not accompanied by an earthquake that can be called upon as an initiating cause. This new island was apparently the result of destabilization of mud at depth and mud upwelling driven by a high gas content. The island disappeared within months due to wave action, but remnants were still detectable by an echolot survey performed in November 1999. Gas exhalation into the water column was noted then. Analysis of this gas indicated methane of predominantly bacterial origin with only traces of admixed components of higher hydrocarbon components (for details see Delisle et al., 2002).

The rise of islands in the near coastal region of Pakistan comes as no surprise considering the fact that the upper slope of the Makran accretionary prism is known for its numerous gas seeps, detected by surveys using a deep towed TV-camera system on the shallow the sea floor. The gas seeps are indicated by white bacterial mats with sizes of typically one to several m^2 (von Rad et al., 2000).

5. THE LOWER SLOPE OF THE MAKRAN ACCRETIONARY COMPLEX AND THE ABYSSAL PLAIN

Extensive marine seismic and bathymetric surveys during BGR cruises in 1995 and 1998 with RV SONNE have not offered conclusive evidence for the existence of mud volcanoes on the lower slope of the Makran accretionary wedge, which, however, is characterized in places by a zone more than 600 m thick with gas hydrates. Delisle and Berner (2002) have actually presented arguments in favour of the gas hydrate zone to act as cap rock for fluids, because extensive gas plumes were observed immediately landward of the boundary, where the gas hydrate zone pinches out. This concept implies extensive fluid flow under the gas hydrate zone towards the shallow waters.

The high sedimentation rates on the abyssal plain seaward of the accretionary front apparently result in undercompacted layers, from which mud diapirs and mud volcanoes rise. Bathymetric surveys of the sea floor show a large number of circular elevations (Wiedicke et al., 2001) near the foot and seaward of the front of the accretionary prism. High-resolution 4 kHz sediment echo-sounding has identified a series of conical mounds, up to 65 m high with diameters of up to 1.5 km. In addition, the well-stratified sediments in the vicinity show repeatedly acoustically transparent zones with 100 to 300 m diameters, rising to near the sea floor. These are interpreted (Wiedicke et al., 2001) as being caused by the rise of gas-rich fluids and mud.

The seismic line SO122-04A of BGR (Roeser et al., 1997) images a slightly bulging sea floor. It appears to represent the early nascent stage of a newly forming accretionary ridge, which contains a buried sediment mound covered and surrounded by flat-lying sediments (Wiedicke et al., 2001). This pile (or possibly a composite feature made up of two neighbouring mounds) is associated with a deep fracture imaged by the seismic record. The base of the pile about 120 m high is approximately 420 m below sea floor. Its diameter is estimated to be ≤ 2.2 km. Wiedicke and others interpret this pile as a former mud volcano on the sea floor that became inactive and was subsequently covered by further sedimentation. The regional sedimentation rate points to a burial age of about 460 ka BP.

Taken all together there is strong evidence for the existence of numerous mud diapirs in the abyssal plain of the Arabian Sea.

6. EMPLACEMENT MODE OF MUD VOLCANOES IN PAKISTAN

Judging from the long record of observations, it appears that two modes of mud volcano activity can be distinguished in Pakistan. Mud flow tongues are conspicuously absent on the flanks of the Chandragup mud volcanoes. Periodic overflow of mud from the muddy waters of the crater lake does not add to the main mass of the volcanoes. The emplacement of the main mass of these volcanoes probably occurred during one event, followed by a long period of low gas exhalation activity. It appears that a continuous leakage of gas from depth prevents the build-up of sufficient pressure for a major new eruption of a mud flow. The same argument might apply for large areas of the upper slope of the Makran accretionary wedge, where we observe abundant gas seeps, but outflow of viscous mud in only a few selected places.

In contrast, the Kandewari mud volcano and apparently some mud volcanoes inland (e.g. the Thok mud volcano) appear to be able to periodically discharge mud flows. In a similar way, some areas offshore appear to possess suitable subsurface conditions to allow the episodic formation of temporary islands.

7. OPEN QUESTIONS

Since the activity mode of mud volcanoes is most likely to be linked to the specific permeability distribution in the subsurface and the physical properties oil/gas reservoirs at depth, studying the gas exhalation periodicity of mud volcanoes in greater detail appears viable in the hope of learning more about the long-term behaviour of oil and gas fields.

To monitor these periodicities in gas flow and variations in the gas composition, monitoring stations should be set up on selected mud volcanoes. Figure 7 shows a first development of a suitable instrument by BGR.

Figure 7. This is a BGR-instrumentation equipped to monitor gas flux and gas composition on mud volcano lakes.

A chamber supported by a float will isolate a gas vent of a mud lake from the atmosphere and force the gas stream past selected sniffers. The gas flow and composition will be continuously recorded. Data are sent via telemetry to a nearby station. In addition, instrumentation for monitoring the mud temperature and variations in the water level of the mud lake are mounted on the float. It is hoped that this instrument will be able to perform its first test at the end of 2003.

REFERENCES

1. Bannert, D., Cheema, A., Ahmed, A., and Schäffer, U., 1992. The structural development of the western fold belt, Pakistan., *Geol. Jahrb. Series B*, **80**, 1-60.
2. DeJong, K. A., 1982. Tectonics of the Persian Gulf, Gulf of Oman, and southern Pakistan region. In: Nain , A.E.M. and Staehli , F.G. (eds) *The Ocean Basins and Margins, v6: The Indian Ocean*, 315-351.
3. Delisle, G., Von Rad U., Andruleit H., Von Daniels CH., Tabrez A. R., and Inam, A., 2002. Active mud volcanoes on- and offshore eastern Makran, Pakistan, *International Journal of Earth Sciences*, **91**, 93-110.
4. Delisle, G., and Berner, U., 2002. Act gas hydrates as cap rock in the Makran accretionary prism? In: Clift, P.D., Kroon, D., Gaedicke,, C. and Craig. J. (eds), *The Tectonic and Climatic Evolution of the Arabian Sea Region*, Geol. Society, London, Special Publications, 195, 137-146.
5. DeMets, C., Gordon, R.G., Argus, D.F., and Stein, S., 1990. Current plate motions:

Geophysical Journal International, **101**, 425-478.

6. Gaedicke, Ch., Prexl, A., Schlüter, H-U., Meyer, H., Roeser, H., and Clift, P., 2002. Seismic stratigraphy and correlation of major regional unconformities in the north Arabian Sea. ? In: Clift , P.D., Kroon, .D, Gaedicke, C., Craig, J., (eds), *The Tectonic and Climatic Evolution of the Arabian Sea Region*, Geol. Society, London, Special Publications, 195, 137-146.
7. Harrison, J. V., 1944. Mud volcanoes on the Makran coast. *Geographical Journal,* **103**, 180-181.
8. Hart, J, 1840. Some account of a journey from Kuzrachee to Hinglaj, in the Lus Territory, descriptive of the intermediate country, and of the port of Soumeanee, *J. Asiatic Soc. Bengal*, **9**, 134-154.
9. Hunting Survey Corporation Ltd., 1960. *Reconnaissance geology of part of West Pakistan.* Maracle Press, Oshawa, Ontario, Canada.
10. Lavrushin, V. U., Polyak, B. G., Prasolov, R. M., and Kamenskii, I. L., 1996. Sources of material in mud volcano products (based on isotopic, hydrochemical, and geological data), *Lithology and Mineral Resources*, **31**, 6, 557-578.
11. Minshull, T. A., and White, R. S., 1989. Sediment compaction and fluid migration in the Makran accretionary prism., *Journal of Geophysical Research*, **94**, 7387-7402.
12. Roeser, H. A., and Shipboard Scientific Party, 1997. The Makran accretionary wedge off Pakistan: Tectonic evolution and and Fluid migration-Part 1. *BGR open file report, Archive No 116 643.*
13. Stiffe, A. W., 1874. On the mud-craters and geological structure of the Mekran coast, *Quarterly Journal of the Geol. Soc. London*, **30**, 50-53.
14. Skrine, C. P., 1936. The Quetta earthquake, *Geographical Journal*, **88**, 414-430.
15. Snead, R. J., 1964. Active mud volcanoes of Baluchistan, west Pakistan, *The Geograpical Review*, **54**, 545-560.
16. Sondhi, V. P., 1947. The Makran Earthquake, 28 November 1945, the birth of new islands, *Indian Minerals* **1**, 146-154.
17. Taponnier, P., Mattauer, M., Proust, F., and Cassaigneau, Ch., 1982. Mesozoic Ophiolites, Sutures, and large scale Tectonic Movements of Afghanistan, *Earth and Planetary Science Letters*, **52**, 355-371.
18. Von Rad, U., Berner, U., Delisle, G., Doose-Rolinski, H., Fechner, N., Linke, P., Lückge, A., Roeser, H. A., Schmaljohann, R., Wiedicke, M., and Sonne 122/130 Scientific Parties, 2000. Gas and fluid venting at the Makran accretionary wedge off Pakistan, *Geo-Marine Letters* **20**, 10-19.
19. White, R. S., and Louden, K. E., 1982. The Makran continental margin: Structure of a thickly sedimented convergent plate boundary. In: Watkins J.S., Drake C.L. (eds) *Studies in continental margin geology*, American Association of Petroleum Geologists Memoir 34, 499-518.
20. Wiedicke, M., Neben, S., and Spiess, V., 2001. Mud volcanoes at the front of the accretionary complex, Pakistan, *Marine Geology*, **172**, 57-73.

MONITORING OF DEFORMATION PROCESSES BY MEANS OF ELECTRIC FIELD OBSERVATION

Igor A. Garagash, Mikhail B. Gokhberg, Nikolat I. Kolosnitsyn
Schmidt United Institute of Physics of the Earth,Russian Academy of Sciences, Moscow, Russia

Abstract: Generation of the electrotelluric field (ETF) by geodynamic processes in the earth crust is considered. It is supposed that the ETF arises due to the electrokinetic effect caused by deformation of the earth crust. Based on this model, a new technique for ETF measurement is proposed. The possibility to use electrical fields caused by the tidal waves for calibration of the measurements is also considered. It is shown that the tidal variations of pore pressure and electrical fields essentially depend on medium permeability, porosity, and viscosity of a liquid. It is shown that by using observations of the vertical electrical field it is possible in principle to determine the mechanical properties of deformable medium. All effects reach a maximum near the earth surface and can be measured.

Key words: electrotelluric field, electrokinetic effect, vertical electric field, tide wave

1. INTRODUCTION

It is well-known that variations in an electromagnetic field formed as a result of fluid flow in deformable saturated near-surface sediment rocks usually attributed to deformations in the earth's crust. For calibration of the variations it is possible to use the changes of porous pressure and their gradients associated with distribution of tidal deformations.

2. DEFORMATION ANOMALIES AND THEIR MIGRATION

The preparation of earthquake is accompanied by abnormal changes in deformations and stresses in a surrounding medium. Fig. 1 shows the changes of one component of tectonic stress (Swolf and Walsh, 1990).

The record of deformations on a slope of the Elbrus volcano (Sobisievich et al., 2001) has shown that after strong earthquake oscillations of deformations with a period of 2-3 days are observed. In Fig. 2 the responses of deformografs on strong earthquakes in Turkey (August 1999) and in Sumatra (June 2000) are shown. The amplitude of deformations is of the order of 10^{-5} .

In addition we note that the abnormal changes in quasi-periodic deformations caused by earthquakes occur during tidal deformations, whose maximum value is about $2 \cdot 10^{-8}$ (Latynina and Karmaleeva, 1978).

G. Martinelli and B. Panahi (eds.), Mud Volcanoes, Geodynamics and Seismicity, 171–179

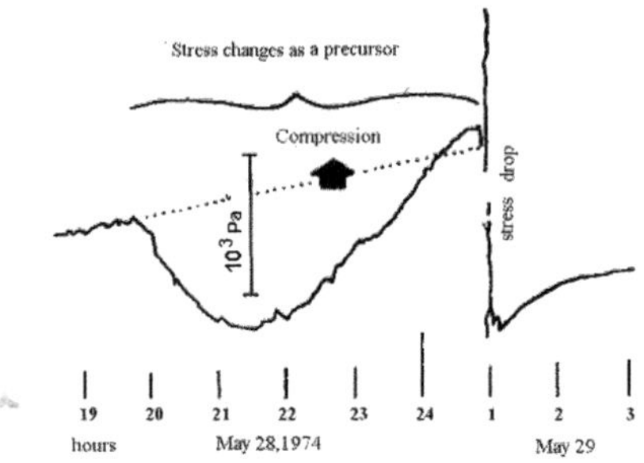

Figure 1. Anomalous tectonic stresses recorded by the hydraulic sensor placed at a crack of a mountain massif.

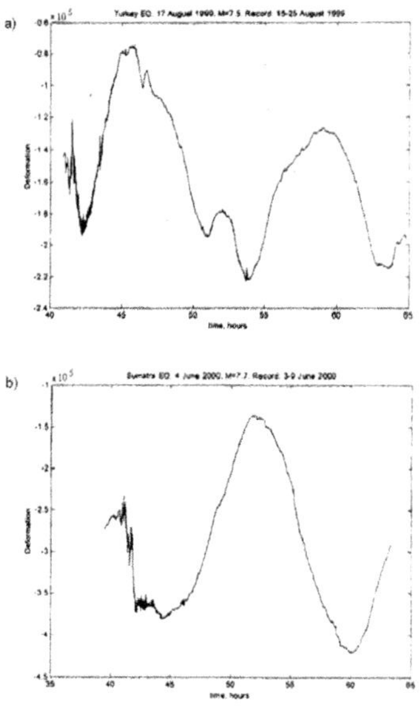

Figure 2. Observed deformations. Duration of the record is equal to 24 hours. The record starts at the moment of the first entrance of the seismic wave from the earthquakes in Turkey (a) and in Sumatra (b).
Sampling rate of the data is equal to 2 s.

Thus, the observations of deformations can provide information about the processes happening in the earth's crust before and after earthquakes. The direct measurements of deformations are rather complex. Therefore, the observations of variations of electromagnetic field generated by deformations and formed as a result of fluid flow in deformable saturated near-surface sediment rocks are of particular interest. For the calibration of such measurements it is possible to take advantage of the analysis of pore pressure and their gradients associated with the distribution of tidal deformations.

3. MODEL OF A SATURATED ELASTIC MEDIUM BY BIOT

Let us consider the behaviour of an elastic isotropic porous medium, saturated by fluid under pressure p. The key relationship proposed by Biot, (Detournay and Cheng, 1993) and connecting effective stresses $\sigma_{ij}^{eff} = \sigma_{ij} + \alpha p \delta_{ij}$ to deformations

$$\varepsilon_{ij} = \frac{1}{2}(u_{i,j} + u_{j,i})$$

can be written in the form

$$\sigma_{ij} + \alpha p \delta_{ij} = 2G\varepsilon_{ij} + \frac{2G\nu}{1-2\nu}\varepsilon\delta_{ij} \tag{1}$$

where

$$\alpha = 1 - \frac{K}{K_s}$$

is the Biot factor,

$$K = \frac{2(1+\nu)}{3(1-2\nu)}G \tag{2}$$

is the bulk modulus of saturated medium under drainage, K_s is the bulk modulus of a matrix material, G is the shear modulus, ν is the Poisson's constant for a drainage material, $\varepsilon=\varepsilon_{kk}$ is the cubic strain, u_k is the component of a displacement vector.

The pore pressure p is connected to the cubic strain of a skeleton ε and change of fluid per unit volume ζ by the relation

$$p = M(\zeta - \alpha\varepsilon) \tag{3}$$

Here

$$M = \frac{2G(\nu_u - \nu)}{\alpha^2(1-2\nu_u)(1-\nu)} \tag{4}$$

is the Biot modulus, ν is the Poisson's constant for a saturated material without drainage.

Substituting Eq. (3) into the expression (1), we can convert it to the form

$$\sigma_{ij} + \alpha p \delta_{ij} = 2G\varepsilon_{ij} + \frac{2G\nu}{1-2\nu}\varepsilon\delta_{ij} \tag{5}$$

For a body being in equilibrium, the equations have to satisfy the condition

$$\sigma_{ij,j} = 0 \tag{6}$$

Besides the equation for mass conservation should be

$$\frac{\partial \xi}{\partial t} + q_{i,i} = 0, \tag{7}$$

where $q_i \rho_f$ is the velocity of a mass transfer through the unit element of erea per unit time.

Using (3), equation (7) becomes

$$q_{i,i} = \frac{1}{M}\frac{\partial p}{\partial t} + \alpha \frac{\partial \varepsilon}{\partial t} \tag{8}$$

Taking into account thr Darcy law,

$$q_i = -\frac{\kappa}{\eta} p_{,i} \tag{9}$$

equation (8), finally, can be represented as (similar equation for the field of pressure in a seismic wave is used by Gershenzon & Bambakidis, 2001)

$$\frac{\partial p}{\partial t} - \frac{\kappa}{\eta} M \nabla^2 p = -\alpha M \frac{\partial \varepsilon}{\partial t} \tag{10}$$

Here κ is the transparency coefficient, η is the dynamic viscosity of fluid,

$$\nabla^2 = \frac{\partial^2}{\partial x^2} + \frac{\partial^2}{\partial y^2} + \frac{\partial^2}{\partial z^2}$$

4. CHANGES OF PORE PRESSURE IN A SATURATED HALF-SPACE UNDER THE ACTION OF A TIDAL WAVE

Let's rewrite the equation (10) as follows:

$$\frac{\eta}{\kappa M}\frac{\partial p}{\partial t} - \nabla^2 p = -\frac{\alpha \eta}{\kappa}\frac{\partial \varepsilon}{\partial t} \tag{11}$$

In the right-hand side of the equation there is a source of periodic stresses, which we relate with the luni-solar tidal waves. The form of the equation (11) coincides with the equation describing a skin effect (see Landau & Lifshits, 1982). By virtue of the boundary condition: p = 0 at z = 0 equation (11) describes an effect that we call "reverse skin effect", i.e. damping of a pressure field when approaching the dayside surface instead of increasing from the surface as in a customary skin effect. In both cases characteristic distance, on which there is the damping, is determined by the frequency and the factor at a temporal derivative in equation (11):

$$\delta = \sqrt{\frac{2\kappa M}{\eta \omega}} \tag{12}$$

We have $\omega = (2\pi / 24)$ $(hour)^{-1}$(a diurnal tidal wave), $\eta = 10^{-3}$,$\kappa = 2{\bullet}10^{-16}\mathrm{m}^2$. Parameter M in view of values of Poisson's constants: $\nu = 0.12$, $\nu_u = 0.31$, factor $\alpha = 0{,}5$ (see Gershenzon & Bambakidis, 2001), Young's modulus $G = 1{,}310^{10}$ Pa and according to the formula (4) is $M = 6{,}842{\bullet}10^{10}$ Pa. From here for a diurnal tidal wave we obtain $\delta = 19.4$ m. For a semidiurnal wave we have $\delta = 13.7$ m.

zIn a saturated half-space along an x-axis the wave of a bulk tidal deformation is propagated. The tidal wavelength is about a

few thousand kilometers, so we neglect a gradient of deformation along the x-axis and take that

$$\varepsilon(t) = \varepsilon_0(1/2)\exp(\omega t) + c.c. \tag{13}$$

where ω is the frequency, ε_0 is the amplitude of bulk strain.

We search for the solution to equation (11) in the form

$$p = A(z)(1/2)\exp(\omega t) + c.c. \tag{14}$$

Inserting (14) and (13) into the (11) we get

$$\left[A''(z) - iA(z)\frac{\omega\eta}{\kappa M}\right]\exp(\omega t) + c.c. = -i\frac{\eta\alpha\varepsilon_0\omega}{\kappa}\exp(\omega t) + c.c. \tag{15}$$

From this we obtain the equation for the function A (z)

$$A''(z) - i\omega\frac{\eta}{\kappa M}A(z) = i\omega\frac{\alpha\eta\varepsilon_0}{\kappa} \tag{16}$$

The constant in the right side of Eq. (16) is eliminated by the substitution

$$A(z) = -\alpha M\varepsilon_0 + \tilde{A}(z) \tag{17}$$

After that the function

$$\tilde{A}(z)$$

is determined by the homogeneous equation:

$$\tilde{A}''(z) - i\omega\frac{\eta}{\kappa M}\tilde{A}(z) = 0 \tag{18}$$

The solution to this equation is searched for in the form:

$$\tilde{A}(z) = A_0\exp(kz) + c.c. \tag{19}$$

From Eq. (18) follows

$$k^2 + i\omega \frac{\eta}{\kappa M} = 0$$

From the latter equation we obtain

$$k = i(1+i)\sqrt{\frac{\omega\eta}{2\kappa M}} = i(1+i)\frac{1}{\delta} \quad (20)$$

The solution looks like:

$$p = \alpha M\varepsilon_0 \left\{ \exp\left[-(1+i)\frac{z}{\delta}\right] - 1 \right\} \frac{1}{2}\exp(i\omega t) + c.c. \quad (21)$$

Taking the real values and requiring fulfillment of the boundary condition $p = 0$ at $z = 0$, we obtain:

$$p = \alpha M \varepsilon_0 \left[\exp\left(-\frac{z}{\delta}\right)\cos\left(\frac{z}{\delta} - \omega t\right) - \cos(\omega t) \right] \quad (22)$$

Another important characteristic of changes of the pore pressure is its gradient. Pursuant to the expression (21) change of the gradient of the pore pressure with depth look like

$$\frac{\partial p}{\partial z} = -\alpha\varepsilon_0 \frac{M}{\delta} \exp\left(-\frac{z}{\delta}\right)\left[\cos\left(\frac{z}{\delta} - \omega t\right) + \sin\left(\frac{z}{\delta} - \omega t\right)\right] \quad (23)$$

The value of the pressure gradient is determined by amplitude $\alpha\varepsilon_0 M / \delta$ equaled to 35.6 Pa/m for diurnal tide and 49,9 Pa/m for a semidiurnal tide (at $\varepsilon_0 = 2\cdot 10^{-8}$, $\nu = 0.19$, $\nu_s = 0.31$).

5. NUMERICAL ANALYSIS OF TIDAL CHANGES IN PORE PRESSURE

The formula (22) allows us to construct the plots of change of the pore pressure on depth at propagation of the tidal wave depending on a permeability of medium and its compliance. We will use the following set of parameters: $\nu = 0.19$, $\nu = 0.31$, $G_s = 1.55\cdot 10^{10}$ Pa, $\omega = (2\pi / 24)\ (hour)^{-1}$, $\eta = 10^{-3}$ Pa/s, $\varepsilon_0 = 2\cdot 10^{-8}$, $\nu = 0.12$, $\nu_u = 0.41$, $\kappa = 2\cdot 10^{-16}$ m^2.

Fig. 3 shows a surface of change of the pore pressure with a depth during 48 hours for a shear modulus $G = 1{,}3\cdot 10^{10} Pa$, relevant to the Ruhr sandstone (Detournay&Cheng, 1993). The values of variations of the pore pressure and its gradient are determined by the parameter

$$\delta^{-1} = \sqrt{\frac{\eta\omega}{2\kappa M}}$$

the Biot parameter M and parameter α, and also the amplitude of a deformation ε_0.

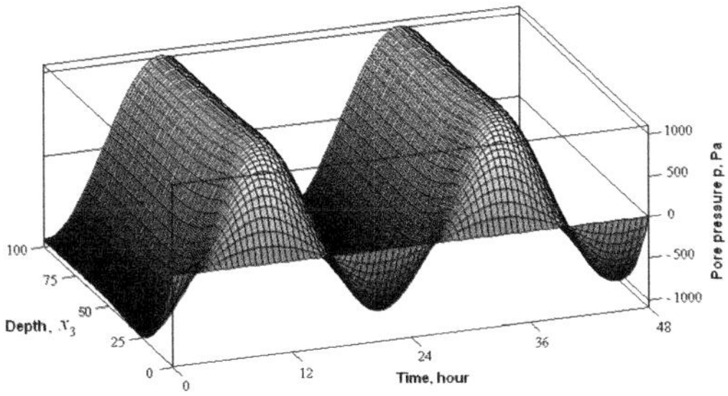

Figure 3. Surface of change of the pore pressure p_i in coordinates "depth"- "time" at ε_0=2·10^{-8}

Fig. 4 shows the dependence of the gradient of the pore pressure near the earth's surface on the Poisson's constant and the depth z in the absence of drainage at $\varepsilon_0 = 2 \cdot 10^{-8}$, $\kappa = 1 \cdot 10^{-17}$ m^2, $\nu = 0.19$, $\nu_s = 0.3$, $G = 1.3 \cdot 10^{10} Pa$, $G_s = 1.55 \cdot 10^{10} Pa$.

Fig. 5 shows the gradient of the pore pressure near the earth's surface as a function of the transparency coefficient κ and the depth z for the Poisson's constant $\nu_u = 0.45$, $\varepsilon_0 = 2 \cdot 10^{-8}$, $\kappa = 1 \cdot 10^{-17}$ m^2, $\nu = 0.19$, $\nu_s = 0.3$, $G = 1.3 \cdot 10^{10} Pa$, $G_s = 1.55 \cdot 10^{10} Pa$.

Finally, Fig. 6 shows the gradient of the pore surface pressure as a function of the Poisson's constant ν_u and the coefficient of transparency κ constructed.

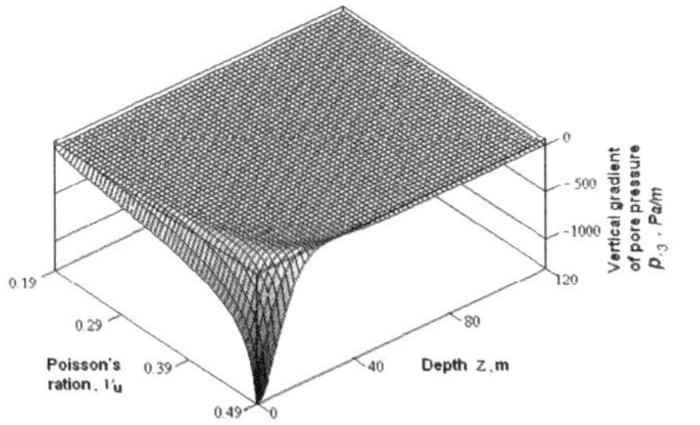

Figure 4. Dependence of change of the vertical gradient of the pore pressure p_z on the depth z and the Poisson constant ν_u

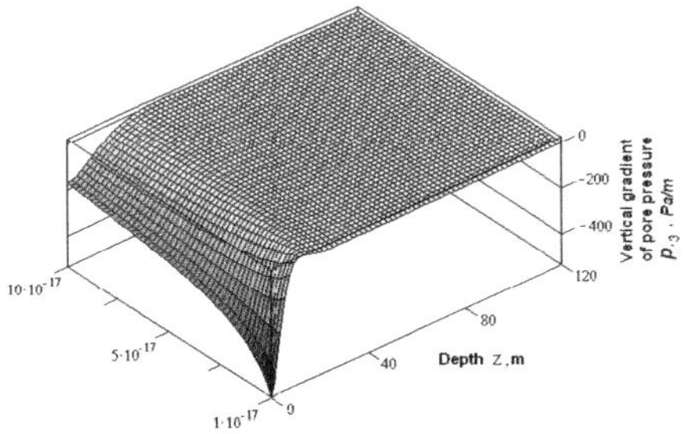

Figure 5. Dependence of the vertical gradient of the pore pressure p_z on the depth z at different values of the coefficient of transparency κ

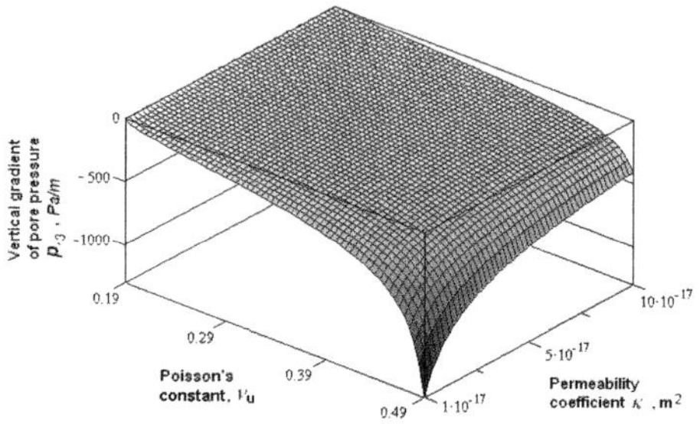

Figure 6. Changes of the vertical gradient of the pore pressure p_z near the earth's surface as a function of the coefficient of transparency κ and the Poisson is constant ν_u

CONCLUSIONS

The accomplished analysis shows that the tidal variations of the pore pressure essentially depend on mechanical properties of medium, its permeability, viscosity of fluid, and pressure gradient reaches a maximum value near the earth surface. According to the calculations, for the characteristic values of parameters of medium, the tidal variations of the pore pressure gradient reach appreciable values, and the electrical field associated with the gradient can be recorded.

Measurements of the vertical component of the electric field can be used to determine medium parameters (permeability, elasticity, and so on). In addition, there is a chance to use the results of calculations in monitoring media in the vicinity of geological structures as oil deposits and mud volcanoes.

REFERENCES

1. Sobisievich, L.E., Milyukov, V.K., Sobisievish, A.L. 2000. Mechanic-mathematical monitoring magmatic of structures of the Elbrus volcano. In: *Geophysics and mathematicians*, Moscow, 223-248.
2. Latynina, L.A., Karmaleeva, P.M., 1978. Deformografic of measurement. Moscow, Nauka.
3. Landau, L.D., Lifshits, E.M,., 1982. Electrodynamics of continuous media. Moscow, Nauka.
4. Gershenzon, N. and Bambakidis, G., 2001. Modeling of seismo-electromagnetic phenomena, Russian Journal of Earth Sciences, **3**, No 4, 247-275.
5. Detournay, E., Cheng, A. H.-D., 1993. Fundamentals of Poroelasticity, in *Comprehensive Rock Engineering: Principle, Practice and Projects*, Vol. II, Analysis and Design Method, Fairhurst, C. (ed.), Pergamon Press, 113-171.
6. Swolf, H.S., Walsh, J.B., 1990. The theory and prototype development of a stress-monitoring system., *Bull. Seism. Soc. Amer.*, **80**, No 1, 197-208.
7. Gershenzon, N. I., Gokhberg, M.B., 1994. On the origin of anomalous ultra low-frequency geomagnetic disturbances prior to Loma Prieta, California, earthquake. Phisics of the Solid Earth, 30, 112-118

MUD VOLCANO: METHODS, APPARATUS - FUNDAMENTAL AND APPLIED ASPECTS OF RESEARCH

O. B. Khavroshkin, V. V. Tsyplakov, E. M. -K. Urdukhanov
and Natalya A.Vidmont
Schmidt United Institute of Physics the Earth, Russian Academy of Sciences, 10 Bolshaya Gruzinskaya, 123995 Moscow, Russia

Abstract: Reliable baro-resistant instruments for mud volcano investigation have been designed for investigation of mud volcano. These instruments and method of using were tested during field evaluation at Krasnodar region.

Key words: baro-resistant geophysical instruments, mud volcano, volcanic channel, eruption phase, seismic field, seismic emission, microseism, seismic noise, deformation fields, active deep faults, earthquake, nonlinear seismology

1. INTRODUCTION

It is well known that of mud volcanoes are unique natural wells. However till now the study of these interesting natural objects is restricted to recording reflected and/or transmitted seismic waves and geochemical analysis of fluid and gas samples from gryphons as well as natural seismic field of a mud volcano (Apanasovich et al., 2003; Godano and Capuano, 1999) when eruption. These methods restrict opportunities of step forward to a new stage of investigations.

Recording and analysis of high-frequency seismic noise and seismic emission (HFSN and SE) are the perspective universal approaches for nonlinear seismology to study complex geophysical objects. Twenty-year authors' experience in this matter (Khavroshkin et al., 1978; Khavroshkin et al., 1997) has been integrated in the first monograph devoted to nonlinear seismology (Khavroshkin, 1999).

The key processes of mud volcanoes are represented in statistical characterizations of HFSN (spectrum, spectral-time analysis, etc.) (Khavroshkin, 1999). So the investigation of HFSN and SE of mud volcanoes will allow to advance qualitatively present knowledge (Khavroshkin et al., 1997). Interesting results of study of Stromboli volcano by means of nonlinear seismology evidence their potential (Urquizi and Correig, 1998; Godano and Capuano, 1999).

On the other hand, the geologic features of mud volcanoes eliminate an opportunity of informative HFSN and SE recording on a dayside due to a strong attenuation of noises at the line of route «deep structures of a mud

G. Martinelli and B. Panahi (eds.), Mud Volcanoes, Geodynamics and Seismicity, 181–186

volcano - dayside». A principally new pressure proof geophysical system (baro-resistant instruments, BRI) allows to solve this problem. It makes possible to investigate wave processes of deep structures of a mud volcano and its volcanic channel up to 12 km and more (Khavroshkin et al., 2001). The BRI was designed (Fig. 1), patented (Khavroshkin et al., 1995) and awarded by Diploma and Gold Medal of exhibition "Brussels Eureka 2001".

Testing the system at descent of the BRI into gryphons of mud volcanoes (Taman peninsula) testified the efficiency of the new approach and revealed some features of microseismic field of a mud volcano in the dying state.

2. EXPERIMENTAL

We investigated Karabetovsky volcano 4-5 kilometers distant from Tamansky settlement and 150 meters above sea level. The volcano is remarkable for a system of weekly acting and dying down gryphons. The gryphons are single or grouped (2-3 ones in a group). The maximum height of the gryphon's cone is 3-5 m, the average height is 0.1-0.3 m. The gryphons have a weak bubbling gas emission with a time interval about 10-15 seconds, the maximum pauses are about 45 seconds.

The mud volcanoes of western Kuban are connected with Indol-Kuban and Kerch-Taman deflections. Their geological structure is represented by Triasic till Quaternary sediments. The foundation is immersed and placed at the depth of 12-14 km. In south-eastern part of the deflection the volcanoes spatially coincide with its southern edge. At the east of Taman peninsula, where are located tectonic elements of Kuban and Kerch-Taman transperyclinal deflections, mud volcanoes are developed at whole peninsula territory. They immersed considerably in Mesozoic-Cainozoic time.

Indol-Kuban deflection is a tectonic element of north-western immersion of the Great Caucasus. Within Taman interpericlinal deflection are revealed anticlinal zones divided by relatively wide synclines formed by Pliocenic and Quaternary sediments. Most of folds are complicated by mud volcanoes and have special shape.

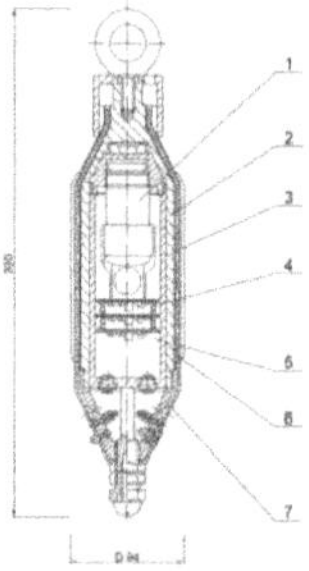

Figure 1. Design of the baro-resistant instruments: 1 - seismic detector units; 2 - outer case; 3 - inner case; 4 - envelope extraction unit and preamplifier; 5 - free space for filling with an inertial unviscous liquid; 6 - cable net; 7 - cover.

We detected and analysed the seismic noises at inner points and at the surface of the mud volcanoes gryphons. The noises measured at the acting gryphons of the cluster of the Karabetovsky mud volcano that have a lot of acting and extinct volcanic manifestations. To detect surface noises were used standard CM-3 seismometers, vertical and horizontal, orientated north and south. Used as detecting instrument at inner points of medium was dipping baro-resistant seismic instruments developed specially for these purposes. Its design is based on the modernized vertical seismic sensor CB-10 with sensitivity about 11 V/m/sec. The three largest of acting gryphons spaced 70 meters apart were chosen. The gryphons were selected subject to absence of any visible communications through mud channels defined by assessment of frequency and periodicity of bubble gas emission at each gryphon.

The technique of seismic noises measurements was following: the BRI self-dipping for a depth to 15-20 meters, and two other CM-3 seismometers were set at the surface next to the gryphon to detect vertical and horizontal components of surface noises. Signals with numbering frequency of F=0.08 sec were recorded simultaneously on the hard medium of the self-contained seismic station.

The signal processing consisted in obtaining the Fourier power spectra of the seismic noises and correlation and cross-correlation functions. To increase reliability of the spectral characteristics signals were smoothed to obtain 10% standard deviation when using square coefficient of distribution χ. According to this technique three gryphons were studied. The first one is located at the bottom of Karabetovsky volcano. The second gryphon is 20-25 meters higher than the first one, 50-60 meters distant from it. And the third one is located beyond the ridge of volcanic ejection at the same height. Presented in Figure 2 are Fourier power spectra of surface noises signals (vertical and horizontal) and power spectra of vertical component of the dipped BRI obtained at three different gryphons.

Analysis showed that maxima of the signal spectra don't coincide at inner points of medium (0.5 Hz) and at surface points (1-1.5 Hz). Recalculations of the seismometers sensitivity at surface and at inner points showed that amplitudes of the oscillations at inner points are greater than at surface approximately by an order of magnitude.

The spectra of the signals of the first and the second gryphons are similar or even coincide for some peaks that probably proves the evidence of their common microseismic field, whereas spectra of the third gryphon differ from the two first gryphons. Cross-correlation spectra for all the three gryphons contain usually the same spectral peaks as surface horizontal ones. Analysis have revealed a considerable frequency set describing a wave pattern of the microseismic field of the three gryphons, as a common interdependent system. The general peculiarity of these interrelations is that the basic spectral peaks correlate as the terms of the one series n = 1,2,3.4. These correlations are sometimes observed with a high accuracy.

On the whole we consider that the exogenic processes play the defining role in a local microseismic field of a mud volcano. Thus using of mud volcanoes to study both local and regional tectonics and deformation processes is perspective.

In accordance with nonlinear seismology we suggest the following noise zone structure for a mud volcano (Fig. 3). The properties of HFSN of the zone I are studied well, zone II is under investigation. For other zones we can note the following. In a stage of complete dying of volcanic activity the noise of Maikopian clays (III) will weaken because emissive component due to fluid transfer is faint but increases with depth in more deep structures (IV).

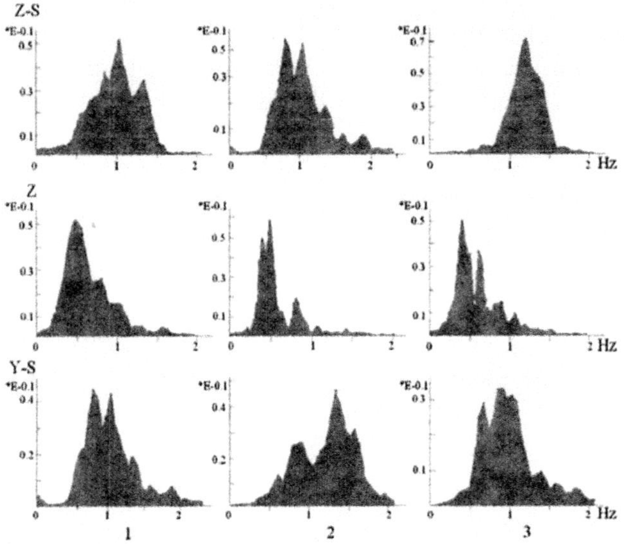

Figure 2. Fourier power spectra of the seismic noises of the three gryphons (1, 2, 3) of Karabetovsky volcano: Y-S, Z-S - horizontal and vertical noise components at the surface; Z - vertical noise component inside gryphon channel.

A sudden increase of the level of the noise with a continuous spectrum will take place in a gas-containing stratum (V), below in Jurassic sediments (VI) will appear some peaks due to deformations and faults motions.

Recording a mud volcano HFSN and SE related to active deep faults (VII) represents the most scientific value, as their genesis is caused by processes in the zone «crust - upper mantle». One of features of the wave field would probably be the mode of seismic auto-generation peculiar to seismic active zones (Khavroshkin, 1999).

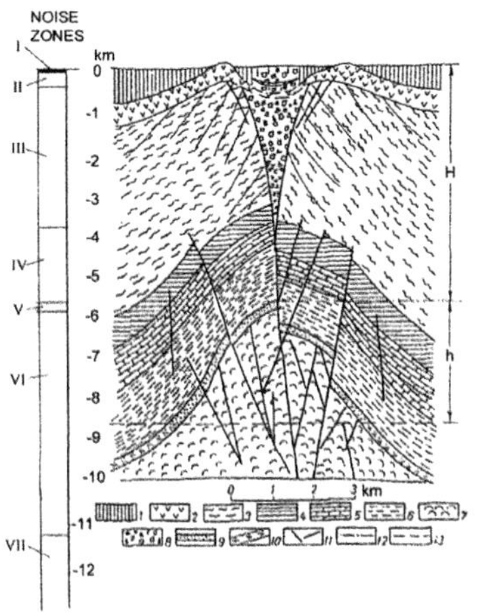

Figure 3. Structure of a mud volcano (Karakin et al., 2001) and vertical section of presumed noise seismic zones.
Sediments: 1 - Quaternary and Pliocenic; 2 - Miocenic; 3 - Maikopian; 4 - Eocenic and Palaeocenic; 5 - Upper Cretaceous; 6 - Lower Cretaceous; 7 - Jurassic; 8 - volcanic or bajada breccia; 9 - gas-containing stratum; 10 - watered stratum; 11 - faults; 12 - location of a gas stratum's roof; 13 - location of a gas-water contact; h - height of a gas deposit's floor; H - depth of the gas deposit. Noise seismic zones: I - not informative seismic noise and microseisms, 1-10 m; II - wave field of gryphones, 10 - 300 m; III - noise of Maikopian clays, 0.3 - 3.5 km; IV - noise of more dense structures (sediments 4-6), 3.5-5.5 km; V - noise of gas-containing stratum, 5.5-6.0 km; VI - noise of Jurassic sediments, 6.0 - 12 km; VII - noise and seismic auto generation of fissile deep faults.

3. CONCLUSIONS

When a mud volcano is in the dying activity a transportation of the BRI into the zone VII (Fig. 3) generally is possible only with the help of a special towboat. At a final activity stage it is real to reach depths of 10-12 km in regime of self-dipping through a volcanic channel opened by a stream of fluids of hyper pressure. We also designed some specialized variants of the BRI represented at the Fig. 1 for different purposes.

In general the experience of nonlinear seismology and the created hardware - methodical basis of mud volcanoes' study allow to advance the following steps:

1. development of wave dynamics of a region and the Earth
2. development of deep faults dynamics
3. details of hydrocarbon deposits shaping
4. realisation of effective short-term forecast of deep-seated earthquake of Vrancea region by using local mud volcanoes (Khavroshkin et al., 2000).

REFERENCES

1. Apanasovich, T.V., Dadashev, F.G., .Milkov A. V., and Sassen, R., 2003. Global gas flux from mud volcanoes: a significant source of fossil methane in the atmosphere and the ocean., *Geophysical Research Letters*, **30** (2), 1037.
2. Godano C., and Capuano P., 1999. Source characterisation of low frequency events at Stromboli and Vulcano Islands (Isole Eolie Italy), *Journal of Seismology* **3** (4): 393-408.
3. Karakin, A.V., Kambarov, N.S. and Kambarova, G.N., 2001. Geodynamics of mud volcano activity and oil-gas-bearing capacity of the Earth interior. Schmidt United Institute of Physics of the Earth Press., Moscow (in Russian).
4. Khavroshkin, O.B., 1999. Some problem of nonlinear seismology, Schmidt United Institute of Physics of the Earth Press., Moscow, 286 (in Russian).
5. Khavroshkin, O.B., Rykunov, L.N., and Tsyplakov, V.V., 1978. The modulation of high frequency microseism, *J. Dokl. Science Section*, **238**. Translated from *Reports of the Academy of Sciences USSR*, **238**, 303-306.
6. Khavroshkin, O.B., Rykunov, L.N., and Tsyplakov, V.V. and Vidmont, N.A., 1997. Modulation of high-frequency seismic noise in the case of low earth seismicity, *J. Doklady Earth Sciences*, **358**, 114-117. Translated from *Doklady Academii Nauk*, **358**, 256-259.
7. Khavroshkin, O.B., Tsyplakov, V.V. and Urdukhanov, R.I., 1995. Method of proofing geophysical equipment against environment. Patent #2110816. Priority: August 30, 1995. Inventor's application:August 30, 1995.
8. Khavroshkin, O.B., Tsyplakov, V.V. and Urdukhanov, R.I. ,1997. Mud Vulcano as an object of Study and an Instrument of Geophysical Reserch. In: Abstracts of 29th General Assembly IASPEI. Thessaloniki, Greece, 18-28 Aug., 1997, 179.
9. Khavroshkin, O.B. Tsyplakov, V.V. and Urdukhanov, R.I., 2001. Pressure resistant deeply-dipped geophysical equipment (GGS): tests of a working model, *Seismic Instruments*, **36**, 38-44, (in Russian).
10. Khavroshkin, O.B., Tsyplakov, V.V., and Vidmont, N.A., 2000. Earthquake prediction: problems and results, *Herald of the DGGGMS RAS: Electr. Sci.-Inf. J.*, **2(12)**, 103-109,
11. http://www.scgis.ru/russian/cp1251/h_dgggms/2-2000/prognosis.engl.htm#begin
12. Urquizi M., and Correig, A.M.,1998. Analysis of seismic dynamical systems, *Journal of Seismology* **2** (2): 159-171.

MUD VOLCANO MONITORING AND SEISMIC EVENTS

Giovanni Martinelli [1]
Andrea Dadomo [2]

1 ARPA Environmental Protection Agency of Emilia-Romagna Region, Via Amendola, 2, 42100 Reggio Emilia, Italy, e-mail: giovanni.martinelli15@tin.it

2 Department of Geological Sciences and Geotechnology, University of Milan, Piazza delle Scienze, 4, 20126 Milan, Italy

Abstract : Mud volcanic activity have been sometimes connected to seismicity but mud-volcanic gases and clayey waters have been poorly monitored. The lack of long-term monitoring data is due to the semi-erratic nature of mud volcanic emissions and to geological and technical constraint factors. A review of geophysical and geochemical available data has been carried out. Data confirm that mud volcanoes are confined fluids accumulations capable to work, in principle, as natural strain-meters. The absence of reliable long-term data sets hampers any definitive conclusions on the sensitivity of mud volcanoes to seismic events and evidence the need for reliable ground-based and satellite-based measuring techniques.

Key words: earthquake precursors, automatic monitoring, confined fluids, mud volcanism.

1. INTRODUCTION

Various methods have been used to explore natural fluid emissions. Geophysical and geochemical parameters have been used to describe the quantity and the chemical features of the fluids expelled from volcanoes, geysers, soil gases, cold and thermal springs, water wells, hydrocarbon wells, and so on. The experiences undertaken in many geological environments have allowed us to state that every natural fluid emission can be characterized by the description of mass flow rate, temperature and chemical composition. Repeated measurements of the flow rate, the temperature and the chemical composition have allowed for the collection of records or data sets. Sampling rate, precision and accuracy adopted in the measurement techniques have revealed unsuspected details concerning natural fluids. Their space distribution, origin and evolution have been explained and every day a large amount of data are collected and stored for exploration purposes, environmental control and other research activities.

G. Martinelli and B. Panahi (eds.), Mud Volcanoes, Geodynamics and Seismicity, 187–199

Mud volcano fluids are connected to deep-seated reservoirs and are not fed by surface fluids. Thus, they can work as natural strain-meters in poroelastic media (Bodvarsson, 1970) and can make a major contribution to a better understanding of seismogenesis. The idea that mud volcanoes could be sensitive to crustal deformations was first proposed by Tamrazyan (1972), who observed eruptive changes linked to moon-induced tides.

Unfortunately, among the natural fluids mud-volcanic gases and clayey waters are the least monitored.

Only a few mud volcanic waters or gases have been geochemically characterized. A review of the available geochemical data has been proposed by Martinelli and Dadomo (this book). Flow rate and temperature have often been inferred and sometimes measured. A review of the available gas-flow rate data has been proposed by Judd in this book. Liquid phase flow-rate data have been measured in only some mud volcanic areas (Brown, 1990) and reviewed by Kopf (2002). Clay dominating flow rates have been estimated in Azerbaijan and reviewed by Aliyev et al. (2002) and commented on by Panahi (this book).

There is a lack of long-term monitoring data but significant efforts have made done in developing technologies applied to monitoring mud volcanic areas by Mellors et al. (this book) and Delisle et al. (this book).

The lack of long-term automatic monitoring data is due to the semi-erratic nature of mud volcanic emissions, to the continuous morphological evolution and to the scarcity of sensors suited to extreme environments, so a significant amount of the available data have been recorded by manual sampling. To better understand the physical and chemical time-behaviour of mud volcanoes a review of the available data sets has been prepared.

2. FINDINGS FROM AVAILABLE GEOPHYSICAL DATA SETS

Many authors have for a long time attributed the flow-rate variations in the total fluid emissions of mud volcanoes to seismic activities. In particular, Guliyev and Feizullayev (1997) have reported in Fig.1 the number of seismic events per month in the period 1669-1982 in line 1, and in line 2 the

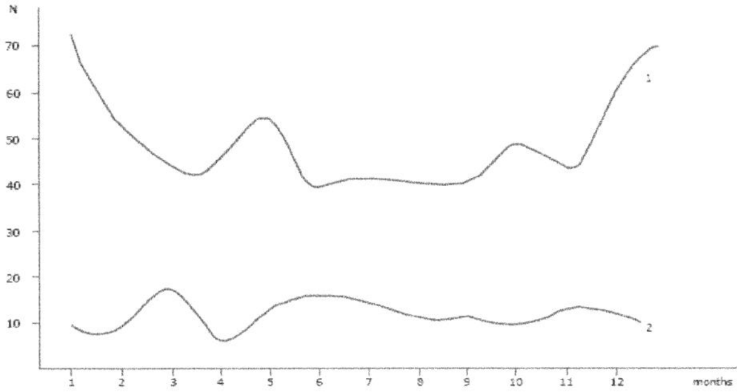

Figure 1. Number of seismic events per month registered in a selected area of Azerbaijan (line 1) and number of eruptions per month (line 2). After Guliyev and Feizullayev, 1997, modified.

monthly number of eruptions recorded in the period 1972-1987. The Authors have concluded that an increasing number of shocks is responsible for an increase in eruptive activity. Although the published graphs do not allow for definitive assessments, a variation in the eruptive activities during the year appear to be confirmed. Meteorological data have not been reported but a relationship with seismic events appears to be partially confirmed. Fig.2 was drawn by Aliyev et al., (2002) and shows that the clayey mud-flow rate varied over time in the period 1887-2001.

Figure 2. Variation of the erupted mud in the period 1887 – 2001 in the Lokbatan mud volcano (Aliyev et al., 2002).

The authors concluded that a strong eruptive period observed also in other mud volcanoes was triggered by strong local seismic events. Strong variations in the gaseous flow rate were detected by Caneva (1958) with an automatic recorder on the Regnano mud volcano located in Northern Italy (Fig.3). The monitoring period lasted six months and strongest variations were retrospectively attributed by Martinelli and Ferrari (1991) to an increase in the local seismic activity. The recorded data do not allow us to definitively state whether the gas flow rate variations preceded or followed the seismic activity. An indirect method for inferring variations in the flow-rate emission involves accurate temperature recording.

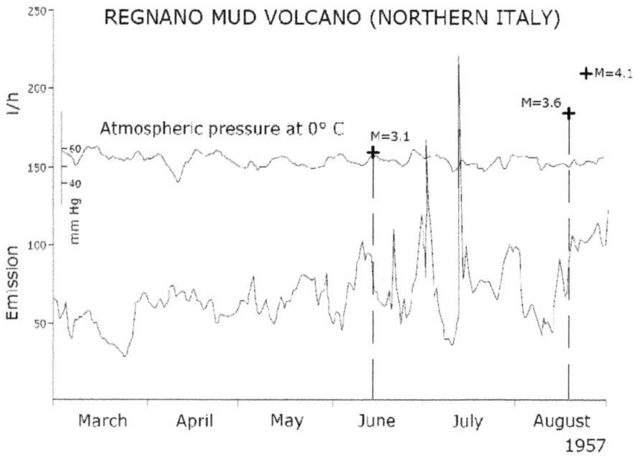

Figure 3. Methane flow-rate, atmospheric pressure and local seismic events recorded at the Regnano mud volcano, North Italy (Caneva, 1958; Martinelli and Ferrari, 1991).

Hasiotis et al.(1996) detected a sharp temperature variation in a pockmark field in the Patras Gulf before a local M=5.4 seismic event (Fig.4).

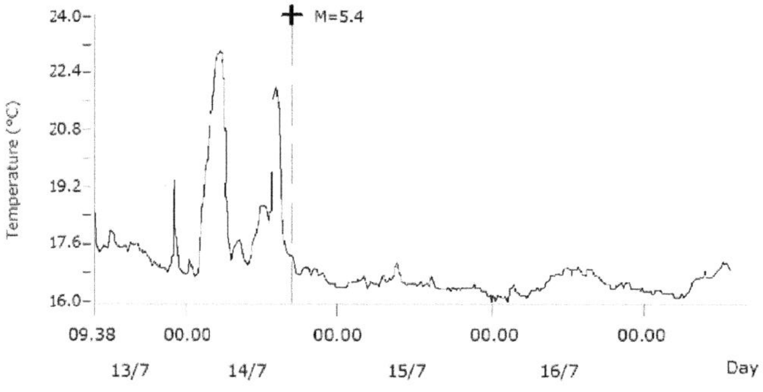

Figure 4. Temperature variation before a M=5.4 seismic event in fluids expelled from a pockmark field in the Patras Gulf, Greece (Hasiotis et al., 1996).

The pockmark cannot be strictly considered to be a mud volcanic structure but some geological features are common to both fluid emissions and suggest that similar phenomena could occur in mud volcanic areas. Foucher et al. (1992) monitored fluid temperature in accretionary wedges located in the Nankai complex. (Fig.5 and Fig.6).

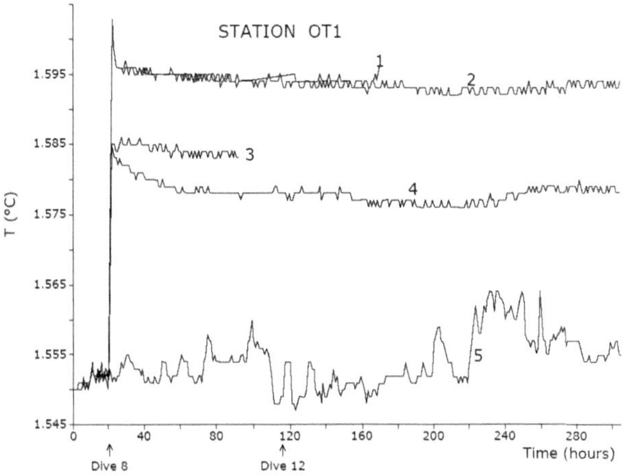

Figure 5. Graph shows temperature data collected at the Nankai accretionary complex. Record 1, 2, 3, 4 were collected at 64 cm depth (1 and 2), 44 cm depth (3) and 34 cm depth (4). Probe 5 was in sea water (Foucher et al., 1992).

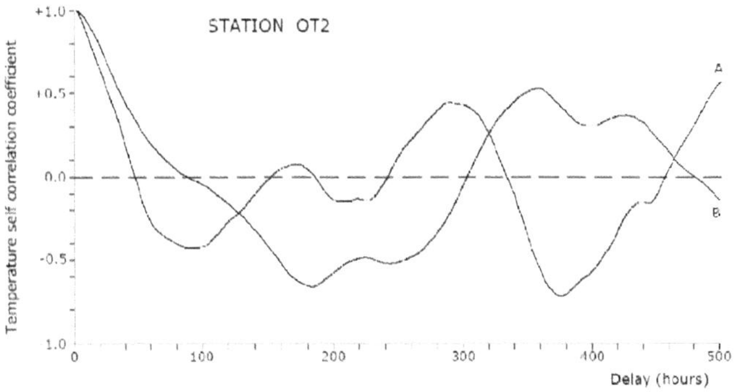

Figure 6. Variation with delay T of the self correlation coefficient for both sea water temperature (line A) and temperature at 34 cm depht in Nankai accretionary complex (line B). Sea water temperature is non periodic while sediment temperature may be affected by a periodicity of about 350 hours (Foucher et al., 1992).

Raw and processed temperature data demonstrated that fluid expulsion variations occurred for reasons independent of the sea currents. Becker et al. (1997) reached similar conclusions after monitoring pressure and temperature in the Barbados accretionary prism (Fig.7).

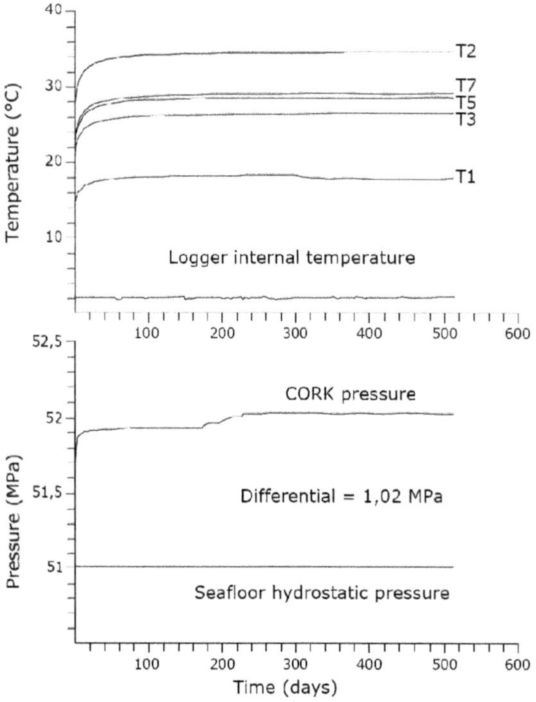

Figure 7. Long term records of temperatures and pressures measured in hole 949 C in Barbados accretionary prism (Becker et al., 1997)

Data are shown to demonstrate that long-term monitoring of temperature and pressure is possible in a mud-volcanic type of geological environment. Streaming potential has also been applied by Segawa and Toh (1992) to track fluid emissions in the Nankai Trough area (Fig. 8).

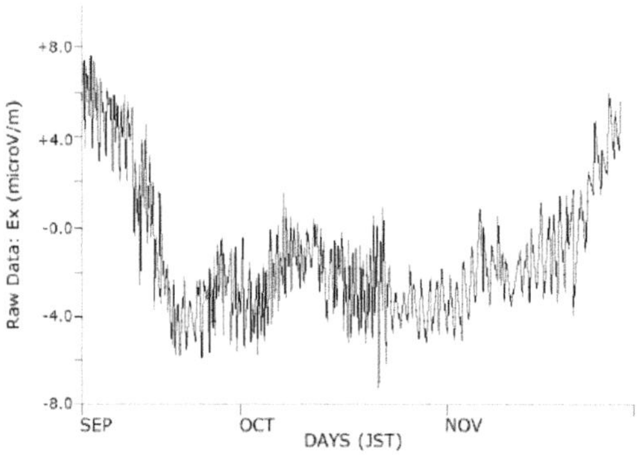

Figure 8. Electric field variations observed at the Nankai Trough in 1989 (Segawa and Toh, 1992).

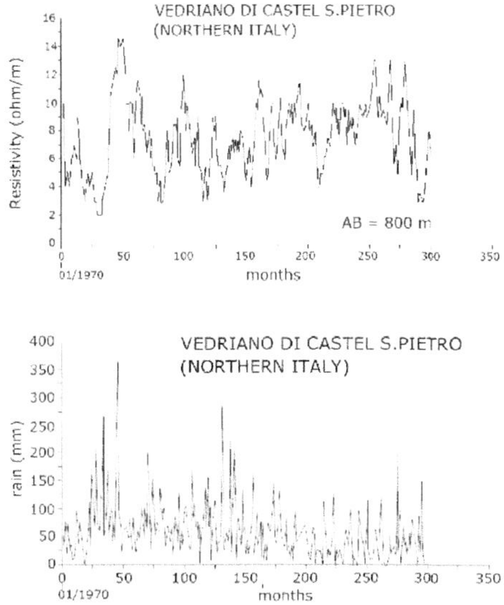

Figure 9. Resistivity values (top) and rains registered close to a small mud volcanic area in North Italy (Albarello et al., 1999).

Long-term monitoring has evidenced variations in the flow-rate not attributable to sea dynamics. A long-term monitoring program was carried out close to a small mud volcanic area of Northern Italy (Fig.9) by Parenti (1999, personal communication). Apparent resistivity was measured in the period 1970-1996 at a monthly sampling rate. Albarello et al. (1999) processed the data and reached the conclusion that strong fluctuations observed in resistivity values were mostly attributable to rain and other meteorological factors.

No further conclusions can be drawn since within a 50 km radius of. the measurement point no seismic events characterized by M> 3.5 occurred during the whole monitoring period. Furthermore, apparent resistivity measurements turned out to be unsuitable in the mud volcanic environment due to the very low electric resistivity of soils impregnated by highly conductive brackish waters. Geophysical records do not allow for a complete understanding of the possible link between seismicity and eruptive activity.

3. FINDINGS FROM THE AVAILABLE GEOCHEMICAL DATA SETS

Geochemical data obtained by manual sampling have been recorded by Capozzi and Picotti (2003) from the Regnano mud volcano located in North Italy. The data evidence a strong variation in the cation ratio during an 8-months time-period (Fig.10).

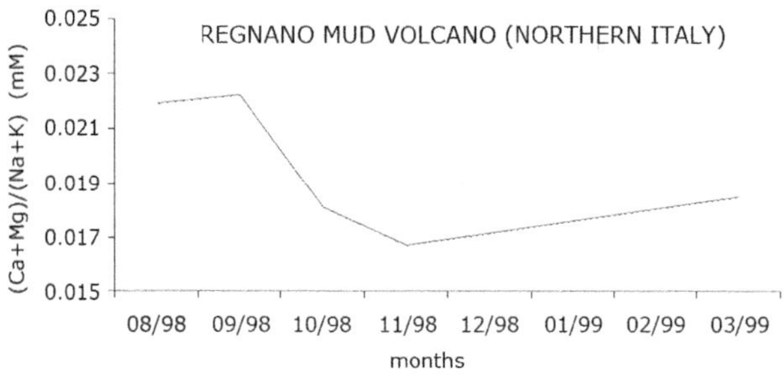

Figure 10. Cations variation observed at the Regnano mud volcano (North Italy) in 1998- 1999 . After Capozzi and Picotti (2003) , modified.

Martinelli and Ferrari (1991) published Radon data obtained by manual sampling in the liquid phase of Nirano (Fig.11) and Regnano (Fig.12) mud volcanoes. Radon data strongly fluctuated coinciding with a local seismic swarm, but a weekly sampling rate turned out to be unsuitable for evidencing possible earthquake precursors.

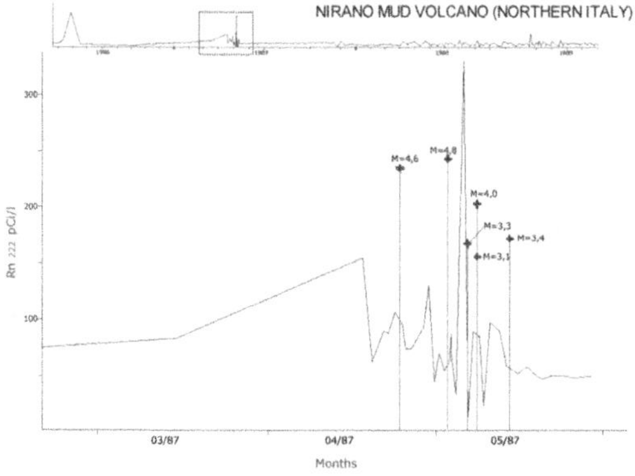

Figure 11. Radon fluctuations observed in Nirano mud volcano (North Italy) in 1986 – 1987 and local seismic events. After Martinelli and Ferrari (1991), modified.

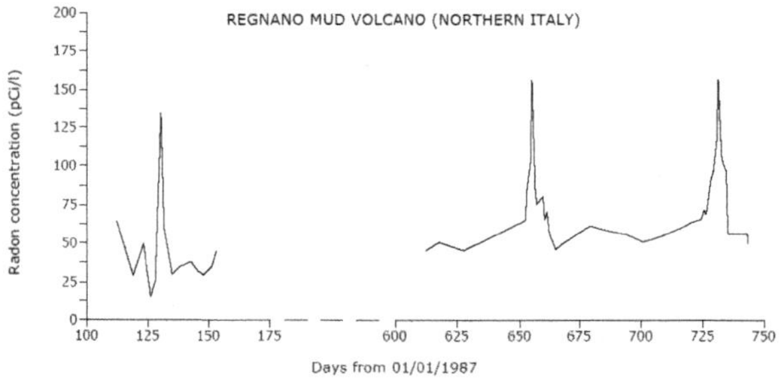

Figure 12. Radon fluctuations observed in Regnano mud volcano (North Italy) in 1987 – 1989 (Martinelli and Ferrari, 1991).

Hence, automatic Radon monitoring was attempted in the gas phase of the Regnano and Pujanello (Fig.13) mud volcanoes. A contemporaneous gas flow rate monitor evidenced a correlation between the gas flow rate and Radon anomalies. Possible relations between the gas flow rate increase and the local seismicity was also observed, although strong dependencies upon meteorological and other unidentified parameters was also identified by Albarello et al. (2003).

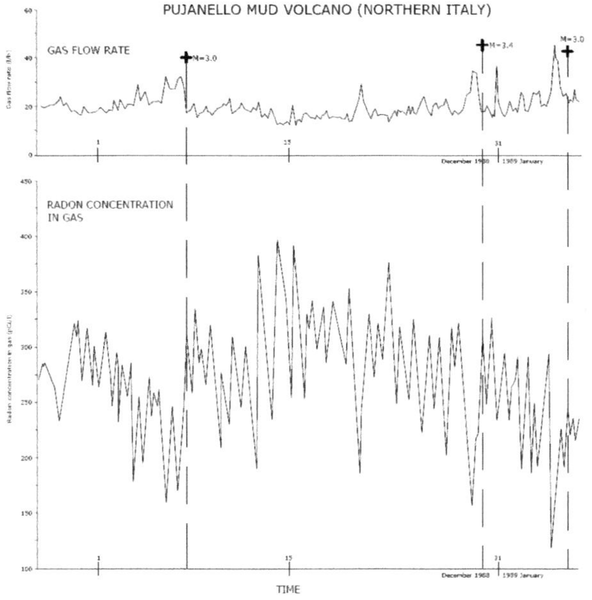

Figure 13. Pujanello mud volcano (North Italy) flow rate and Radon fluctuations compared with local seismic events (Martinelli and Ferrari, 1991).

The manual sampling of chemical and isotopic components of gases evidenced a significant variation in the gas composition (Fig. 14) in concomitance with local seismic events in a mud volcano located in Turkmenistan (Voytov et al., 1989). In particular, Voytov et al.(1989) started the geochemical monitoring after observing an increased eruptive activity following increased local seismicity.

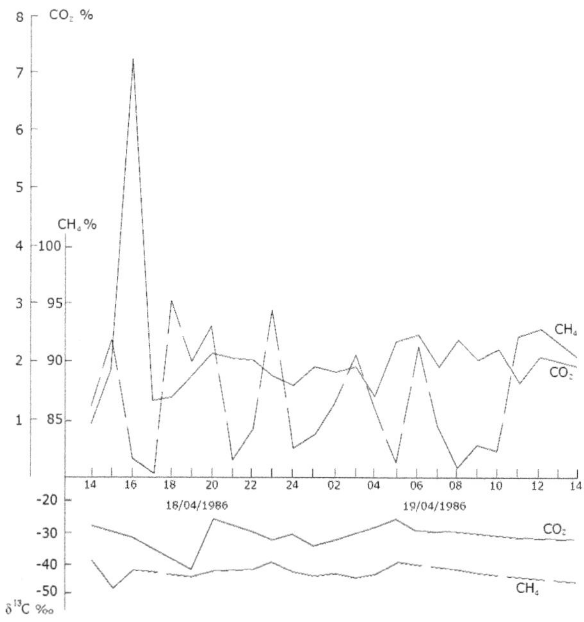

Figure 14. Geochemical fluctuations recorded in Kipyashchiy Bugor mud volcano (Turkmenistan) in 1986. After Voytov et al.(1989), modified.

Experiments in automatic geochemical methane monitoring (Fig.15) were attempted by Guliyev and Feyzullayev (1997). Strong gas composition variations were detected in a less than one hour period.

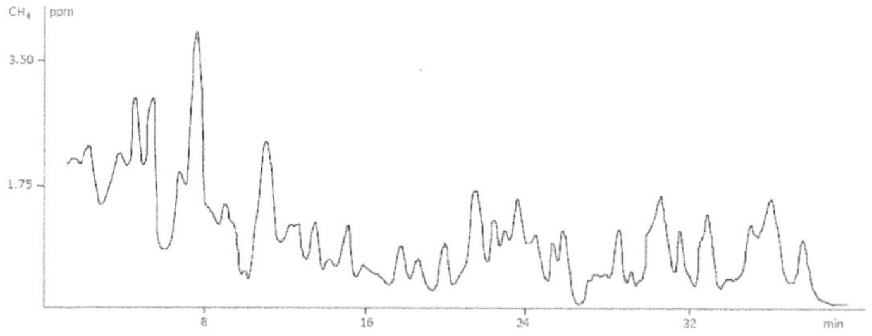

Figure 15. Methane fluctuations automatically registered at the Astrakhanka mud volcano, Azerbaijan (Guliyev and Feyzullayev, 1997).

Fission track Radon passive sensors were also utilized by Nevinsky et al. (2001) in monitoring a mud volcano in the Taman area (Fig.16). A positive relation Radon- flow rate was evidenced while further isotopic data were also reported for a better knowledge of the area.

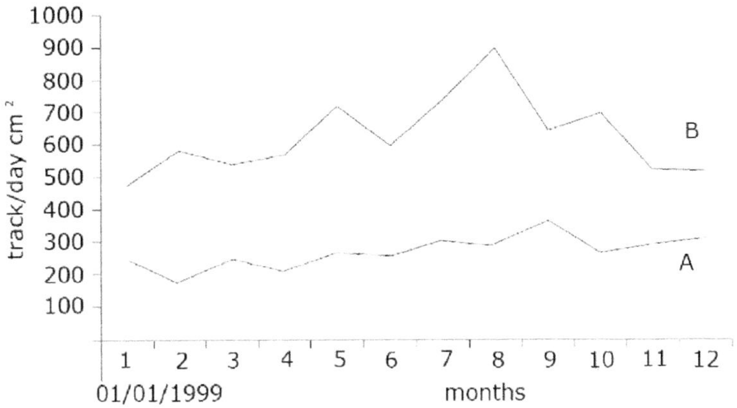

Figure 16. Radon fluctuations in the gas phase recorded in a mud volcano of the Taman area (Russia) in 1999 (Nevinsky et al., 2001).

All the available geochemical data sets show data compatible with the models proposed by Kopf (this book) and by Albarello (this book). Although important attempts have been made by many researchers no definitive conclusions can be drawn from the processing geochemical data sets in order to make out the possible links between fluid emissions and seismicity.

4. FINDINGS FROM THE AVAILABLE GEOCHEMICAL DATA SETS

The review of the available data sets recorded during geophysical and geochemical monitoring has evidenced that not all sensors are suitable in extreme environments. Data is often missed by the complex equipment utilized in mud volcanic monitoring. Temperature and pressure transducers are probably the most reliable devices (see also Khavroshkin et al., in this book). Advanced technology robots, rather like those used in space missions, have also been proposed (Delisle, in this book) for monitoring geochemical and geophysical parameters. Local seismic networks and geophones have also been proposed for long-term monitoring by Panahi (this book) and Albarello (this book). Mellors has proposed some new, satellite-based techniques that have produced promising results (Mellors et al., in this book).

Tramutoli et al. (2001) has used satellite techniques to monitor methane gas emissions expelled by mud volcanoes or carbon dioxide-dominated gas vents.

CONCLUSIONS

The mud-volcanic environment is characterized by the permanence of extreme chemical and physical conditions. Such conditions strongly limit the collection of long-term data sets. The obtained data allow us to consider mud volcanoes as confined fluid accumulations capable of working as natural strain-meters. If a pre-seismic compression occurs in a mud-volcanic area the possible flow-rate variations are, in principle, measurable. Most of the observed data show that crustal deformations have followed the local seismic events and influenced the fluid emissions. The absence of reliable long-term data sets strongly hampers any definitive conclusions to be drawn on the sensitivity of mud volcanoes to seismic events, as well as evidencing the need for reliable ground-based or satellite-based measuring techniques.

REFERENCES

1. Albarello, D., Mud volcanoes as natural strainmeters: a working hypothesis, in Martinelli, G., Panahi, B. (Eds), *Mud Volcanoes, Geodynamics and Seismicity, Proceedings of NATO Advanced Research Workshop, Baku, May 20-22,2003.*
2. Albarello, D., Lapenna, V., Martinelli G., Telesca, L., 2003. Extracting quantitative dynamics from 222Rn gaseous emission of mud volcanoes, *Environmetrics*, **14**, 63-71.
3. Albarello, D., Martinelli, G., Parenti, R., 1999. *Analisi dei dati di resistività registrati a Vedriano di Castel San Pietro*. Università di Bologna, Dipartimento di Fisica, Settore Geofisica, technical report.
4. Aliyev, Ad.A., Guliev, I.S., Belov, I.S., 2002. *Catalogue of recorded eruptions of mud volcanoes of Azerbaijan,* Nafta Press, Baku, 89 pp.
5. Becker, K., Fisher, A.T., Davis, E.E., 1997. The CORK experiment in hole 949C: long-term observations of pressure and temperature in the Barbados accretionary prism, in Shipley,T.H., Ogawa,Y., Blum, P., Bahr, J.M.(Eds.) *Proceedings of the Ocean Drilling Program, Scientific Results*, 156, 247-252.
6. Bodvarsson, G., 1970. Confined fluids as strain meters, *J. Geophys. Res.*, **75**, 2711-2718.
7. Brown, K.M., 1990. Nature and hydrogeologic significance of mud diapirs and diatremes for accretionary systems, *J. Geoph. Res.*, **95**, 8969-8982.
8. Caneva, A., 1958. Ricerche geofisiche sulla Salsa di Regnano (Prov. di Reggio Emilia), *Geofis. Meteorol.*, **6**, 71-84.
9. Capozzi, R., Picotti, V., 2002. Fluid migration and origin of a mud volcano in the Northern Apennines (Italy): the role of deeply rooted normal faults, *Terra Nova*, **14**, 363-370.
10. Delisle, G., Mud volcanoes of Pakistan – an overview: a report on three centuries of historic and recent investigations in Pakistan, in Martinelli, G., Panahi, B. (Eds), *Mud Volcanoes, Geodynamics and Seismicity, Proceedings of NATO Advanced Research Workshop, Baku, May 20-22,2003.*
11. Foucher, J.-P., Henry, P., Le Pichon, X., Kobayashi, K. 1992. Time – variations of fluid expulsion velocities at the toe of the eastern Nankai accretionary complex, *Earth and Planetary Science Letters*, **109**, 373-382.
12. Guliyev, I.S., Feizullayev, A.A., 1997. *All about mud volcanoes*, Nafta Press, Baku, 52 pp.

13. Hasiotis, T., Papatheodorou, G., Kastanos, N., Ferentinos, G. 1996. A pockmark field in the Patras Gulf (Greece) and its activation during the 14/7/93 seismic event, *Marine Geology*, **130**, 333-344.
14. Khavroshkin, O.B., Tsyplakov, V.V., Urdukhanov, E.M.-K., Vidmont, N.A., Mud volcano : methods, apparatus-fundamental and applied aspects of research, in Martinelli, G., Panahi, B. (Eds), *Mud Volcanoes, Geodynamics and Seismicity, Proceedings of NATO Advanced Research Workshop, Baku, May 20-22,2003.*
15. Kopf, A.J., 2002. Significance of mud volcanism, *Rev. Geophys.*, **40** (X), 1005, doi:10.1029/2000RG000093.
16. Judd, A., Gas emissions from mud volcanoes: significance to global climate changes, in Martinelli, G., Panahi, B. (Eds), *Mud Volcanoes, Geodynamics and Seismicity, Proceedings of NATO Advanced Research Workshop, Baku, May 20-22,2003.*
17. Martinelli, G., Albarello, D., Mucciarelli M., 1995. Radon emissions from mud volcanoes in Northern Italy: possible connection with local seismicity, *Geoph. Res. Lett.*, **22**, 1989-1992.
18. Martinelli, G., Dadomo, A. Geochemical model of mud volcanoes from reviewed worldwide data, in Martinelli, G., Panahi, B. (Eds), *Mud Volcanoes, Geodynamics and Seismicity, Proceedings of NATO Advanced Research Workshop, Baku, May 20-22,2003.*
19. Martinelli, G., Ferrari, G., 1991. Earthquake forerunners in a selected area of Northern Italy: recent developments in automatic geochemical monitoring, *Tectonophysics,* **193**, 397-410.
20. Mellors, R. J., Bunyapanasarn, T., Panahi, B.M., INSAR analysis of the Absheron Peninsula and nearby areas, Azerbaijan, in Martinelli, G., Panahi, B. (Eds), *Mud Volcanoes, Geodynamics and Seismicity, Proceedings of NATO Advanced Research Workshop, Baku, May 20-22,2003.*
21. Nevinsky, I., Nevinsky, V., Panyushkin, V., Ferronsky, V., Tsvetkova, T., 2001. An attempt to determine the tritium, 22Na, 36Cl and Radon in territory of mud volcano in Taman, *Radiation Measurements*, **34**, 349-353.
22. Panahi , B., Mud volcanism, geodynamics and seismicity of Azerbaijan and the Caspian Sea region, in Martinelli, G., Panahi, B. (Eds), *Mud Volcanoes, Geodynamics and Seismicity, Proceedings of NATO Advanced Research Workshop, Baku, May 20-22,2003.*
23. Segawa, J., Toh, H., 1992. Detecting fluid circulation by electric field variations at the Nankai Trough, *Earth and Planetary Science Letters*, **109**, 469-476.
24. Tamrazyan, G. P., 1972. Peculiarities in the Manifestation of Gaseous-Mud Volcanoes, *Nature*, **240**, 406-408.
25. Tramutoli, V., Di Bello, G., Pergola, N., Piscitelli, S., 2001. Robust Satellite Techniques for Remote Sensing of Seismically Active Areas, *Ann. Geofis.*, **44**, 295-312.
26. Voytov, G.I., Belikov, V.M., Orlova, T.G., Nosik, L.P., Kucher, M.I., Krivomazova, N.G., Nikulina, I.V., 1989. Chemical and carbon isotope instability in the natural gases of the Kipyashchiy Bugor mud volcano (southwestern Turkmenistan), *Transactions (Doklady) of the USSR Academy of Sciences*, **309**, 68-73.

INSAR ANALYSIS OF THE ABSHERON PENINSULA AND NEARBY AREAS, AZERBAIJAN

Robert J. Mellors and Ted Bunyapanasarn
San Diego State University
San Diego, CA
rmellors@geology.sdsu.edu

Behrouz Panahi
Geological Institute of Azerbaijan
bpanahi@gia.ab.az

Abstract: A set of Synthetic Aperture Radar (SAR) data covering the Absheron Peninsula is processed using interferometic synthetic aperture radar (InSAR) to generate a detailed digital elevation map of the region and to measure possible surface deformations. The dataset includes eight ERS-1 and ERS-2 scenes from 1996 though 1999 and images Radarsat images from 1999. The radar has a wavelength of 5.66 cm allowing theoretical resolution along the line of sight to the satellite of surface deformations on the order of mm. Possible causes of surface deformation are earthquakes, mud volcanism, groundwater changes, and hydrocarbon withdrawal. We also use optical satellite images (Advanced Spaceborne Thermal Emission and Reflection - ASTER) and ground truth data. Initial In-SAR processing used a digital elevation model (DEM) based on the combination of a global DEM with a high resolution DEM derived from ASTER stereo optical data. This DEM was then refined with radar pairs possessing short temporal baselines. Correlation over the area was fair with moderate to low correlation over a time period of 2 years. Preliminary results show no clear indications (> 10 cm line-of-sight) of large deformation over known mud volcanoes during the periods spanned by the interferograms. Preliminary modeling of likely fluid movement associated with the mud volcanoes indicates that it should be possible to estimate the depth of increased pressure and determine constraints on chamber size using InSAR. Modeling of the two large (M_W 6.8 and 6.5) earthquakes near Baku in November 2000 using global catalog locations and focal mechanisms indicates that the deformation from these events should be observable onshore using InSAR.

Key words: Azerbaijan, interferometric synthetic aperture radar, mud volcanoes

G. Martinelli and B. Panahi (eds.), Mud Volcanoes, Geodynamics and Seismicity, 201–209

1. INTRODUCTION

The mud volcanoes of the Absheron Peninsula, Azerbaijan are the most spectacular examples of onshore mud volcanism in the world [Jakubov et al., 1971; Hovland et al., 1997]. As well as being fascinating and occasionally dangerous structures, these volcanoes are of interest for the clues they provide to sub-surface fluid flow and stress. In general, sub-surface fluid flow is poorly understood but important in many applications, including hydrocarbon extraction. Mud volcanism can also pose a natural hazard, not only to onshore facilities and people but also to offshore facilities and drilling. With increases in population and urban development in areas of mud volcanism, a better understanding of the mechanics underlying eruptions is needed.

Recent studies based on 3D seismic data [Cooper, 2001] suggest that at least some offshore mud volcanoes in the South Caspian region possess shallow mud chambers which refill prior to eruption. This raises the possibility that inflation of the chamber may produce measurable surface deformation. Similar techniques have been used successfully to detect precursors to magmatic volcanism (e.g. [McGuire et al., 1995]). Some data from leveling lines also supports deformations related to mud volcanism [Sinelnikov and Svistun, 1980] Recently, interferometric synthetic aperture radar (InSAR) has been used to monitor magmatic volcanoes. In this work we explore the use of In-SAR for investigating onshore mud volcanism and in particular the potential for monitoring subtle changes due to sub-surface fluid flow.

2. METHOD

InSAR relies on measuring the phase difference between two (or more) images from satellite based radar to construct an image of the surface deformation between satellite passes. It provides very accurate measurements (mm scale resolution with pixel size on the order of 10's of m) of the relative surface deformation over a wide area (typical scenes are 100 km square). The exact measurement is along the satellite line-of sight, which for the ERS satellites is at an average (scene center) angle of 23°. Consequently, the technique is most sensitive to vertical motion. The method has been used successfully to image a wide range of deformations including those due to earthquakes, groundwater extraction, and volcanic activity (see [Hanssen, 2001] or [Burgmann et al., 2000] for a review).

However, limitations exist with InSAR data. These limitations include limited data coverage, decorrelation, and artifacts due to topography or atmospheric propagation. For the South Caspian region, perhaps the most substantial difficulty is the limited data. Only a few radar scenes suitable for interferometry currently exist, which make a thorough data analysis difficult.

However, sufficient data does exist to assess the potential use. Another difficulty is decorrelation, which refers to the loss of phase stability between pixels and can be caused by surface slope or changes in the ground surface. Decorrelation increases with time but the rate of temporal decorrelation varies greatly from region to region [Zebker and Villasenor, 1992]. Urban areas tend to retain correlation well while vegetated areas rapidly decorrelate. Surface topography also creates a phase change between images that depends on the distance between the satellites (baseline) at the time of imaging and which requires an accurate knowledge of the topography to remove. Excessive topography also creates features such as layover or shadowing due to the radar imaging geometry. In general, these effects are best compensated for with the use of an accurate digital elevation model. Another potential difficulty is phase changes due to variations in atmospheric water content, either laterally or vertically in combination with topography [Delacourt et al., 1998; Hanssen, 2001], which can easily distort observed signals [Rigo and Massonnet, 1999]. The proximity to the Caspian Sea suggests that such effects could easily occur and create substantial gradients in atmospheric water vapor. Groundwater or extraction of hydrocarbons changes can also cause surface elevation changes [Bawden et al., 2001; Galloway et al., 1998]. The best way to identify and compensate for these possible artifacts is through use of multiple data sets over the same area.

Modeling of possible deformation. A crude estimate of the expected deformation prior to an eruption can be made by assuming a spherical pressure source embedded in a homogenous half-space [Mogi, 1958]. The depth of the chamber must be much greater than the radius. Figure 1 shows the amount of vertical deformation expected from a sphere with radius of 250 m and a pressure change equivalent to 82,000 m^3 at depths of 3 km and 1 km. This roughly corresponds to the amount of mud expelled from the recent eruption of Lokbatan. Larger quantities have been recorded during other eruptions [Jakubov et al., 1971]. This should be considered as a lower bound, as it does not include the large quanities of gas that escape.

Consequently, any measured deformation can be used to place constraints on the depth, size, and shape of the chamber (and pressure change). Shallow chambers produce a high amplitude signal while deeper chambers will produce a longer wavelength but low amplitude signal.

Remote sensing data. A suite of radar imagery as well as selected optical imagery was collected (Table 1). The optical data was from the Advanced Spaceborne Thermal Emission and Reflection Radiometer (ASTER) satellite, which has a resolution of 15 meters at optical and near-infrared wavelengths [Abrams et al., 2001]. The ASTER satellite also collects stereo pairs, which are useful for the construction of elevation models. The DEM had a nominal resolution of 30 m.

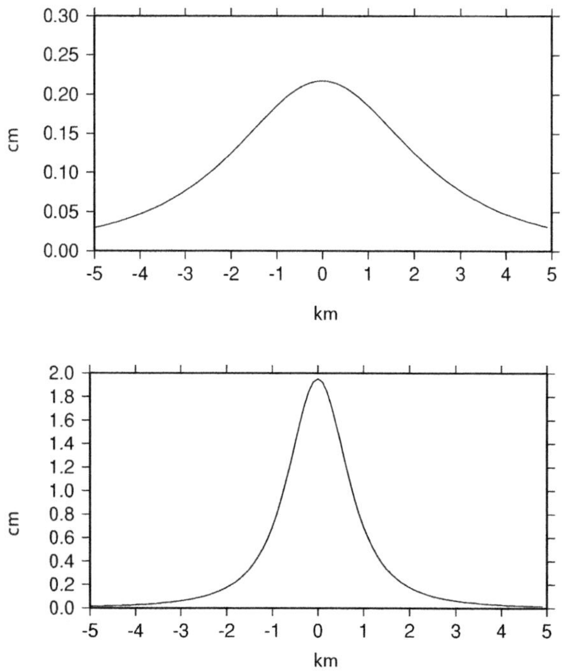

Figure 1. Cross-sections of expected deformation assuming a spherical pressure source with a radius of 250 m at a depth of 1 and 3 km. A pressure change equivalent to a volume change of 82,000 m^3 is used.

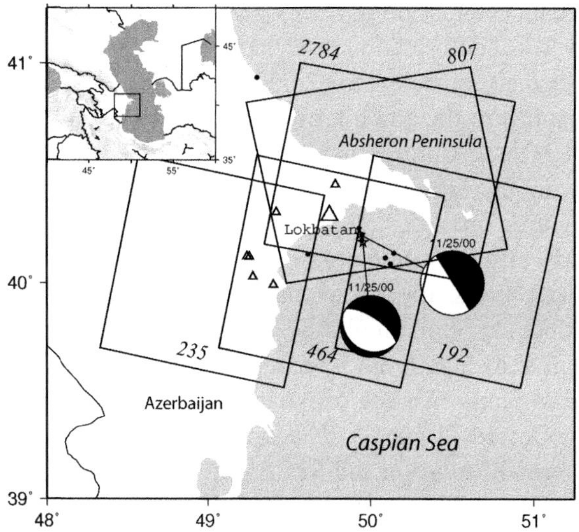

Figure 2. Map of data showing locations of radar scenes and focal mechanisms and locations of November 2000 events.

For optimal use in interferometry, the two pairs of radar images should be within several hundred meters of each other in the respective orbits, which places considerable constraints on the available data (Table 1). Data from both the ERS and Radarsat satellites was collected. The general procedure used the pairs with longer spatial baselines but shorter temporal span to generate the topographic corrections. Pairs with short baselines were used to observe deformation.

Table1. Radar data used in this study. The radar data shows the pairs and baselines. Pairs marked with an asterisk were used to generate DEMs.

Sat.	*Track*	*Frame*	*Date 1*	*Sat.*	*Date 2*	*Perp. Base*	*Days*	*Corr.*
ERS-1	192	2781	1996/05/12	ERS-2	1996/05/13	-156	1	good
ERS-2	192	2781	1996/05/12	ERS-2	1998/10/05	124	876	poor
ERS-2	192	2781	1998/10/05	ERS-2	2000/12/18	-217	805	poor
ERS-2	464	2799	1998/08/15	ERS-2	1999/03/15	36	210	good
ERS-1	235	2799	1998/08/15	ERS-2	1999/05/16	-528	934	poor
ERS-2	235	2799	1996/05/16	ERS-2	1999/05/06	234	1120	poor
ERS-2	235	2799	1996/05/16	ERS-2	1999/06/10	-137	970	poor
ERS-2	235	2799	1996/05/06	ERS-2	1999/06/10	-378	35	good
RSAT	807	26083	2000/10/09	RSAT1	2000/11/02	63	24	good
RSAT	2784	25690	2000/10/06	RSAT1	2000/10/30	488	24	poor

Track 192. This track covered the Absheron Peninsula and provided the only data that spanned the earthquakes of 25 November 2000. A pair of scenes with a time gap of 1 day (tandem pair) was available for elevation correction, which is minimal for the relatively flat peninsula. Unfortunately the data had relatively poor correlation due to a relatively long time span and does not cover many mud volcanoes. Nevertheless, fringes possibly due to the earthquakes of 25 November 2000 are apparent on the pair spanning October 1998 to December 2000.

Track 235/464. These scenes cover the Absheron Peninsula as well as much of the coast south of Baku. Similar to Track 192, the length of time covered make deformation determinations difficult but is excellent for elevation models. Track 464 has only one pair but with a short baseline and spans 210 days.

Track 807/2784. These data were recorded by the Radarsat satellite in October 2000 and covered only a short time period (24 days) immediately prior to the earthquakes. No obvious signs of deformation are visible. Track 807 has a short baseline (67 m) and therefore well suited to measure deformation. Track 2784 had a much longer baseline and showed considerable decorrelation.

3. RESULTS

The combination of the various data sets has allowed the construction of a DEM, which is a necessary first step. The DEMs have been compared with topographic maps and with independently derived topography (Figure 1). Correlation is fair and appears to be retained for up to roughly two years, with isolated patches of longer-term coherence in urban areas. The offshore platforms also retain coherence, which allows the potential for limited measurement offshore as well.

Nov. 25, 2000 earthquakes. On November 25, 2000 two large (M_W 6.8 and 6.5) earthquakes occurred within minutes of each other offshore Baku. One pair (track 192 1998-2000) spanned the time of the events. Although coherence was low, some fringes were observed on the Absheron Peninsula although it is possible that these may be contaminated by orbital or atmospheric errors.

Due to the poor coverage and multiple number of events, it was impossible to determine a seismic mechanism from the InSAR data alone. However the InSAR does serve to place constraints on the depth and location.

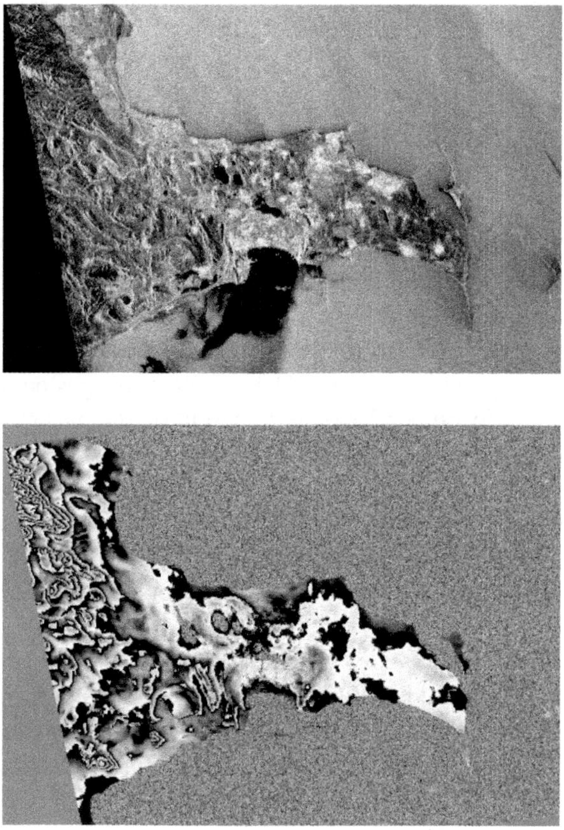

Figure 3. Examples of amplitude (top) and phase data (bottom) for radarsat (track 807).

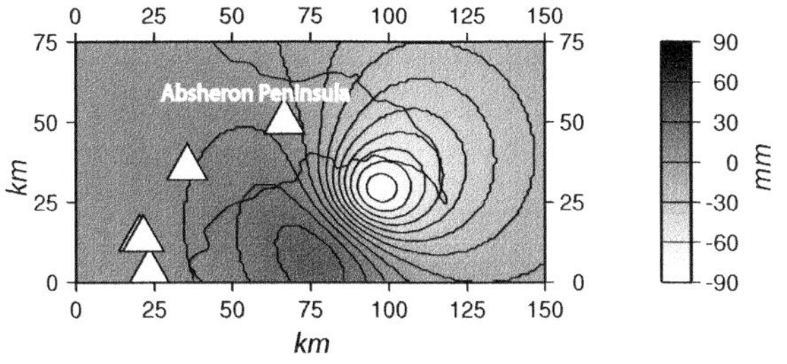

Figure 4. Model of estimated deformation from the November 2000 earthquakes. A finite fault in an elastic half-space is assumed, with slip scaled to fault dimensions. Focal mechanism is from Harvard CMT (e.g. Dziewonski and Woodhouse, 1983) with depth and location from the ISC (ISC, 2001).

Modeling was performed to estimate the amount of expected surface deformation using the available seismic data (Figure 4). The modeling was performed using a finite fault in an elastic half space [Okada,1995], as implemented by [Fiegl and Dupre, 1999]. Fault size was scaled to the moment and slip scaled to length using standard relationships. The fault was assumed to be square. This model predicted a high spatial deformation along the south side of the Absheron Peninsula, which was not observed. Therefore the model was revised to use a deeper depth that was consistent with the body-wave solution. Even increasing the depth to 50 km still predicted more fringes than observed, and it may be that the ISC (global) location is in error. Shifting the location farther offshore would help resolve the differecnes.

Mud volcanoes. Observation of mud volcanoes has been focused on the area around Lokbatan due to the possibility of observing deformation prior to the 2001 eruption (Figure 3). No large-scale movement is observed so far but work is still in progress.

Conclusions and future work. Preliminary results suggest that InSAR is feasible for longterm monitoring of mud volcano activity but more work needs to done on estimating errors. Coherence in the area is good for periods of up to two years at the wavelengths used for the ERS and Radarsat satellites (5.66 cm). The data was suitable to construct a good quality DEM, which will be useful for future analysis and monitoring. No clear signals have been observed yet associated with the mud volcanoes however fringes representing deformation from the November 2000 earthquake may be visible on pairs spanning the time of the earthquake. Coherence is fairly good for up to two years and longer in urban areas.

To reduce the error, two suggestions are made. Combining InSAR with GPS monitoring would enhance the results of both, as has been shown in other areas [Bawden et al., 2001]. While GPS is spatially limited to a few points, it is relatively immune to variations in atmospheric water vapor due to the use of two wavelengths. The GPS sites could also be used to identify fixed points with known elevations for use in orbital corrections. In additional, the collection of a long series of InSAR data is possible with the new Envisat satellite which would be highly useful. This would allow careful monitoring of the possible signals.

REFERENCES

Abrams, M., S. Hook, and B. Ramachandran,(2001), ASTER User handbook downloaded from http://asterweb.jpl.nasa.gov/, 135 pg.

Bawden, G.,W. Thatcher, R.S. Stein, K.W. Hudnut, and G. Peltzer, (2001), Tectonic contraction across Los Angeles after removal of groundwater pumping effects, *Nature*, 412, 812-8151.

Burgmann, R., P. A. Rosen, and E. Fielding, (2000), Synthetic aperture radar interferometry to measure earth's surface topography and its deformation, *Ann. Rev. Earth, Planet. Soc.*, 28, 169-209.

Cooper, C., (2001), Mud Volcanoes of the South Caspian Basin - Seismic Data and Implications for Hydrocarbon Systems, *AAPG Annual Meeting*, Denver, Colorado, June 3-6.

Delacourt, C., P. Briole, and J. Achache, (1998), Tropospheric correction of SAR interferograms with strong topography: Application to Etna, *Geophys. Res. Lett.*, 25, 2849-2852.

Dvorak, J., and D. Dzurisin,(1997), Volcano geodesy: the search for magma reservoirs and the formation of eruptive vents. *Rev. Geophys.* 35, 343-384.

Dziewonski, A. M., and J. H. Woodhouse, (1983), Studies of the seismic source using normalmode theory, in Kanamori, H. and E. Boschi, eds., Earthquakes: observation, theory, and interpretation: notes from the International School of Physics "Enrico Fermi" (1982: Varenna, Italy), North-Holland Publ. Co., Amsterdam, pp. 45-137.

Fiegl, K., and Dupre, E., (1999), RNGCHN: a program to calculate displacement components from dislocations in an elastic half-space with applications for modeling geodetic measurements of crustal deformation, *Computers and Geosciences*, 25, 695-704.

Galloway, D. L., K. W. Hudnut, S. E. Ingebritsen, S. P. Phillips, G. Peltzer, F. Rogez, and P. A. Rosen, (1998), Detection of aquifer system compaction and land subsidence using interferometric synthetic aperture radar, Antelope Valley, Mojave Desert, California, *Water Resources Research*, 34, 2573-2585.

Hanssen, R. F., Radar interferometry: Data interpretation and error analysis, (2001), Kluwer Academic Press, pp. 308.

Hovland, M., A. Hill, and D. Stokes, The structure and geomorphology of the Dashgil mud volcano, (1997), Azerbaijan, *Geomorphology*, 21, 1-15.

International Seismological Centre, (2001), *On-line Bulletin*, http://www.isc.ac.uk/Bull, Internatl. Seis. Cent., Thatcham, United Kingdom, 2001.

Jakubov, A. A., A. A. Ali-Zade, and M. M. Zeinalov, (1971) Mud volcanoes of the Azerbaijan SSR Atlas, (1971), Publishing House of Academy of Sciences of AzSSR, Baku.

McGuire, B., C. Kilburn, and J. Murray, (1995), Monitoring Active Volcanoes: Strategies, Procedures, and Techniques: University College London Press, 421 pp..

Mogi, K., (1958), Relation between eruption of various volcanoes and the deformations of the ground surfaces around them, *Bull. Earthquake res. Inst.*, Univ. Tokyo, 36, 99-134.

Okada, Y., (1995), Surface deformations due to shear and tensile faults in a half-space, *Bull. Seism. Soc. Am.*, 75, 1135-1154.

Rigo, A., and D. Massonnet, Investigating the 1996 Pyrenean earthquake (France) with SAR Interferograms heavily distorted by atmosphere, *Geophys. Res. Lett*, 26, 3217-3220.

Sinelnikov, D. A. and V. I. Svistun, (1980), The effect of mud volcanoes on leveling lines, *Geodesy, Mapping, and Photgrammetry*, 19, 253-254.

Zebker, H. A., and J. Villasenor, (1992), Decorrelation in interferometric radar echoes, *IEEE Trans. Geo. Rem. Sensing*, 30, (5), 950-959.

Chapter 6
GEOCHEMICAL FEATURES OF MUD VOLCANOES

GEOCHEMICAL MODEL OF MUD VOLCANOES FROM REVIEWED WORLDWIDE DATA
Short review

Giovanni Martinelli[1] Andrea Dadomo[2]

1 ARPA Environmental Protection Agency of Emilia Romagna Region, Via Amendola 2, 42100 Reggio Emilia - Italy, e-mail: giovanni.martinelli15@tin.it

2 Dept. of Geological Sciences and Geotechnology, University of Milan, Piazza delle Scienze 4, 20126 Milan, Italy

Abstract: Geochemical data from mud volcanic fluids obtained from various geological environments have been reviewed and reprocessed. The chemical and isotopic components of the liquid and gas phases have been studied. Notwithstanding the geographical distance between the mud volcanic areas and the differences between the geological environments, a common originating fluid derived from seawater has been recognized. Diagenetic processes due to sediment compaction can be considered to be responsible for the evolutionary patterns observed in the liquid phase and in the associated gas-emissions. Geochemical data fit the currently available physical models.

Key words: mud volcanoes, geochemical modelling

1. INTRODUCTION

Accretionary prisms undergoing deformative stresses due to subductive processes host sediments that are subjected to compaction and mineral dehydration phenomena. In particular, subduction zones host fluids that are often expelled from thick sedimentary layers accumulating over millions of years (Carson and Screaton, 1998). These pore waters still show the geochemical signature of seawater as the starting material and are often enriched with hydrocarbons. The flow-rate at which pore waters are expelled may be affected by local permeability coefficients and pressure gradients strongly linked to the tectonic setting. A significant part of the localized fluid flow is represented by mud volcanoes (Kopf, 2002).

The available geochemical data obtained in the mud volcanic areas of Alaska, Azerbaijan, Barbados, Italy, Russia, Taiwan and Trinidad were

G. Martinelli and B. Panahi (eds.), Mud Volcanoes, Geodynamics and Seismicity, 211–220

compared. Data inter-comparison can be of help in better understanding the genetic features of fluids expelled by mud volcanic areas in question. The present review tracks the main cations, anions, stable isotopes of waters, Sr isotopes and the main gases sampled at the considered sites.

1.1 Water origin and classification

All the analyzed fluids show some kind of relation with the meteoric waters. The geological setting and the chemical composition indicate that sea water and clay minerals are the main starting materials of the sampled fluids. The strong prevalence of Cl-Na water geochemical *facies* allow us to consider the sampled fluids as a sort of oilfield waters (Fig. 1, Fig. 2, Fig. 3, Fig. 4 and Fig. 5)

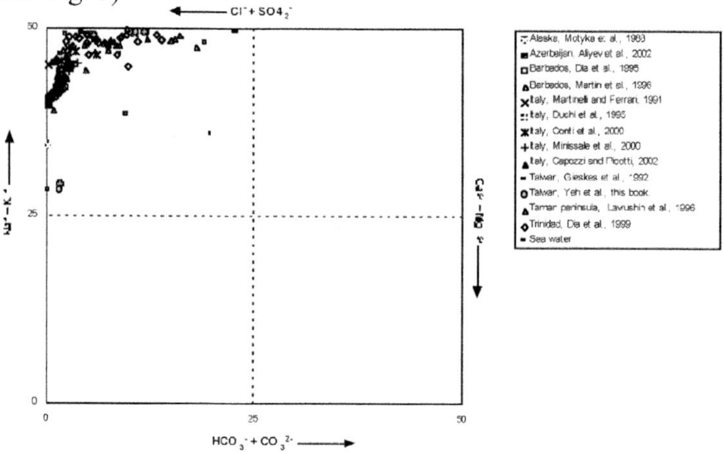

Figure 1. Langelier-Ludwig diagram of considered mud volcanic waters

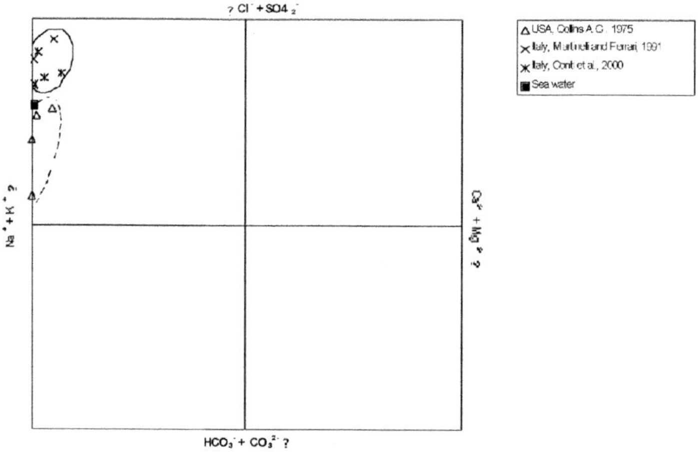

Figure 2. Oil field character of mud volcanic waters sampled in Italy

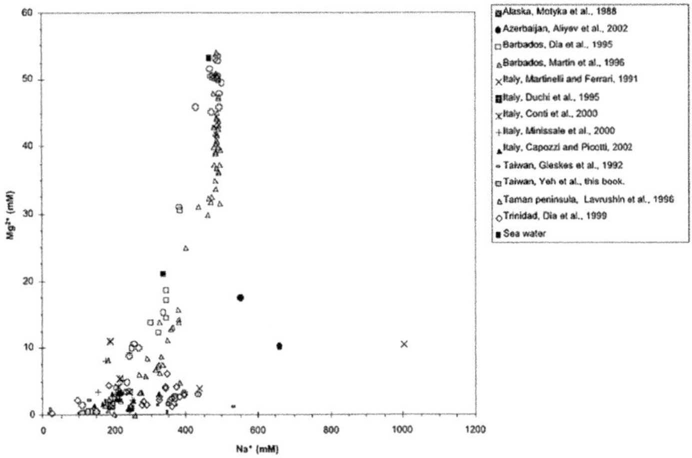

Figure 3. Na/Mg ratio in considered mud volcanic waters

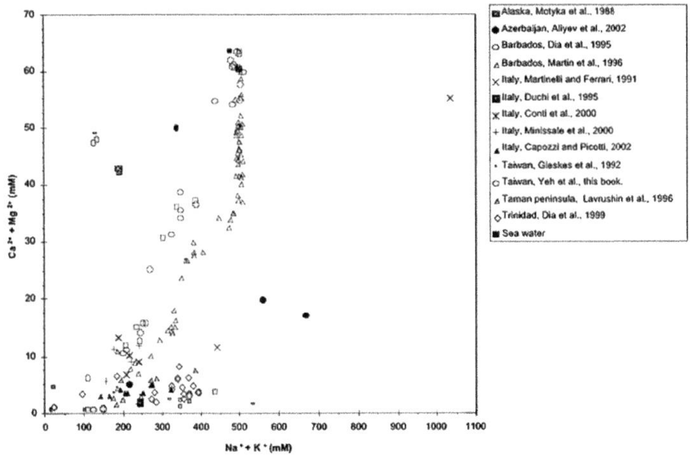

Figure 4. (Ca + Mg) / (Na + K) ratio in considered mud volcanic waters

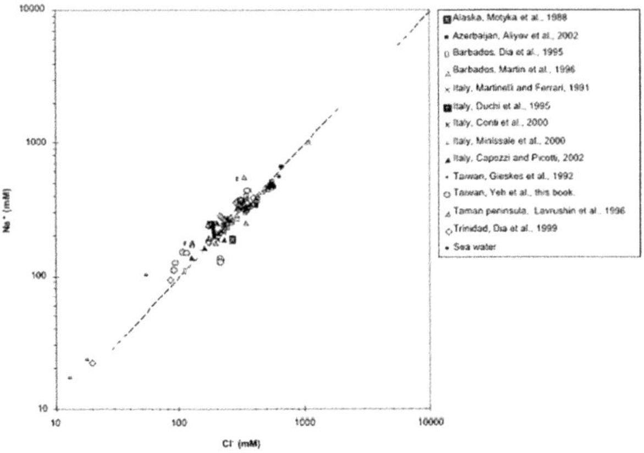

Figure 5. Cl/Na ratio in considered mud volcanic waters

Most of the geochemical alterations observed in the samples is due to compaction and dehydration phenomena of the marine sediments. Oxygen and hydrogen isotope ratios confirm the marine origin of the fluids under examination (Fig. 6).

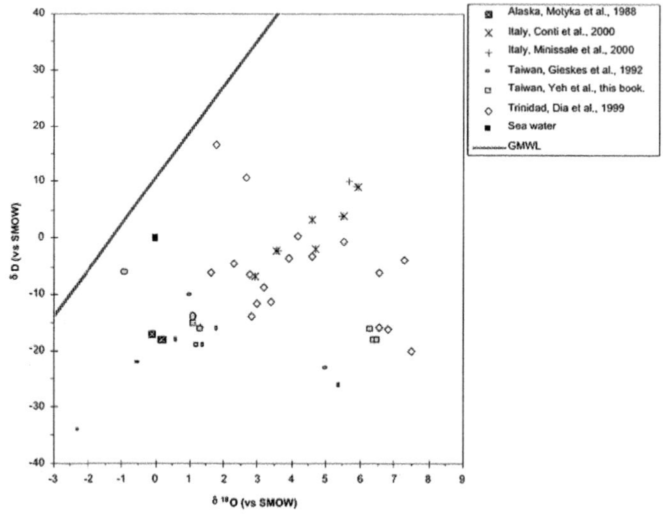

Figure 6. δ^{18} O / δ D ratio in considered mud volcanic waters. Isotopic data are available only for some localities. Global Meteoric Water Line is also indicated.

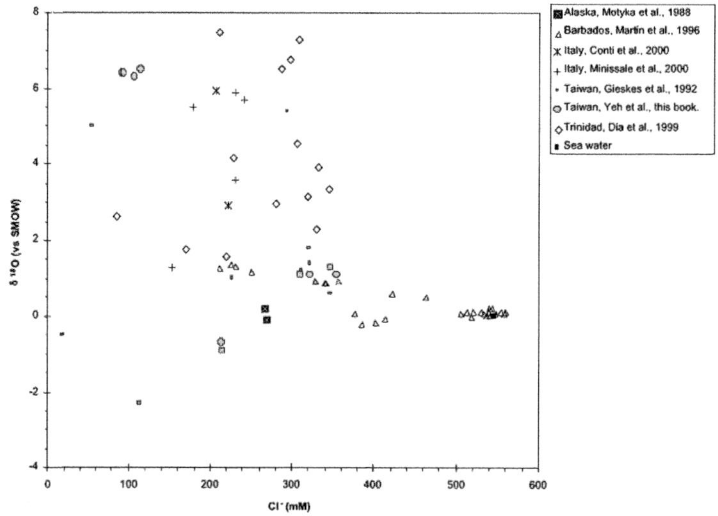

Figure 7. Cl / δ^{18} O ratio in mud volcanic waters.

Barbados and Trinidad confirm the common marine origin of the considered fluids (Fig. 7). In spite of their well-constrained marine origins some chemical differences relative to modern sea water have been observed. Clay behaviour under compaction is subjected to a variety of diagenetic processes, as they can release water enriched or depleted in electrolytes as compared with seawater (Bredehoeft et al, 1963). When water and solutes are driven by hydraulic head gradients across semi-permeable membranes, the ionic solute passage through the membranes is restricted in relation to the water. The solute concentration on the input side of the membrane thus increases in regard to the output concentrations. The ion-exclusion effects are described as salt-filtering or ultrafiltration processes (Kharaka and Smalley (1976). Water that accrues from possible gas hydrate dissociation in the clay layers can be considered responsible for significant dilution phenomena (Martin et al., 1996). Further amounts of water can be released during possible conversion from smectite to illite (Bekins et al., 1994).

1.2 The evolutionary trend of water

Fluids sampled in the Barbados area have been interpreted by Martin et al. (1996) as the result of the mixing of two end-member fluids, seawater and water derived from gas hydrate dissociation. Fluids sampled in Trinidad have been interpreted by Dia et al. (1995) and Dia et al. (1999) as the result of diagenetic processes affecting the original sea water. Similar low- temperature seawater diagenetic processes have been proposed by Conti et al.(2000), Minissale et al., (2000) and Capozzi and Picotti (2002), to explain the origin of fluids sampled in the mud volcanic areas of Italy. Fluids sampled in Taiwan have been interpreted by Gieskes et al.(1992) and by Yeh et al., (this volume) as the result of diagenetic processes affecting seawater. Aliyev et al. (2002) have considered mud volcanic fluids as oilfield waters syngenetic to extruded clay sediments. Over the past ten years geochemists have paid particular attention to Strontium isotopes as indicators of diagenetic processes. Indeed, Dia et al.(1995), Martin et al. (1996), Dia et al. (1999), Conti et al. (2000) have analyzed Sr86/Sr87 ratio in fluids expelled by mud volcanoes located in Barbados, Trinidad and Italy. Analyses were carried out on the basis of findings described by Burke et al. (1982 and references therein) and Dia et al. (1992 and references therein) as to the sensitivity of the Sr86/87 ratio to global climatic changes during the geological eras. A close correlation between Sr 86/87 and O16/18 was found in the geological samples by Dia et al. (1992) through the past 300 kyr. Figure 8 shows the available strontium and oxygen isotopic analytical data accruing from the mud volcano fluids.

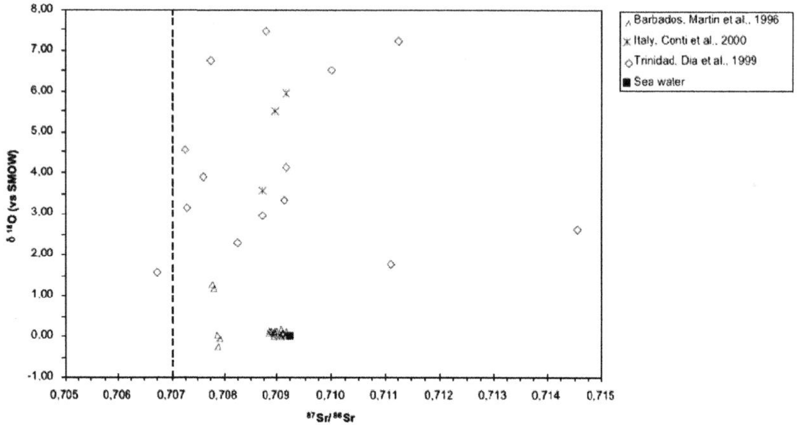

Figure 8. Relation δ^{18} O vs.^{87} Sr / ^{86} Sr. A slight scatter affects Strontium isotopic data, while a relatively wide interval affects Oxygen isotopic ratio.

The available data show that seawater is a sort of common building material for all the sampled mud volcanic fluids. A slight scattering in 87Sr/86Sr confirms the conservative character of the Strontium isotopes. The wider scattering observed in 18 O/16 O confirms that diagenetic processes may be responsible for the evolutionary geochemical water trends observed, but seawater is confirmed as the fundamental primary matter. Paroxystic eruptive periods have at times caused slight fluctuations in water chemical composition and have been detected in Italy (Martinelli et al.,1995 and references therein; Capozzi and Picotti, 2002) and Azerbaijan (Aliyev et al., 2002 and references therein). The absence of significant thermal anomalies in the waters sampled in Italy during the paroxystic period indicate that the ejected liquid phase came from depths at which thermal re-equilibration processes had occurred (<50 m.). Slight chemical fluctuations observed during the paroxystic periods are probably due to the feeding of the main conduit by the different sub-reservoirs characterized by a different clay-water suspension ratio and a different water-rock interaction time.

The low magnitude of the observed geochemical fluctuations is consistent with the existence of a main reservoir and the possible existence of some secondary reservoirs within the mud volcano structure. Significant thermal anomalies over time were recorded in Italy during the 19th and 20th centuries in the course of exceptionally strong eruptive events. Recorded thermal anomalies fit with the waters coming from layers located at depths of 500-600m (Fig. 9).

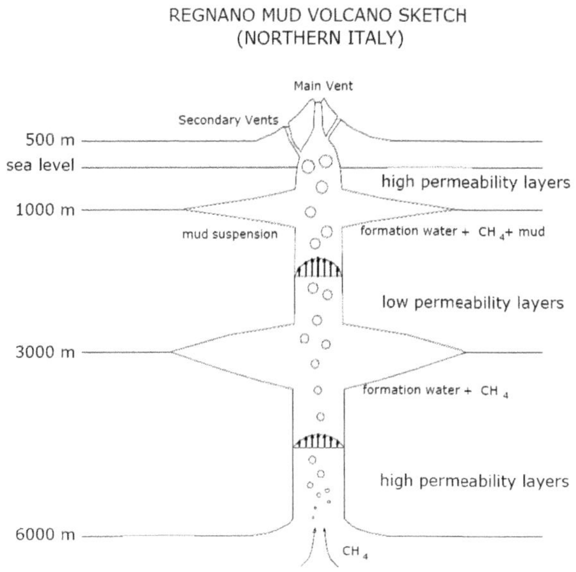

Figure 9. Sketch of the Regnano mud volcano (North Italy). After Caneva (1958), Martinelli and Ferrari (1991), Capozzi and Picotti (2002) modified.

1.3 Gas phases

Methane is the most common gas released in virtually all of the mud volcanic areas listed by Kopf (2002). The data available from the geochemically surveyed areas come from Barbados, Trinidad, Italy, Azerbaijan and Taiwan.

The isotopic methane composition can be used to distinguish between biogenic or thermogenic sources (Schoell, 1980) enabling the identification of the approximate depth at which the methane had formed. Methane sampled in the Barbados mud volcanoes ranges between –60 and –65 delta 13 C and can be considered a mix of biogenic and thermogenic gases. Little is known about the isotopic data, although Dia et al. (1999) report that the methane sampled in Trinidad should be considered as a deep originating gas. Fluids sampled in Taiwan show that a thermogenic methane source is mixed with a biogenic one (Gieskes et al., 1992; Yeh et al., this book, and references therein). Methane analyzed in the Italian mud volcanoes show carbon isotopic data typical of mixed origin (-40 >delta 13C>-50). Slight fluctuations in carbon isotopes were observed in a Northern Italian mud volcano during the eruptive phases, showing the existence of at least two different gas sources (Capozzi and Picotti, 2002). Similar phenomena were observed in Azerbaijan by Aliyev et al.(2002). Methane sampled in Azerbaijan mud volcanoes is characterized by –40>delta 13C> -50 and a multi-layered gas source has been hypothesized by Guliev and Feizullaev (1996). The small magnitude of fluctuations over time in the methane carbon isotopes recorded in Azerbaijan and Italy lead us to confirm that the main reservoir is significantly larger than the secondary ones. Carbon isotope values obtained in methane and upper hydrocarbons in the mud volcanoes of Azerbaijan and Italy confirm that the gases originated at about 6-8 Km in depth (Guliyev et al., 2001., Capozzi and Picotti, 2002). Helium isotopes analyzed in the gases sampled in the mud volcanic areas of North Italy, Sicily and Azerbaijan show data in the range of the crustal values. Some relatively anomalous data found in the Helium isotopes show that mantle- contaminated deep fluids can reach shallow rock layers through enhanced crustal permeability rock volumes (Aliyev et al., 2002; Etiope et al., 2003; Martinelli and Judd., 2004 and references therein). Despite the link with the deep-mantle connected fluids, no chemical or physical differences were observed in mantle-contaminated mud volcanoes as compared with the pure crustal mud volcanic structures.

1.4 Transient fluid geochemical phenomena

Transient geochemical phenomena were detected in Italy and Azerbaijan. Indeed, radon anomalies were detected in the Northern Apennine mud volcanoes in the liquid phase in the period 1986-1987 by Martinelli et al., (1995 and references therein). Further geochemical fluctuations were detected in the major stable water isotopes, and in the major cations and anions analyzed in a Northern Italian mud volcano. Other geochemical fluctuations affected the gas phase of a mud volcano in Northern Italy in 1988-1992 and in the period 1998-1999. These were revealed by means of radon monitoring (Albarello et al., 2003 and references therein) and by carbon isotope methane monitoring (Capozzi and Picotti, 2002). The observed geochemical fluctuations revealed the sensitivity of mud volcanoes to possible crustal deformative processes and to local climatological conditions. In actual fact, geochemical anomalies detected in the liquid phase in the 1986-1987 period turned out to be principally related to geodynamic factors, whilst anomalies detected in the period 1998-1999 followed an extraordinarily warm climatic phase that dried the ejected mud and sealed the main vents. The overpressure due to vent-sealing generated a paroxystic eruption and followed the fluid flow anomalies recognizable within the chemical parameter fluctuations. Similar anomalies have been observed in Azerbaijan and Romania (Baciu and Etiope, this book) without reaching definitive conclusions, although many paroxystic eruptive phenomena in Azerbaijan have been found to be triggered by local earthquakes. The observed behaviour have confirmed that confined fluids in the crust-like mud volcanoes can act as natural strain-meters.

CONCLUSIONS

Mud volcanic fluids are mostly syngenetic to extruded sediments. Ocean water is the common raw material utilized by the "subduction factory" to build up mud volcanoes. Clay compaction and organic matter diagenesis originates the liquid and gas phases observed in the vents. Diagenetic processes and possible mixing phenomena with seawater can account for the observed data-scattering. The gases collected in the mud volcanic areas are consistent with a mixed syngenetic origin with hosting sediments contaminated by deeper originating gases. This phenomenon is probably due to fact that gases can migrate more easily than water in geological environments. Geochemical data confirm that mud volcanoes are not linked to meteorological environment or to very deep geological strata. Fluid circulation tracked by geochemical parameters is characterized by a "mainstream fluid current", fed by a main reservoir. Slight data scattering accounts for near-surface fractionation processes affecting fluid fluxes and chemical parameters. Mud extrusion is confirmed to occur chiefly as a result of the compaction phenomena generated by the lithostatic load. The peculiar confined fluid nature of mud volcanism can be considered responsible for virtually all the observed tectonically-related phenomena.

REFERENCES

1. Albarello, D., Lapenna, V., Martinelli, G. and Telesca, L., 2003. Extracting quantitative dynamics from 222Rn gaseous emissions of mud volcanoes, *Environmetrics*, **14**, 63-71.
2. Aliyev, Ad.A., Guliyev, I.S., Belov, I.S., 2002. *Catalogue of recorded eruptions of mud volcanoes of Azerbaijan for period of years 1810-2001*, Publishing House Nafta-Press, Baku.
3. Baciu, C., Etiope, G., Mud volcanoes and seismicity in Romania, in Martinelli, G., and Panahi, B. (Eds.), *Mud volcanoes,Geodynamics and Seismicity.Proceedings of NATO Advanced Research Workshop, Baku, May 20-22, 2003.*
4. Bekins, B., Mc Caffrey, A.M., Dreiss, S.J. , 1994. The influence of kinetics on the smectite to illite transition in the Barbados accretionary prism., *J.Geophys. Res.*, **99**, 145-158.
5. Bredehoeft, J.D., Blyth, C.R., White, W.A., Maxey, G.M., 1963. Possible mechanism for concentration of brines in subsurface formations, *Bull. Amer. Assoc. Petrol. Geol.*, **47**, 257-269.
6. Burke, W.H., Denison, R.E., Hetherington, E.A., Koepnick, R.B., Nelson, H.F., Otto, J.B, 1982. Variation of seawater 87 Sr/ 86 Sr throughout Phanerozoic time, *Geology*, **10**, 516 - 519.
7. Capozzi , R., Picotti ,V. , 2002. Fluid migration and origin of a mud volcano in the Northern Apennines (Italy): the role of deeply rooted normal faults, *Terra Nova*, **14**, 363-370.
8. Carson, B., Screaton, E.J., 1998. Fluid flow in accretionary prisms: evidence for focused, time-variable discharge, *Reviews of Geophysics*, **36**, 329-351.
9. Conti, A., Sacchi, E., Chiarle, M., Martinelli, G., Zuppi , G.M., 2000. Geochemistry of the

formation waters in the Po plain (Northern Italy): an overview, *Applied Geochemistry*, **15**, 55-65.

10. Dia, A.N., Castrec, M., Boulègue, J.,Boudou, J.P. , 1995. Major and trace element and Sr isotope constraints on fluid circulations in the Barbados accretionary complex. Part 1: Fluid origin, *Earth and Planetary Science Letters*, **134**, 69-85.
11. Dia, A.N., Castrec-Rouelle, M., Boulègue, J., Comeau, P. ,1999. Trinidad mud volcanoes: where do the expelled fluids come from ?, *Geochimica et Cosmochimica Acta*, **63**, 1023-1038.
12. Dia, A.N., Cohen, A.S., O'Nions, R.K., Shackleton, N.J., 1992. Seawater Sr isotope variation over the past 300 kyr and influence of global climate cycles, *Nature*, **356**, 786 - 788.
13. Etiope, G., Caracausi, A., Favara, R., Italiano, F. , Baciu, C. , 2003. Reply to comment by A. Kopf on "Methane emission from the mud volcanoes of Sicily (Italy)", and notice on CH 4 flux data from European mud volcanoes, *Geophys. Res. Lett.* , **30**, no. 2, 1094, doi : 10.1029/2002GL016287.
14. Gieskes, J.M., You, C.F., Lee, T., Yui, T.F., Chen, H.W., 1992. Hydro-geochemistry of mud volcanoes in Taiwan, *Acta Geol. Taiwan..*, **30**, 79-88.
15. Guliev, I.S., Feizullayev, A.A. , 1996. Geochemistry of hydrocarbon seepages in Azerbaijan. In *Hydrocarbon migration and its near-surface expression*, Schumacher D. and Abrams M.A. (eds), AAPG Memoir 66,Tulsa, Okla, 63-70.
16. Guliev, I.S., Feizulayev, A.A., Huseynov, D.A. , 2001. Isotope geochemistry of oils from fields and mud volcanoes in the South Caspian Basin, Azerbaijan, *Petroleum Geoscience*, **7**, 201 – 209.
17. Kharaka, Y.K., Smalley, W.C., 1976. Flow of water and solutes through compacted clays, *Bull. Amer. Assoc. Petrol. Geol.*, **60**, 973-980.
18. Kopf, A.J., 2002. Significance of mud volcanism, *Reviews of Geophysics*, **40** (X), 1005, doi: 10.1029/2000RG000093.
19. Martin, J.B., Kastner, M., Henry, P., Le Pichon, X., Lallemant, S., 1996. Chemical and isotopic evidence for sources of fluids in a mud volcano field seaward of the Barbados accretionary wedge, *J. Geophys. Res.*, **101**, 20325-20346.
20. Martinelli, G., Albarello, D. and Mucciarelli, M., 1995. Radon emission from mud volcanoes in Northern Italy: possible connection with local seismicity, *Geophys. Res. Lett.*, **22**, 15, 1989-1992.
21. Martinelli, G., Judd, A., 2004. Mud volcanoes of Italy, *Geological Journal* , **39**, 49-61.
22. Minissale, A., Magro, G., Martinelli, G., Vaselli, O., Tassi, G.F. , 2000. Fluid geochemical transect in the Northern Apennines (central-northern Italy): fluid genesis and migration and tectonic implications, *Tectonophysics*, **319**, 199-222.
23. Schoell, M., 1980. The hydrogen and carbon isotopic composition of methane from natural gases of various origins., *Geochimica et Cosmochimica Acta* , **44**, 649 – 661.
24. Yeh, G.H., Yang, T.F., Chen, J.C., Chen, Y.G., Song, S.R., Fluid geochemistry of mud volcanoes in Taiwan, in Martinelli, G. and Panahi, B. (Eds.), *Mud volcanoes, Geodynamics and Seismicity, Proceedings of NATO Advanced Research Workshop, Baku, May 20-22, 2003.*

BIOSENSOR CONTROL OF ACUTE TOTAL TOXICITY OF WATER AND SOIL POLLUTED BY POLYCYCLIC AROMATIC HYDROCARBONS

Nikolay F. Starodub
Department of Biochemistry of Sensory and Regulatory Systems of A.V. Palladin Institute of Biochemistry of National Academy of Sciences, 9 Leontovicha Str., 01030 Kiev, Ukraine, Tel.: 380-44-229 47 43, e-mail: nstarodub@hotmail.com

Abstract: In this article a short analysis of total toxicity of different samples of water and soil (obtained from some Azerbaijan mud bobbling volcanoes, region of oil production, railway lines, Dnieper river and prepared by experimental way) is given. The analysis was fulfilled on the basis of the determination of bioluminescence of bacteria of *Vibrio fisheri* and the registration of level of chemiluminescence of *Daphnia magna* staying medium.

Key words: Polycyclic aromatic hydrocarbons, total toxicity, determination, bioluminescent bacteria, chemiluminescence of *Daphnia*.

1. INTRODUCTION

As a rule polycyclic aromatic hydrocarbons (PAHs) including naphthalene, phenanthrene, benzo(a)pyrene are in environment as result of petroleum production and other high-temperature industrial processes connected with oil treatment. These products are very dangerous for people health and their concentration in the environment should be controlled and for this case a lot of different methods are proposed. These methods based on trad itional approaches of biological monitoring. Unfortunately these approaches are very complicate, expensive and time consumable. Early [1, 2], we have proposed some biosensors for control of acute total toxicity of environment. They are based on:

a) the determination of the intensity of bioluminescence of bacteria of *V. fisheri*, which are immobilized on the surface of fiber optics

and

b) the registration of level of chemiluminescence of *D. magna* staying medium with the use of fiber optics device.

In this report the experimental results about the examination of efficiency of application of the developed biosensors for the control of acute total toxicity of water and soil polluted by PAHs are discussed.

G. Martinelli and B. Panahi (eds.), Mud Volcanoes, Geodynamics and Seismicity, 221–226

2. EXPERIMENTAL

Samples of water were analyzed direct without any preliminary treatment. To obtain samples from soil we have used two ways:

a) extraction of soluble PAHs by water

and

b) by ethanol.

The level of total toxicity was expressed in toxic units (TU) according to the following formula:

$$TU = 100/EC50Extr - 100/EC50Blank \tag{1}$$

where EC50Extr is the effective concentration causing a 50% decrease for the measured endpoint (in%), EC50Blank – effective ethanol concentration causing a 50% decrease for the measured endpoint (in %). The samples of water were collected in Dnieper river near technical port and from mud volcanoes near Baku (Azerbaijan). The samples of soil were taken in region of petroleum production (Ukraine), near railway lines and in regions of mud volcanoes (Azerbaijan). In additional to it was analyzed experimental samples of water and soil which where special polluted by oil, namely: 3g of oil/kg of water or dry soil.

In this investigations *V. fischeri* F1 (IMB B-7070) and *V. fischeri* Sh1 purified from Black sea (Katzev, 2001) were used. The cultivation of bacteria and their preparation for the analysis are accomplished according to method described early (Katzev, 2002). The bioluminescent (BL) analysis was fulfilled by the two different ways: with the use of the stationary chemiluminometer (ChML-3, Ukraine) and with the portable device based on the fiber optics.

In first case the device included signal amplification block on the basis of photoelectric counter (PhEC-176, Ukraine), block for the injection of chemical reagents and control block with digital microprocessor of signals treatment, which allowed us to have maximally automated process. Signals from chemiluminometer were registered by personal computer switched trough special interface. The samples contented 0.8 ml of the tested samples in 2,5% solution of NaCl, 0,1 ml of 0.5 M phosphate (pH 7.0) or phosphate citrate (pH 5.5) buffers and 0.2 ml of bacterial suspension including 5x105 cells/ml. In second case BL bacteria (105 cells) were immobilized in sepharose gel (about 0.1 ml) deposited at the end of fiber optics. In both case the BL intensity (I) was registered through 30-120 min. The level of toxicity was presented as the concentration, which caused 50% decrease of the intensity of BL (EC50) (Azur Environmental, 1995). The BL signal was recorded and processed with the help of chemiluminometer.

For the estimation of toxicity of the analyzed samples with the use of *D. magna* it was kept the following conditions. *Daphnia* was cultivated according to the International (ISO 6341-82) and Ukrainian (RD 211.1.7.054-97) standards (ISO 6341, 1996; KND 211.1.4, 1997). For experiments it was taken animals, which leaved 24 h. As a rule 5 animals were introduced in 10 ml of analyzed sample. The incubation of them accomplished at the temperature of 28 0C during 24 h. For determination of activated chemical luminescence it was added 0,2 mM of luminol and 1% of hydrogen peroxide (50 µl of each) to 500 µl of the medium of *Daphnia* cultivation. The standard medium (ISO 6341-82) served as control.

Samples from mud bubbling volcanoes in regions of Dashgil and Kechaldag (Azerbaijan) were preliminary treated by the following way. In case of the analysis of the total toxicity of water the samples were centrifuged and supernatant was used in the experiments. The total scheme of experiments with soil was organized according to recommendation of (Bispo et al., 1999) and it was presented in Figure 1 (see below).

3. RESULTS AND DISCUSSION

In result of investigations it was established:

1. water samples were less toxic than samples obtained after ethanol extraction;
2. experimental samples had most toxic effect;
3. water samples from region of petroleum production had the same toxicity as samples from Dnieper river.

According to bioluminescent test the level of toxicity of the last samples may be expressed in 5-10 TU. Ethanol extracts from soil of petroleum production region had toxicity about 15-30 TU in dependence on place of obtaining samples. The soil from places near railway lines after extraction of ethanol had toxicity about 650-870 TU. Of course water extracted samples obtained from this place had toxicity much less (about 100-150 TU). At the same time the experimental samples (with 3g of oil/kg of soil) after ethanol extraction have characterized by about 4200 TU. The toxicity of water of these experimental samples was considerably less (about 20-30 times). It is necessary to pay attention that the level of the toxicity of the analyzed samples was in 3-5 times higher if the *Daphnia* chemiluminescent test was used (Table 1). The difference in level of revealed toxicity of the investigated samples by bioluminescent and *Daphnia* chemiluminescent tests stipulated by the differences in the sensitivity of these biological organisms to above indicated substances. Early (Levkovetz et al., 2003; Katzev and Starodub, 2003) we demonstrated that the *Daphnia* chemiluminescent test is more sensitive to different environment pollutants in comparison with the bioluminescent one based on the use of *V. fischeri* and other luminescent bacteria, including Photobacterium phosphoreum K3 (IMB B-7071).

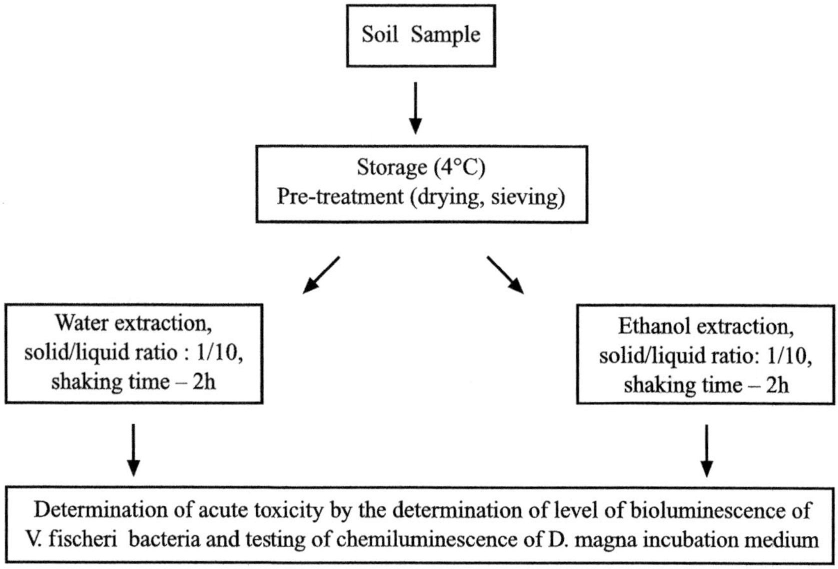

Figure 1. The overall scheme of the fulfillment of experiments with determination of total acute toxicity of soil

The level of toxicity of water from both above indicated mud bubbling volcanoes is not much high in comparison with oil polluted samples. The same was observed in case of the investigation of both soil samples, extracted by water and ethanol. Moreover it is necessary to notice that the differences between toxicity of mud volcano soil samples extracted by water and ethanol are much less than soil polluted by oil (experimental samples or samples obtained near railway lines or region of oil production). All these data may testify that in samples of mud bubbling volcanoes the content of organic carbohydrates is not high and their toxicity may be caused by some non-organic components. This effect will be special analyzed in some additional experiments.

At the some time the differences in toxicity of water and ethanol extracted samples are connected with that ethanol dissolves more effective some organic substances in comparison with water. The method obtaining samples with the use of ethanol is very simple and allows preparation of material for analysis very rapid. We have demonstrated that the developed biosensor approaches are very effective for the determination of acute toxicity of environment since it is very simple, fast and cheap. The developed biosensors can be used in stationary conditions and in field. They could be completely automated with special program for control of limited toxicity of water and soil. There is possibility to develop multi-biosensor system for simultaneous analysis of number of samples.

Table 1. The level of acute toxicity of different analyzed samples of water and soil

Item	Type of samples	Type of treatment of samples	Level of toxicity (in TU) according to boluminescent test	*Daphnia* chemiluminescent test
1.	Water from Dniper riva	None	5-10	20-45
2.	Soil from region of oil production	Ethanol extraction	15-30	75-150
3.	Soil from railway lines	Ethanol extraction	650-870	3000-4600
4.	Soil from railway lines	Water extraction	100-150	500-700
5.	Experimental samples of polluted soil (3g of oil/kg of solid phase)	Water extraction	140-200	400-1000
6.	Experimental samples of polluted soil (3g of oil/kg of solid phase)	Ethanol extraction	~ 4200	~ 12200
7.	Water from mud bubbling volcanoes	Supernatant after centrifugation	~30	~120
8.	Soil from mud bubbling volcanoes	Water extraction	~ 70	~230
9.	Soil from mud bubbling volcanoes	Ethanol extraction	~ 90	~ 300

AKNOWLEDGEMENTS

This work has been supported by the grant from INCO-Copernicus Programs (Grants N: ICA2-CT-2001-10007 and ICA2-CT-2000-10033).

REFERENCES

1. Levkovetz, I.A., Starodub, N.F., Nazarenko, V.I., Ivashkevich, S.P., Goncharuk V.V., Klimenko, N.A., Vakulenko, V.F., Shvadchina, Ju.O. , 2003. The estimation of toxicity of water solutions of SAS in process of their oxidative treatment, *Chemistry and Technology of Water*, N1.
2. Katzev, A.M., Starodub, N.F. , 2003. Investigation of the influence of some surface active substances on the intensity of luminescence of bacteria, *Ukr. Biochem.J.*, **75**, N2, 50-54.
3. Katzev, A. M., 2001. The characteristics of luminescent bacteria form Black see and the perspective of their application for the analysis of water toxicity, *Tavricheskij Medical Biological Zh.*, 4, 112–116.
4. Katzev , A.M., 2002. Utilities of luminous bacteria from the Black Sea, *Applied Biochem. and Microbiol.*, **38**, 189-192.
5. Azur Environmental, 1995. *Microtox Acute Toxicity Basic Test Procedures.* Carlsbad, C.A.
6. ISO 6341:1996 (E), 1996. *Water quality. Determination of the inhibition of the mobility of Daphnia magna Straus (Cladocera, Crustacean) – Acute toxicity test.*

7. KND 211. 1. 4. 054-97, 1997. The method of determination of acute toxicity of water based on the use of cladocera of Daphnia magna Straus.
8. Bispo, A., Jourdain, M.J., Jauzein , M. , 1999. Toxicity and genotoxicity of industrial soils polluted by polycyclic aromatic hydrocarbons (PAHs), *Organic Geochemistry*, **30**, 947-952.

FLUID GEOCHEMISTRY OF MUD VOLCANOES IN TAIWAN

Yeh, Gau-Hua[1], Yang, Tsanyao Frank[1], Chen, Ju-Chin[2], Chen, Yue-Gau[1] and Song, Sheng-Rong[1]

1 Department of Geosciences, National Taiwan University No.1, Sec.4, Roosevelt Road, Taipei 106, Taiwan, R.O.C.

2 Institute of Oceanography, National Taiwan University No.1, Sec.4, Roosevelt Road, Taipei 106, Taiwan, R.O.C.

** Corresponding author T.F. Yang Department of Geosciences, National Taiwan University No. 1, Sec. 4, Roosevelt Road, Taipei 10617, Taiwan, R.O.C.*
Tel: +886-2-3366-5874 Fax: +886-2-2363-6095, e-mail: tyyang@ntu.edu.tw

Abstract: Mud volcanoes (MVs) are commonly found in southern Taiwan. They are believed to be the products related to the accretionary prism due to the on-going arc-continent collision between Eurasian plate and Luzon arc. For better understanding their fluid compositions, samples of representative MV fluids were collected for geochemical analyses. Compared to the hot spring waters in Taiwan, all the MV fluids show much higher content of chlorine concentration. This indicates that marine pore waters play an important role for the sources of the MV fluids. Interestingly MV fluids from eastern Taiwan show unique geochemical characteristics of sodium-deficit and calcium-excess, which may be explained by the process of albitization during water-rock interaction.The H-O isotopic data of the MV fluids samples from eastern Taiwan, in contrast to those from western Taiwan, fall along the local meteoric water line. It implies that meteoric water is an important component for the source of MV fluids from eastern Taiwan but not for those from western Taiwan. Meanwhile, samples collected from the region near deep structures in western Taiwan usually exhibit higher $\delta^{18}O$ value, which suggests that these fluids may contain dehydrated waters of clay minerals at deep source.

Key words: fluid geochemistry, mud volcanism, neotectonic

1. INTRODUCTION

Worldwide distributions of MVs are mostly related to the convergent plate boundaries (Milkov, 2000). Taiwan locates on the arc continent collision zone between Eurasia Plate and Philippine Sea Plate, and develops a

G. Martinelli and B. Panahi (eds.), Mud Volcanoes, Geodynamics and Seismicity, 227–237

typical accretionary prism. Hence MVs are common landscapes in southern Taiwan.

Early MV studies in Taiwan are mainly focused on the exploration of natural gas and petroleum. However, most of the investigation stopped because no valuable reservoirs were found from MV related research. Shih (1967) pointed out that there were 64 MVs in 17 MV active zones in Taiwan. He divided the MVs into five types based on their geomorphologic shapes. Furthermore, he concluded that the shapes of MVs are closely related to the viscosity of the fluids but not related to the erupted materials, which were mainly clay minerals.

Gieskes et al. (1992) analyzed the fluid compositions of selected MVs in Taiwan and concluded that seawater is an important component for their sources. The δD-δ18O plot of the fluids can be explained by the fluid-mineral interaction during the dehydration of clay mineral at high temperature. They inferred that the MV fluids might be derived from décollement at bottom of the accretionary prism.

Recently, Yang et al. (2004) analyzed the gas compositions of representative MVs and showed that methane is the major gas component for most MVs in Taiwan, although some MVs exhibit unusual carbon dioxide dominated and/or nitrogen excess compositions. Helium isotopic data demonstrated that those MV gases were mainly derived from crustal component in southwestern Taiwan (Yang, 2002; Yang et al., 2003b). Nevertheless, MV gases from eastern Taiwan exhibit significant mantle signature (Yang et al., 2003a).

In addition to the fluids from MVs, some hot springs from non-volcanic areas in Taiwan show MV-like geochemical characteristics, i.e., high chlorine and other ion concentrations (Chen, 1975). In this study we collected fluid samples from representative MVs and MV-like hot springs in different tectonic domains in Taiwan to compare their fluid compositions and sources.

2. GEOLOGICAL BACKGROUND OF SAMPLING SITES

Present active MVs can be grouped into five zones in Taiwan based on their distribution in different tectonic domains (Yeh, 2003; Yang et al., 2004). They are:

1. Chu-kou Fault zone,
2. Gu-ting-keng anticline zone,
3. Chi-shan Fault zone,
4. Coastal Plain zone,
and
5. southern Coastal Range zone, respectively.

Gu-ting-keng Formation is one of the most dominant bedrocks in southwestern Taiwan. It mainly consists of muddy materials. According to previous studies, its lower stratum is considered to be deposited in the environment of continental shelf to slope, and the upper stratum is defined as offshore to sub-offshore sedimentary faces (Chou, 1971; Covey, 1984). Many MVs are distributed along the axis of Gu-ting-keng anticline. However, some isolated MVs erupted in Coastal Plain of SW Taiwan. Although covering with Quaternary deposits, their eruptive fossils were recognized to be derived from Gu-ting-keng Formation (Shih, 1967). Hence, MVs from Gu-ting-keng anticline zone and Coastal Plain zones may have the same source in origin.

Chi-shan Fault and Chu-kou Fault are two major active faults in southwestern Taiwan and many MVs are found along these two fault zones. Fluid samples were collected from representative MVs and MV-like hot springs, which are Wu-shan-ding MV (WSD) and Hsin-yang-nyu-hu MV (HYN) in Chi-shan Fault zone; Chung-lun hot spring (CL) and Kuang-tze-ling hot spring (KZL) in Chu-kou Fault zone. Meanwhile, we also collected samples from Gung-shuei-ping MV (GSP) in Coastal Plain region, and Hsiao-gun-shuei MV (HGS) in Gu-ting-keng anticline zone of southwestern Taiwan (Fig. 1).

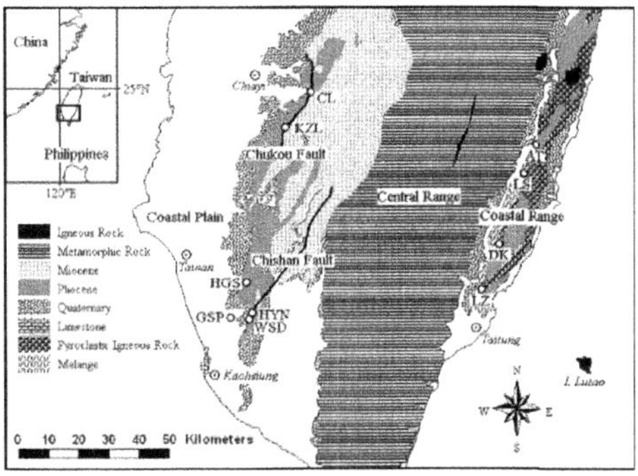

Figure 1. The distributions of major MVs and sampling sites in this study.

MVs and hot springs in the Coastal Range, eastern Taiwan are all distributed in the Formation of Lichi Mélange, which is composed of mudstone matrix mixed with blocks of ophiolites, andesites and sedimentary rocks. Most people believed that they were the obductive products of the accretionary prism of the Luzon arc system during arc-continent collision (e.g., Page and Suppe, 1981; Teng, 1990; Huang et al., 1997).

Gas compositions of MVs and hot springs in Coastal Range, eastern Taiwan show different sources in origin from those from southwestern Taiwan (Yang et al., 2003b; 2004). Hence, fluid samples from representative MVs and hot springs were also collected and analyzed for comparison. They included Diang-Kuang MV (DK), Luo-shan MV (LS), Li-chi hot spring (LZ) and An-tong hot spring (AT) (Fig. 1).

3. ANALITICAL METHODS

Fluids of MVs and hot springs were collected by 50 ml centrifuge tubes from the depth of ca. 30cm of sub-surface water to prevent the evaporation effect which may change the H-O isotopic compositions of the fluids. After collecting, the fluid samples were tightly capped in the PP tubes and carried back to laboratory for analysis. First, the sample was centrifuged for 30 minutes with the speed of 5000 rpm, and then the fluids could be separated from the sediments and suspensions with the 0.45 μm filters. The separated water samples were then stored as two bottles. One was for major ionic composition analysis; and another (adding few drops of mercury chloride to prevent bacteria effect) was for hydrogen-oxygen isotopic analysis.

The concentrations of chlorine and bicarbonate were determined by the titration of silver nitrate and hydrochloric acid. The other major anions were then analyzed by ion chromatography, and major cations were analyzed by inductively coupled plasma-atomic emission spectrometry (ICP-AES). Meanwhile, water samples were prepared for hydrogen isotopic measurement by zinc reduction method (Coleman et al., 1982; Kendall and Coplen, 1985) and for oxygen isotopes by carbon dioxide balance method (Cohn and Urey, 1938; Epstein and Mayeda, 1953), respectively. Utilizing a gas type mass spectrometer (Finnigen Delta Plus), hydrogen and oxygen isotopic compositions were measured. Detailed procedures and method for isotopic analysis have been described by Chan (2001).

4. RESULTS AND DISCUSSIONS

The analysis results indicate that the fluid compositions of MVs and hot springs did not show significant seasonal variations (Table 1). Except for AT, all the fluid samples exhibit very high chlorine concentration. However, MV fluids in Coastal Plain and Gu-ting-keng anticline zones have higher chlorine concentrations than those from Chi-shan and Chu-kou Fault zones. In addition to the isotopic compositions, MV fluids from Coastal Range, eastern Taiwan have distinct geochemical characteristics from those of southwestern Taiwan. Hence, we will divide the geochemical results of analyzed samples into three groups for further discussion in the text and figures. They are: (1) Group of Chi-shan and Chu-kou Fault zones, (2) Group of Coastal Plain and Gu-ting-keng anticline zones, and (3) Group of Coastal Range, eastern Taiwan.

It is worthy to note that GSP, HGS and DK fluids can reach up to two third of the chlorine composition of seawater. Concentration of total cations of the samples shows a good linear correlation with their chlorine content. Furthermore, the regression line can be extrapolated to the composition of seawater component (Fig. 2), indicating that seawater is a major component for those studied fluid samples.

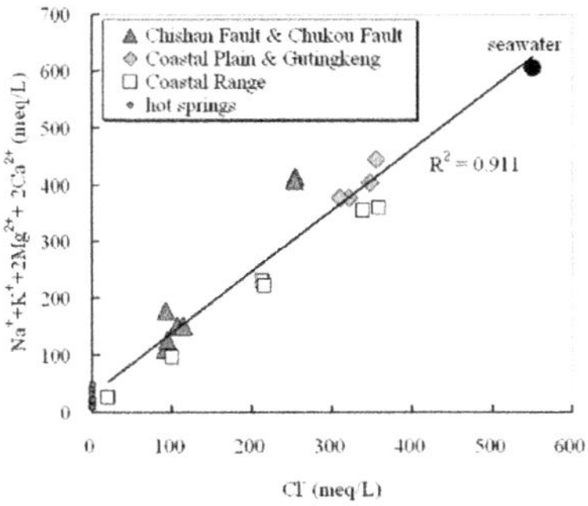

Figure 2. Plot of chlorine concentration vs. total cations of the MV fluids and hot springs in southern Taiwan. Data of normal non-volcanic related hot springs in Taiwan are also plotted for comparison (Chen, 1975). It is clear that the MVs possess much higher concentrations than the normal hot springs. A good linear correlation between chlorine and major cations can be obtained, and extended to the composition of seawater. This implies that seawater was a major component of these fluids.

Other cases at the convergent plate boundary, the fluids from marine formation in non-volcanic areas also showed similar geochemical characteristics, i.e., seawater component was the major source and then mixed with other sources with different proportions (e.g., Martin et al., 1995). Because most erupted materials from MVs in Taiwan are marine sediments, the seawater component could be considered as the residues of the sedimentary pore water, and hence becoming an important source for the fluids of MVs.

The oxygen-hydrogen isotopic compositions of fluids from Costal Range are close to the meteoric water line of local monsoon rain. However, fluids from southwestern Taiwan are far away the meteoric water line (Fig. 3), which generally can be explained by the processes of evaporation, fluid-rock alteration, or mixing with other fluids.

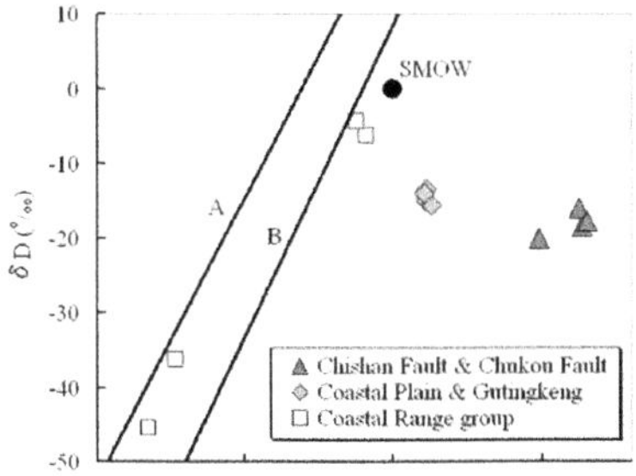

Figure 3. Plot of δD-δ18O isotopic compositions of the MV fluids in Taiwan. Fluid samples from Coastal Range fall close to the local meteoric water line; however, those from southwest Taiwan are far away the meteoric water line. It indicates that meteoric water plays an important role for the source of the fluids in Coastal Range, but not for those from southwest Taiwan. Line A and B are local meteoric water line. Line A: the meteoric water line of northeast monsoon rain; line B: the meteoric water line of southwest monsoon rain (Peng, 1995).

However, evaporation is unlikely to be an important process to account for the isotopic composition of MV fluids. Because those samples exhibiting heavier oxygen isotopes usually also have lower chlorine concentrations (Table 1), which is opposite to the trend induced by the process of evaporation.

Table 1. Composition of fluids from representative mud volcanoes and related hot springs in southern Taiwan.

Loca-tions	date	Na (mM)	K (mM)	Mg (mM)	Ca (mM)	Cl (mM)	HCO_3^- (mM)	SO_4^{2-} (mM)	NO_3^- (mM)	δD (‰)	$\delta^{18}O$ (‰)
WSD	02/02/26	110	0.614	0.329	0.200	92.0	30.0	0.365	1.63	-	6.4
	02/07/25	125	0.629	0.364	0.221	94.2	29.5	0.375	1.61	-18	6.4
HYN	02/02/26	150	0.716	0.411	0.274	107	29.5	0.094	1.40	-16	6.3
	02/07/25	149	0.767	0.436	0.299	115	29.5	0.083	1.45	-18	6.5
CL	02/07/02	404	8.02	1.18	0.204	255	170	1.39	0.452	-23	4.7
	02/07/11	399	7.56	1.13	0.165	254	164	1.35	0.484	-23	4.5
KZL	02/07/24	171	6.14	0.411	0.162	92.2	82.0	2.15	1.63	-20	4.9
GSP	02/02/26	369	2.73	2.79	0.423	322	36.6	0.104	3.48	-14	1.1
	02/07/25	369	2.72	2.46	0.544	310	37.7	0.094	3.39	-15	1.1
HGS	02/02/26	395	1.99	3.16	0.46	347	44.9	0.182	3.05	-16	1.3
	02/07/25	435	2.05	2.98	0.74	355	45.9	0.177	3.07	-14	1.1
DK	02/02/24	147	1.46	4.16	99.3	339	6.95	0.021	2.50	-	-1.6
	02/07/12	147	1.39	3.61	103	358	6.89	0.031	2.58	-4	-1.2
LS	02/02/24	134	0.598	0.482	47.4	214	7.13	0.042	1.40	-	-0.7
	02/07/11	126	0.588	0.505	46.9	215	7.05	0.052	1.37	-6	-0.9
LZ	02/07/12	86.8	0.424	0.102	5.12	99.9	2.80	0.052	1.37	-45	-8.2
AT	02/07/11	22.1	0.168	0.019	2.38	19.4	1.64	4.17	0.016	-36	-7.3

Hence other process must be taken into consideration for the variations of MV fluid compositions in Taiwan. Gieskes et al. (1992) suggested that fluid mineral exchange during the process of clay dehydration at elevated temperatures could account for the heavier isotopic composition of MV fluids in southwestern Taiwan. Recent studies showed that fluids from the hinterlands in southwest Taiwan may have undergone a higher grade of diagenesis and hence exhibit more signature resulting from water-rock reaction and dehydration of clay minerals (Yeh, 2003; Yeh et al., 2002). The authors also further suggested another alternative that deeper thrusts may exist in the hinterland, which consequently could provide a pathway for deep dehydration water addition. The present study supports these conclusions.

MV and hot spring fluids in Coastal Range show distinct compositions from that in southwestern Taiwan. They not only have significant amounts of meteoric water component, in terms of H-O isotopes; but also possess remarkable calcium compositions (Table 1).

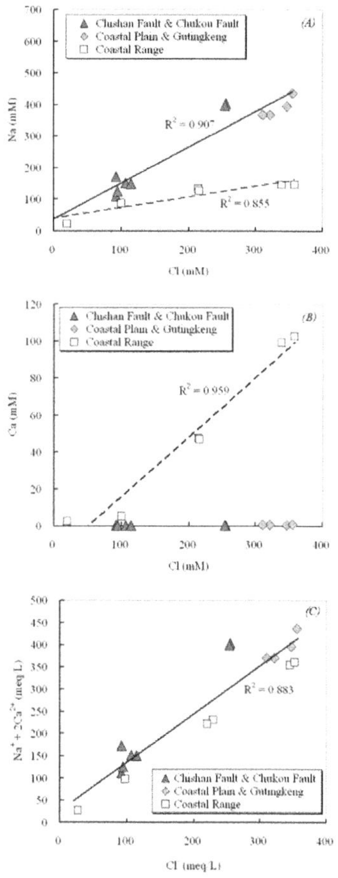

Figure 4. Plot of chlorine vs. sodium, calcium and sodium plus calcium concentrations of the studied fluids. Compared with the MV fluids from southwestern Taiwan, those from the Costal Range exhibit characteristics of deficit-sodium (A) and excess-calcium (B). It indicates that a process of sodium-calcium replacement may have occurred (C).

MV fluids in Coastal Range exhibit the phenomenon of sodium-deficit and calcium-excess compared to that in southwestern Taiwan (Fig. 4A and 4B).

However, all the MV fluids data display a good linear correlation in the plot of alkaline (Na+Ca) vs. chlorine concentration (Fig. 4C), indicating that the deficit of sodium is equal to the excess of calcium. If the sodium and calcium of the fluids mainly came from seawater, we can calculate the concentrations of deficit-sodium and excess-calcium according to the ion content of seawater (Davisson et al., 1994):

$$Na_{deficit} = 1\times [(Na/Cl)_{seawater} \times [Cl]_{sample} - [Na]_{sample}] \quad (1)$$
$$Ca_{excess} = 2\times [(Ca/Cl)_{seawater} \times [Cl]_{sample} - [Ca]_{sample}] \quad (2)$$

The deficit-sodium and excess-calcium in fluids show a very good leaner correlation (R2 = 0.988) (Fig. 5), meanwhile, the slope of the correlation is close to 1. This implies that the sodium-calcium replacement has occurred in these fluid samples.

Sodium in the fluids can replace calcium in feldspar during the process of albitization, and change the fluid compositions (Land and Milliken, 1981; Boles, 1982). The reaction is:

$$CaAl_2Si_2O_8 + 4SiO_2 + 2\ Na^+ \rightarrow 2NaAlSi_3O_8 + Ca^{2+} \quad (3)$$

Consequently, calcium enrichment of the fluids is usually found in a deep silica cataclasis basin (White, 1965; Kharaka and Thordsen, 1992). MVs and hot springs in the Coastal Range all are distributed in the Formation of Lichi Mélange, which is composed of marine clay mineral matrix with large amount of various igneous blocks. These rocks range from rhyolitic to ultrabasic composition and possess plagioclase dominant phenocrysts (Chen, 1986; 1988; Tsao, 1987). Hence, under the circumstance of silica-saturated environment, the sodium-rich fluids of MVs would react with the plagioclase (Ca-feldspar) of the igneous blocks, and hence, to enrich the calcium concentrations in the fluids. Because of lacking the igneous rocks with Ca-feldspar phenocrysts in mud-matrix, such kind of the process would not occur in southwest Taiwan.

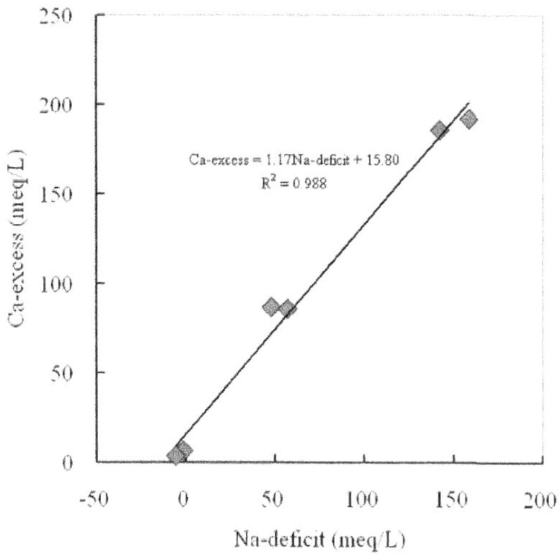

Figure 5. The concentrations of deficit-sodium and excess-calcium of the fluids from Coastal Range display a good linear correlation. It infers that sodium may replace calcium in the fluids. The mechanism of replacement can be explained by the process of albitization (see the further discussion in the text).

5. CONCLUSIONS

Fluids from MVs in Taiwan usually contain a very high chlorine content, which dominates the concentrations of major anions in the samples. The cation compositions are dominant with sodium in all studied MV fluids; nevertheless, significant amount of calcium is also present in samples from Coastal Range, eastern Taiwan.

Fluid samples of MVs from southwestern Taiwan can be divided into two types.

1. MV fluids from Chi-shan Fault and Chu-kou Fault zones.
   They exhibit lower ion concentrations but heavier oxygen isotopes.
2. Fluids from Coastal Plain and Gu-ting-keng anticline zones possess higher ion concentrations but lighter oxygen isotopes.

Sedimentary pore waters originated from seawater are considered to be the major components for the studied MV fluids. In addition, dehydrated waters of clay minerals also play an important role for the fluids of MVs in southwestern Taiwan. According to the H-O isotopic data, nevertheless, meteoric water would be also an important source for the MV fluids in Coastal Range, eastern Taiwan.

Compared to the compositions of MV fluids from southwestern Taiwan, those from eastern Taiwan exhibit the characteristics of sodium-deficit and calcium-excess, which can be explained by the process of albitization.

ACKNOWLEDGEMENTS

The authors wish to thank the help of Hsiao, D.R., Fu, C.C., Wang, C.H., Lan, D.F. and Hung, W.L. during sampling in the field; Lin, B.W., Liu, C.M. in sample handling and analysis. Helpful discussions with Dr. You, C.F. of National Cheng-Kung University are also appreciated. This study was financially supported by Central Geological Survey and National Science Council of Taiwan, R.O.C.

REFERENCES

1. Boles, J.R., 1982. Active albitization of plagioclase, Gulf Coast Tertiary, *Amer. J. Sci.*, **282**, 165-180.
2. Chan, P.S., 2001. $\delta^{18}O$ and δD of spring waters from active structure zones, southwestern Taiwan. MS thesis, Inst. Geosciences, National Taiwan Univ., (In Chinese with English abstract).
3. Chen, C.H., 1975. Origin of hot springs and geothermal exploration in Taiwan. *Ti-Chih*, **1**(2), 107-117, (in Chinese)
4. Chen, C-H., 1986. Petrology and genesis of cognate plutonic in andesites of east Coastal Range, Lutao and Lanhsu, Taiwan., *Mem. Geol. Soc. China*, **7**, 259-282.
5. Chen, C-H., 1988. Mineral compositions of primary phases in ultramafic rocks in the Coastal Range, eastern Taiwan and implication of source region of these rocks, *Acta Geol. Taiwanica*, **26**, 193-222.
6. Chou, J.T., 1971. A preliminary study of the stratigraphy and sedimentation of the mudstone formations in the Tainan area, southern Taiwan, *Petrol. Geol. Taiwan*, **8**, 187-219.
7. Cohn, M. and Urey, H.C., 1938. Oxygen exchange reactions of organic compounds and water, *J. Am. Chem. Soc.*, **60**, 679.
8. Coleman, M.L., Shepherd, T.J., Durham, J.J., Rouse, J.E. and Moore, G.R. , 1982. Reduction of water with zinc for hydrogen isotope analysis, *Anal. Chem.*, **54**, 993-995.
9. Covey, M., 1984. Lithofacies analysis and basin reconstruction, Plio-Pleistocene western Taiwan foredeep, *Petrol. Geol. Taiwan*, **20**, 53-83.
10. Davisson, M.L., Presser, T.S. and Criss, R.E. , 1994. Geochemistry of tectonically expelled fluids from the northern Coast range, Rumsey Hills, California, USA, *Geochim. Cosmochim. Acta*, **58**, 1687-1699.
11. Epstein, S. and Mayeda, T., 1953. Variations of ^{18}O content of waters from natural sources, *Geochim. Cosmochim. Acta*, **4**, 213-224.
12. Friedman, I., 1953. Deuterium content of nature waters and other substances, *Geochim. Cosmochim. Acta*, **4**, 89-203.
13. Gieskes, J.M., You, C.F., Lee, T., Yui, T.F. and Chen, H.W., 1992. Hydro-geochemistry of mud volcanoes in Taiwan., *Acta Geol. Taiwan.*, **30**, 79-88.

14. Huang, C.Y., Wu, W.Y., Chang, C.P., Tsao, S., Yuan, P.B., Lin, C.W., Xia, K.Y. , 1997. Tectonic evolution of accretionary prism in the arc-continent collision terrane of Taiwan, *Tectonophys*. **281**, 31-51.
15. Kendall, C. and Coplen, T.B., 1985. Multisample conversion of water to hydrogen by zinc for stable isotope determination, *Anal. Chem*., **57**, 1437-1440.
16. Kharaka, Y.K. and Thordsen, J.J., 1992. Stable isotope geochemistry and origin of waters in sedimentary basins. In: *Isotopic Signatures and Sedimentary Records*, Clauer , N. and Chaudhuri S. (eds.), Berlin, Springer Verlag, 411-466.
17. Land, L.S. and Milliken, K.L., 1981. Feldspar diagenesis in the Frio Formation, Brazoria County, Texas Gulf Coast, *Geology*, **9**, 314-318.
18. Martin, J.B., Kastner, M. and Egeberg, P.K., 1995. Origins of saline fluids at convergent margins, In: *Active Margins and Marginal Basins of the Western Pacific*, Taylor, B. and Natland, J. (eds.) Geophys. Monogr. Ser., 88, 219-239, Washinghton, American Geophysical Union.
19. Milkov, A.V. , 2000. Worldwide distribution of submarine mud volcanoes and associated gas hydrates., *Marine Geology*, **167**, 29-42.
20. Page, B.M. and Suppe, J. , 1981. The Pliocene Lichi mélange of Taiwan: its plate-tectonic and olistostromal origin, *Am. J. Sci*., **281**, 193-227.
21. Peng, C.R., 1995. Environmental isotopic study (^{13}C, D, ^{18}O, ^{14}C, T) on meteoric water and ground waters in I-Lan area. Ph.D. dissertation, Inst. Geosciences, National Taiwan Univ., (In Chinese with English abstract)
22. Shih, T.T., 1967. A survey of the active mud volcanoes in Taiwan and a study of their types and character of the mud, *Petrol. Geol. Taiwan*, **5**, 259-311.
23. Teng, L.S., 1990. Geotectonic evolution of late Cenozoic arc-continent collision in Taiwan., *Tectonophys*. **183**, 57-76.
24. Tsao, S., 1987. The trace element geochemistry and mineral composition of the exotic basic rocks enclosed in Lichi Mélange, east Coastal range, Taiwan. MS thesis, Inst. Geology, National Taiwan Univ., (In Chinese with English abstract).
25. White, D.E., 1965. Saline waters of sedimentary rocks. In: *Fluids in Subsurface Enviroments*, Young , A. and Galley, E. (eds.), *AAPG Mem*., **4**, 342-366.
26. Yang, T.F. ,2002. $^{3}He/^{4}He$ ratios of fluid samples around Taiwan, The 2002 Goldschmidt Conference, Davos, Cambridge Publication, Cambridge (abstr. A859).
27. Yang, T.F., Chen, C-H., Tien, R.L. Song, S.R. and Liu, T.K. , 2003a. Remnant magmatic activity in the Coastal Range of East Taiwan after arc-continent collision: fission-track data and $^{3}He/^{4}He$ ratio evidence, *Radia. Meas*., **36**, 343-349.
28. Yang, T.F., Chou, C.Y., Chen, C-H., Chyi, L.L. and Jiang, J.H., 2003b. Exhalation of radon and its carrier gases in SW Taiwan, Radia. Meas., **36**, 425-429.
29. Yang, T.F., Yeh, G.H., Fu, C.C., Wang, C.C., Lan, D.F., Lee, H.F., Chen, C-H., Walia, V. and Sung, Q.C. , 2004. Composition and exhalation flux of gases from mud volcanoes in Taiwan, *Environ. Geology* (in press).
30. Yeh, G.H., 2003. Geochemistry and sources of mud-volcano fluids in Taiwan. MS thesis, Inst. Oceanography, National Taiwan Univ., (In Chinese with English abstract).
31. Yeh, G.H., You, C.F., Chen, J.C., Yang, T.F., Chen, Y.G. and Song, S.R. , 2002. Fluid geochemistry of mud volcanoes at the accretionary prism in southern Taiwan, The 2002 Goldschmidt Conference, Davos, Cambridge Publications, Cambridge (abstr. A862).

Chapter 7
PHYSICAL MODELS OF MUD VOLCANOES

MUD VOLCANOES AS NATURAL STRAINMETERS
A working hypothesis

Dario Albarello
Dept. of Earth Sciences, University of Siena, Via Laterina 8, Siena, 53100, Italy, e-mail: alberello@unisi.it

Abstract: Due to the sub-critical status of the seismogenic crust, relatively small strain fluctuations (>0.1 μstrain) induced by post-seismic stress redistribution could significantly affect the seismic hazard on a regional scale (tens to thousands of km) in the medium-term (months to tens of years). The physical feasibility of mud volcano monitoring for the detection of crustal strain field fluctuations is discussed. Simple physical arguments are considered to evaluate the response of the mud volcanic system, in terms liquid-outflow or gas discharge variations, to slow pore pressure changes in the reservoir. The application of a new remote sensing technique for large-scale/medium-term monitoring is also suggested.

Key words: Mud volcanoes, Strain field, Geodynamics, Earthquakes, Seismic hazard

1. SEISMIC HAZARD AND STRAIN/STRESS FIELD VARIATIONS

Seismic hazard assessment is commonly performed by using a "standard" probabilistic procedure (e.g., McGuire, 1993a,b). Basic hypotheses underlying this approach are that the seismogenic process is stationary and that seismicity is the consequence of mutually independent activations of a relatively small number of seismogenic structures whose geometry can be assumed as stable in time (e.g. Cornell, 1968). The generality of these assumptions allows the widespread applicability of the procedure also where detailed information about seismogenic processes and geodynamic conditions are lacking. This is why such an approach has been chosen as the basis of the GSHAP/UN program dedicated to the determination of a worldwide seismic hazard map (Giardini and Basham, 1993).

However, recent seismological research suggests that the actual reliability of hazard estimates performed on this basis is debatable, to say the least.

G. Martinelli and B. Panahi (eds.), Mud Volcanoes, Geodynamics and Seismicity, 239–249

In particular, recent analyses (see, for example, the reviews by Kagan, 1994; Main, 1995, 1996) suggest that assumptions of stationary seismic activity and an independence between seismic sources conflict with growing evidence about the widespread scale invariance of faulting and the time/space clustering of seismicity. These properties are shared by a number of different "critical" processes (fluid turbulence, magnetisation, percolation, stock exchange dynamics, forest fires, etc.) whose macroscopic behaviour can be interpreted as the effect of local interactions between a large number of simple elements (e.g. Kagan, 1992; Turcotte, 1992). In the case of seismicity, such elements are ubiquitous seismogenic structures (faults or possible crustal weaknesses present over a wide range of dimensional scales) whose mechanical interactions can be responsible for the observed phenomenology (e.g. Sacks and Rydelek, 1995; Castellaro and Mulargia, 2001).

A common feature of such "critical" or "sub-critical" systems is their strong sensitivity to small variations in the physical environment, which can result in dramatic macroscopic changes to the system status. Thus, notwithstanding a nearly constant energy supply (e.g. tectonic loading in the case of earthquakes), energy dissipation can occur in a highly intermittent, "bursting", manner. In particular, due to the dynamic interactions between potential seismic sources, the system is dramatically influenced by relatively small variations in the stress/strain field (to the order of tens of kPa or 0.1 μstrain), which can significantly modify the seismic hazard level (e.g. Rydelek and Sacks, 1999).

The presence of strain field variations on the time-scale relevant for seismic hazard assessment (years/tens of years) has been widely recognised also on the basis of empirical evidence (e.g. Kasahara, 1979; Viti et al., 2003 and references therein).

A possible origin of such a perturbation is the regional-scale (10^3-10^5 m) redistribution of tectonic loads after the occurrence of significant earthquakes. Two processes can account for this phenomenon. The first is associated to the elastic rebound of the crust after the earthquake. This is a typical short-range (a few times the fault-length) short-term (hours/days) effect (see, e.g., Okada, 1992; Harris, 1998). The second one is related to the medium-term rheological relaxation of lower crust and asthenosphere after the earthquake (e.g. Ranalli, 1995). This is a typical long-range (several times the fault length) medium/long-term (years, tens of years) effect (e.g., Anderson, 1975). Both processes are considered to be responsible for relevant perturbations in the seismogenic process and capable of inducing significant variations in the local seismic hazard level (e.g. Rydelek and Sacks, 1990; Albarello and Bonafede, 1990; King et al., 1994; Stein et al., 1997; Piersanti et al., 1997; Mantovani and Albarello, 1997; Belardinelli et al., 1999; Pollitz et al., 2000).

2. MONITORING STRESS/STRAIN FIELD PERTURBATIONS BY PIEZOMETRIC OBSERVATIONS RELATIVE TO DEEP-SEATED AQUIFERS

Despite the fact that large agreement exists concerning the importance of reliable evaluations of strain field fluctuations as an important element for new-generation seismic hazard estimates, direct and affective experimental tools for this purpose are still lacking. To be effective, these kinds of monitoring apparatus should be sensitive to small strain variations (to the order of 10^{-7}-10^{-5} strain units or 10-1000 kPa), to operate at relatively large space/time-scales (tens to hundreds km, from months to tens of years). Furthermore, in order to make such monitoring feasible even when the economic resources available for seismic hazard assessment are scarce, it should be cost-effective.

The most direct approach to this problem is space geodesy. However, these methodologies are , in many cases, unable to achieve sufficient accuracy, at least as concerns the time-space scale of concern. An attractive alternative seems to be the monitoring of deep-seated confined fluids.

In the presence of low strain values, solid/fluid interactions within the crust can be modelled in the frame of poroelastic linear approximation (e.g. Wang, 2000). When confined aquifers are of concern, both theoretical considerations and experimental data (e.g. Roeloffs, 1996) suggest that volumetric strain fluctuations can dramatically affect pore pressure. In particular, in undrained conditions, typical of deep-seated confined aquifers, it holds that

$$\delta P \cong \theta \delta \varepsilon \qquad (1)$$

where δP and $\delta\varepsilon$ are the pore pressure and volumetric strain variations, respectively. θ is a constant parameter whose value, in typical crustal situations, is in the order of 5-50 GPa. As an example, this value implies that the piezomeric level of a confined aquifer will vary by 1m if the volumetric strain change is one microstrain. The macroscopic response of confined geofluid reservoirs to microscopic strain variations makes deep geofluid monitoring a useful tool in the study of strain field perturbations.

A number of studies have confirmed the effectiveness of this possibility (see,for example, Bodvarsson, 1970; Kumpel, 1992; Roeloffs, 1996 and references therein). As concerns the local co-seismic perturbations, for example, Muir-Wood and King (1993) discuss several situations where clear-cut hydrological changes (increases in spring and river discharge, variations in well piezomeric levels, etc.) have been observed in association with strain perturbations induced by strong earthquakes. As concerns medium-term/long range strain fluctuations, an analysis of piezometric levels monitored during a 17-year time-span at 400 deep water wells in Northern Italy (Albarello and Martinelli, 1994) has revealed the presence of multi-annual piezometric fluctuations with a typical wavelength of several tens of km, slowly migrating along the monitored area at a velocity in the order of few cm/years. The presence of such migrating perturbations in the area has been also confirmed by independent tilt-metric measurements (Zadro and Rossi, 1991; Rossi and Zadro, 1996).

However, a major drawback to the use of piezometric observations as an inexpensive monitoring tool for strain perturbations is the difficulty in identifying confined fluid reservoirs (water or oil) not previously exploited for anthropic use. The monitoring of mud volcano activity could represent a valid alternative, allowing such limitations to be overcome.

3. MUD VOLCANOES AS NATURAL STRAINMETERS

The association of mud volcano activity to strain field variations has been suggested by Tamrazyan (1982) on the basis of an observed correspondence between mud eruptions and earth tide maxima. Several authors (e.g. Wakita et al., 1988) have hypothesized a sensitivity of mud volcano gaseous emissions to seismogenic processes. Encouraging results in this sense have been obtained in Italy by the monitoring of medium-term Radon fluctuations in the liquid phase of fluids erupted at a mud volcano in Northern Italy (Martinelli et al., 1995). Less significant results have been obtained for the gaseous phase (Albarello et al., 2003).

In order to evaluate the feasibility of mud volcano monitoring to detect strain field variations, a simple physical model can be considered. The mud volcano complex is assumed to be the surface expression of a cylindrical vertical conduit (diapir) open at one edge, connecting a deep-seated reservoir of over-pressurized fluids (mud and dissolved gases) with the surface.

It is reasonable to assume that fluid flow in the conduit is driven by the pressure difference ΔP between the top and the bottom of the conduit of length h. The average pressure gradient can be roughly estimated by

$$\frac{\Delta P}{h} = (\rho_{ovb} - \rho_{mud})g \tag{2}$$

where g is gravity acceleration, ρ_{ovb} and ρ_{mud} are overburden and mud densities, respectively (Brown, 1990).

As long as the gaseous phases are dissolved in the mud, the fluid can be assumed to be incompressible. Reynolds number Re, which characterises the flow of such a fluid in this situation is given by

$$\mathrm{Re} = \frac{\rho_{mud}\bar{u}D}{\mu} \quad (3)$$

where, u is the average fluid velocity, D is the diameter of the conduit, μ and ρ fluid viscosity and density, respectively (e.g., Turcotte and Schubert, 2002). For Re<2300, the flow can be assumed as laminar. In this case, the average fluid velocity throughout the conduit results to be

$$\bar{u} = -\frac{D^2}{32\mu}\frac{\Delta P}{h} \quad (4)$$

corresponding to a flow rate Q given by

$$Q = -\frac{\pi D^4}{132\mu}\frac{\Delta P}{h} \quad (5)$$

Equations (4) and (5) imply that both the flow rate and average fluid velocity dramatically depend on the conduit diameter (Kopf, 2002).

By assuming a density difference between overburden and mud to the order of 10^2 kg/m^3, the pressure gradient results to the order of 10^3 Pa/m (Brown, 1990). Since μ is to the order of 10^6 Pa sec (Kopf, 2002), the ascent velocity provided by equation (3) results to the order of 10^3 m/y for a 1 meter conduit. Corresponding flow rates are to the order 10^3 m^3/y. By using these figures, the relevant Reynolds number results to be very low (<<1) and the laminar flow assumption can be considered reliable.

Pore pressure variations at depth induced by strain field perturbations should reflect in flow rate changes at the surface. However, due to the viscosity of the involved fluids, these variations will take time to develop. It can be demonstrated (Batchelor, 1967, section 4.3) that, a fully developed response of an incompressible fluid to a sudden pressure variation at one end of the rigid conduit requires times τ such that

$$\tau > \frac{\rho D^2}{4\mu} \quad (6)$$

By using the figures considered above, τ results to be to the order of 10-3 sec, which implies a very fast response by the system to hydraulic head variations.

All these considerations hold in the case that mud can be considered an incompressible fluid, i.e., that fluids characterised by very low compressibility (such as water) control the mechanical behaviour of the system. However, several gaseous phases (mainly methane) are ubiquitously observed in association with sub earial mud volcanoes (e.g. Kopf, 2002). These gaseous phases are originated at depth in the mud reservoir or below and are present all along the conduit at least in dissolved form. In general, methane solubility above 4 km is low. Above this depth, as up-rising mud is decompressed, methane will evolve from solution and it will expand together with existing bubbles of free gas. In the event that such gaseous phases remain trapped inside the mud, the mechanics of mud up-rise will be significantly affected.

Due to the very low permeability of mud (10^{-16}-10^{-20} m^2 by following Brown, 1990), the present gaseous phases cannot escape by percolation. Sometimes high permeability scaly fabrics can develop in the marginal shear zone of the conduit due to the peculiar un-loading path (Brown, 1990). In these cases, gas can escape from the mud towards the surface by actually decoupling the mechanical behaviour of liquid and gaseous phases.

When this particular situation does not occur, gas escape from up-rising mud could occur due to fast up-rise of gas bubbles driven by their high buoyancy at a velocity higher than that of the mud. For low Reynolds numbers (typical for the considered situation), the kinematics of bubbles (in the assumption of a spherical shape) is controlled by the average radius of bubbles and by the mud viscosity. In particular, it holds that the velocity u_{bubble} of a bubble of radius r driven by buoyancy with respect to the moving fluid of density ρ_{mud} is

$$u_{bubble} = \frac{2(\rho_{mud} - \rho_{gas})gr^2}{9\mu} \cong \frac{2\rho_{mud}gr^2}{9\mu} \qquad (7)$$

where g is the gravitational acceleration and μ is the fluid viscosity (e.g. Brennen, 1995; Turcotte and Schubert, 2002). In order to evaluate the actual possibility that bubbling may significantly produce gas escape from the mud, the ratio between average mud velocity and bubble up-rise velocity can be computed from (2, 4 and 7) as

$$\frac{\bar{u}}{u_{bubble}} = -\frac{9}{64}\left(\frac{\rho_{ovb}}{\rho_{mud}} - 1\right)\frac{D^2}{r^2} \qquad (8)$$

It results that, for maximum bubbles radius r to the order of 0.1 m and by using the figures adopted above for the relevant parameters, the ratio in equation (7) is to the order of one and rapidly increases by decreasing r. This implies that only relatively big bubbles can escape from the mud while smaller ones remain trapped in the mud and participate in the mud flow. This implies that, at least as concerns the shallowest part of the mud path from the reservoir to the surface, the assumption of fluid incompressibility cannot be considered as fully realistic.

Gas expansion will result in a substantial increase of mud porosity and a decrease in density (Brown, 1990). Since this density variation affects the driving pressure gradient (see equation (2)), the mud out-flow could result faster than the one deduced above.

If the expansion is constrained (e.g. due to the obstruction of the conduit by solidified mud), the fluid pressure will remain at high levels, which must be limited by the strength of the surrounding rocks. In this situation, the pressure gradient can gradually decrease as far as producing a significant slow-down in the mud up-rise. However, even in this case, bubbles continue to move upwards due to their high buoyancy. This produces a gradual accumulation of gaseous phases in the upper part of the conduit. When pressure overcomes the strength of the rocks obstructing the conduit, a typical gas-dominated eruption will occur. Actually, in many situations the volume of released gases during such eruptions may be enormous in relation to the mud component (e.g. Kopf, 2002).

In the case of free outflow, the gas expansion will increase the compressibility of the up-rising fluid. This could modify the dynamic response of the system to pore pressure variations in the reservoir. Modelling the mechanical behaviour of such a high viscosity/compressible fluid requires somewhat complex numerical modelling which, to my knowledge, has not yet been performed. In any case, although the mechanical response of such a dynamic system remains an open problem, it is reasonable to expect that in this situation additional hydraulic loads induced at depth by pore pressure variations in the reservoir will, at least partially, propagate along the conduit in the form of compressional waves. As a first order approximation, propagation of such waves will be characterised by velocities to the order of a few hundred to thousand m/sec (p-wave velocity is 480 m/sec and 1400 m/sec in air and water, respectively).

Thus, both in the case that gaseous phases are actually present (compressible fluid) or nearly absent (incompressible fluid), the response of the system can be considered (at least partially) quasi-static in response to strain perturbations characterised by relatively slow rise times (to the order of days or larger), such as those associated to post-seismic strain diffusion phenomena. This implies that the flow rate relative variations δQ and pore pressure perturbations δP can be linearly related

$$\frac{\delta Q}{Q} \approx \frac{\delta P}{\Delta P} \approx \frac{\theta \delta \varepsilon}{\Delta P} = 10^4 \delta \varepsilon \tag{9}$$

from (1) and assuming ΔP and θ of the order of few MPa and 10 Gpa, respectively. These figures imply that a strain variation to the order of 10 µstrain will result in a 10% variation in the emission flow rate. As far as gas and fluid phases are at least partially mechanically coupled, such a relative flow rate fluctuation can be detected both in the fluid and the liquid phase. At least when major mud volcanic structures are of concern, such variations should be easily detectable.

4. DISCUSSION AND CONCLUSIONS

Fluctuations in the crustal strain field can significantly affect ongoing seismogenic processes. Thus, monitoring of such fluctuations can be used to r constrainseismic hazard evaluations better. Physical considerations suggest that mud volcanoes could be sensitive to volumetric strain fluctuations at the reservoir. Thus, monitoring their activity can be considered as a promising new tool for the detection of volumetric strain fluctuations.

While long-term variations in the fluid discharge of mud volcanoes has been widely explored (see, for example, Carson and Screaton, 1998 and references therein) medium-short term fluctuations have been less extensively analysed. Despite the fact that a number of questions remain open in terms of the actual dynamic response of mud volcano activity to pore pressure variations in the reservoir when free gases are significantly present in the conduit, the considerations reported in this paper and some very preliminary experimental results (see above) seem to encourage experimental and theoretical studies in this direction. In particular, some indications useful to addressing the future experimental activity can be obtained.

In most cases, due to the mechanical coupling of liquid and gaseous phases, monitoring could regard relative variations in the flow rates of gaseous or liquid emissions both being representative of pore pressure variations at depth. Continuously emitting structures should be preferred in the monitoring since eruptive activity can be conditioned by a number of factors not necessarily dependant on pressure variations at depth (e.g. the strength of structures obstructing the conduit). Of course, the monitoring of any mud volcanic activity will result easier and effective where distributed expressions of sub-aerial mud volcanism exist (as in the case of Azerbaijan).

Strain field variations induced post-seismic stress redistribution will involve wide areas (tens to hundreds of km) and develop over relatively long timescales (days to years). Thus, extensive ground-based monitoring of mud volcanoes would be characterised by discouragingly high costs. Remote sensing procedures could represent a valid cost-effective alternative to direct ground observations. Tramutoli et al. (2001) have recently proposed an interesting approach, which can be effectively applied to the monitoring of mud volcanoes. This approach can be used to detect relatively small variations in greenhouse gas emissions (methane, carbon dioxide, etc.) from underground towards the atmosphere, which can be detected in the form of apparent ground temperature anomalies measured by satellite images in the thermal infrared radiances. Due to the relatively large space-scale, characterising the phenomena under investigation , the relatively low-resolving power of such an approach in the space-scale will not represent an obstacle to affective application. Furthermore, aerial averaging of emission anomalies could make observations less sensitive to "spurious" local anomalies due to the variations in the partitioning of fluid and gaseous emissions between secondary structures (gryphons) associated to the same mud volcano. Some encouraging results produced by this approach for the detection of earthquake-related gaseous ground emissions have been recently obtained in Southern Italy (Tramutoli et al. 2001).

ACKNOWLEDGEMENTS

Many thanks are due to Dr. Martinelli and Prof. Panahi for their efforts in the organization of the meeting and in making it so enjoyable and stimulating.

REFERENCES

1. Albarello, D. Bonafede, M., 1990. Stress diffusion across laterally heterogeneous plates. Tectonophysics, **179**, 121-130.
2. Albarello, D., Martinelli, G., 1994. Piezometric levels as a possible geodynamic indicator: analysis of the data from a regional deep waters monitoring network in Northern Italy. Geophys. Res. Lett., **21**, 1955-1958.
3. Albarello, D., Lapenna, V., Martinelli, G. and Telesca, L., 2003. Extracting quantitative dynamics from 222Rn gaseous emissions of mud volcanoes. Environmetrics, **14**, 63-71.
4. Anderson, D., 1975. Accelerated plate tectonics. Science, **17**, 1077-1079.
5. Batchelor, G.K., 1967. Fluid dynamics: an introduction. University Press, Cambridge, 615 pp.
6. Belardinelli, M.E., Cocco, M., Coutant, O., Cotton, F., 1999. Redistribution of dynamic stress durino coseismic ruptures: evidence for fault interaction and earthquake triggering. J.Geophys. Res., **104**, B7, 14925-14945.
7. Bodvarsson, G., 1970. Confined fluids as strain meters. J.Geophys.Res., 75, 14, 2711-2718.
8. Brennen, C.E., 1995. Cavitation and bubble dynamics. University Press, Cambridge.
9. Brown, K.M., 1990. The nature and significance of mud diapirs and diatremes for accretionary systems, J.Geophys.Res., **95**, 8969-8982.
10. Carson, B., Screaton , E.J., 1998. Fluid flow in accretionary prisms: evidence for focused, time-variable discharge. Rev. Geophys., **36**, 3, 329-351.
11. Castellaro, S., Mulargia, F., 2001. A simple but effective cellular automation for earthquakes. Geophys. J. Int., **144**, 609-624.
12. Cornell, C.A. 1968. Engineering seismic risk analysis, Bull. Seism. Soc. Am., **58**, 1583-1606.
13. Giardini, D., Basham, P. (eds.) "Global seismic hazard assessment program", ILP publication 209, Ann.Geofis., **36**, 3-4, 181-200.
14. Harris, R.A., 1998. Introduction to special section: stress triggers, stress shadows, and implications for seismic hazard. J. Geophys. Res., **103**, B10, 24347-24358.
15. Kagan, Y.Y., 1992. Seismicity: turbulence of solids. Nonlin. Sci. Today, **2**, 1, 2-13.
16. Kagan, Y.Y., 1994. Observational evidence for earthquakes as a non linear dynamic process. Physica D, **77**, 160-192.

17. Kasahara, K., 1979. Migration of crustal deformation. Tectonophys., 52, 329-341.
18. King, G.C.P., Stein, R.S., Lin, J., 1994. Static stress change and the triggering of earthquakes. Bull.Seism.Soc.Am., **84**, 3 , 935-953.
19. Kopf , A., 2002. Significance of mud volcanism. Rev.Geophys., **40**, 2, 1-52.
20. Kümpel, H.-J., 1992. About the potential of wells to reflect stress variations within inhomogeneous crust. Tectonophys., **211**, 317-336.
21. Main, I., 1995. Earthquake as critical phenomena: implications for probabilistic seismic hazard analysis. Bull.Seism.Soc.Am., **85**, 5, 1299-1308.
22. Main, I., 1996. Statistical physics, seismogenesis and seismic hazard. Rev.Geophys., **34,** 4, 433-462.
23. Mantovani, E., Albarello, D., 1997. Middle-term precursors of strong earthquakes in southern Italy. Phys. Earth Planet. Int., **101**, 49-60.
24. Martinelli, G., Albarello, D. , Mucciarelli , M., 1995 - Radon emission from mud volcanoes in Northern Italy: possible connection with local seismicity. Geophys. Res. Lett., **22**, 15, 1989-1992.
25. McGuire, R.K., 1993a. Computations of seismic hazard. In Giardini , D. and Basham, P. (eds.) "Global seismic hazard assessment program", ILP publication 209, Ann.Geofis., **36,** 3-4, 181-200.
26. McGuire, R.K. (Ed.), 1993b, The practice of earthquake hazard assessment, IASPEI, Denver, 1-284.
27. Muir-Wood, R. and King, G.C.P., 1993. Hydrologic signatures of earthquake strain. J.Geophys. Res., **98**, B12, 22035-22068.
28. Okada, Y., 1992. Internal deformation due to shear and tensile faults in a half-space. Bull.Seism.Soc.Am., **78**, 1907-1929.
29. Piersanti, A., Spada, G. , Sabadini , R., 1997. Global post-seismic rebound of a viscoelastic earth: theory for finite faults and application to the 1964 Alaska earthquake. J.Geophys. Res., **102**, 477-492.
30. Pollitz, F.F., Peltzer, G., Burgmann, R., 2000. Mobility of continental mantle: evidence form post-seismic geodetic observations following the 1992 Landers earthquake. J. Geophys. Res., **105**, 8035-8054.
31. Ranalli, G., 1995. Rheology of the Earth. Chapman & Hall, London, 413 pp.
32. Rydelek, P.A., Sacks, I.S., 1990. Asthenospehric viscosity and stress diffusion: a mechanism to explain correlated earthquakes and surface deformation in NE Japan. Geophys .J.Int., **100**, 39-58.
33. Rydelek, P.A., Sacks, I.S., 1999. Large earthquake occurrence affected by small stress changes. Bull.Seism.Soc.Am., **89**, 822-828.
34. Roeloffs, E., 1996. Poroelastic techniques in the study of Earthquake related phenomena. Dmowska R. and Saltzman G. (eds.), Advances in Geophysics, **37**, 135-195.
35. Rossi, G., Zadro, M., 1996. Long-term crustal deformations in NE Italy revealed by tilt-strain gauges. Phys. Earth Planet. Int., **97**, 55-70.
36. Sacks, I.S., Rydelek, P.A., 1995. Earthquake "quanta" as an explanation of observed magnitudes and stress drops. Bull.Seism.Soc.Am., **85**, 3, 808-813.
37. Stein, R.S., Barka, A.A., Dietrich, J.H., 1997. Progressive failure on the North Anatolian fault since 1939 by earthquake stress triggering. Geophys. J.Int., **128**, 594-604.
38. Tamrazyan, G.P., 1982. Peculiarities in the manifestation of gaseous-mud volcanoes. Nature, **240**, 406-408.
39. Tramutoli ,V., Di Bello , G., Pergola, N., Piscitelli ,S., 2001. Robust satellite techniques for remote sensing of seismically active areas. Ann. of Geophys., **44**, 2, 295-312.
40. Turcotte, D.L., 1992. Fractals and Chaos in Geology and Geophysics. University Press, Cambridge, 398 pp.

41. Turcotte, D.L., Schubert, G., 2002. Geodynamics, 2nd edition, University Press, Cambridge, 456 pp.
42. Viti, M., D'Onza, F., Mantovani, E., Albarello, D. , Cenni, N., 2003. Post-seismic relaxation and earthquake triggering in the Southern Adriatic region. Geophys .J. Int., **153**, 645-657.
43. Wang, H.F., 2000. Theory of linear poroelasticity. Princeton Univ.Press,, Princeton, 287 pp.
44. Wakita, H., Nakamura, Y. , Sano, Y., 1988. Short term and intermediate term geochemical precursors. Pure Appl.Geophys., **126**, 2657-278.
45. Zadro, M., Rossi, G., 1991. Long term strain variations in Friuli (NE Italy) seismic area. Proc. of the Intern. Conf. "Earthquake prediction: state–of-the art", Strasbourg, 435-441.

MUD VOLCANO MODEL RESULTING FROM GEOPHYSICAL AND GEOCHEMICAL RESEARCH

Akper A. Feyzullayev, Fakhretdin A. Kadirov, Chingiz S. Aliyev
Geology Institute of Azerbaijan National Academy of Sciences, H. Javid Prospekt, 29A, 370143 Baku, Azerbaijan, e-mail: akperf@yahoo.com

Abstract: Measurements have been made of the gravitational field, the geodetic uplift together with a radioactive survey in the general area of the Lokbatan mud volcano after its explosion on 25 October 2001, as well as in the crater itself. In addition, geochemical measurements of vitrinite reflectance with depth and isotopic composition of carbon in methane and ethane have been performed. This massive compendium of information represents the most detailed investigation ever into the deep structural effects of a mud volcano and also of the mud and gas sources at outflow time. The data have been inte-grated into a combined picture showing the roots of both the mud outflow and the gas causing the flaming eruption, several km deep inside the sedimentary pile. The overall behavior is best served by a model in which a relatively thin jet of liquefied mud is extruded from depth following the action of the varying tectonic stresses in the region, as adduced from the GPS tectonic movement data [6]. The variation of Bouguer gravity across a profile including the Lokbatan mud volcano, and combined with the geodetic vertical motion straight after and a long time after (10 months) the explosion, confirms this basic model. The geochemical evidence, showing a low vitrinite maturity (<0.6%) down to a depth of around 6 km, indicates the origin of gas from greater depths, as do the isotopic carbon measurements of methane and ethane. To conclude, it would seem that the tectonic "squeezing" of a low strength plastic mud layer from depth through a narrow vent with entrained gas and mud is the primary driver for mud volcano explosions.

Key words: Model, Mud Volcanoes, Azerbaijan.

1. INTRODUCTION

In the South Caspian Basin, the general region of onshore and offshore Azerbaijan is home to over 300 mud diapirs and/or mud volcanoes. These mud structures are associated with a copious production of oil and gas. This association is not merely a coincidence but is related to the dynamic development of mud diapers; the generation, migration, and accumulation of hydrocarbons in the South Caspian Basin has been detailed elsewhere (Dadashev, 1963; Gubkin and Fedorov, 1938; Lerche and Bagirov, 1999; Rakhmanov, 1987). The genetic relationship between mud volcanoes and oil and gas formation has been confirmed by oil field investigations, in particular

G. Martinelli and B. Panahi (eds.), Mud Volcanoes, Geodynamics and Seismicity, 251–262

when a well drilled at the cover of breccia 1.5 km from the Lokbatan mud volcano crater in 1933 yielded an oil gusher with a production rate of up to 20,000 t/day. Since then, oil geologists have considered that mud volcanism manifestations are a direct indication of the presence of oil and gas at depth (Peive, 1956). In addition, by implication the earlier hypothesis that mud-volcanic processes are evidence for the last stages in the destruction of oil and gas deposits penetrated by mud volcanoes could be rejected. The volcano has erupted 23 times and is still active but has never disrupted oil production. However, problems so far not adequately studied are related to the formation mechanism of mud volcanoes, their role in the origin and/or accumulation of hydrocarbons, the role of deep faults, and the regime of fluid motion in volcanoes. Unfortunately, in spite of complex long-term studies of the Azerbaijan mud volcanoes, no special geophysical investigations have been undertaken. Only available are the results of geophysical studies obtained serendipitously whilst resolving other oil-field problems; these investigations are insufficient to compile a mud volcano model or to study the dynamics of its activity. Such a capability is possible only with special, very detailed geophysical and geodetic investigations. This paper contains the results of such investigations carried out for the first time into volcanoes in the southwest Absheron region (Lokbatan, Akhtarma-Puta and Gushkhana), all located within a single tectonic zone (Figure1).

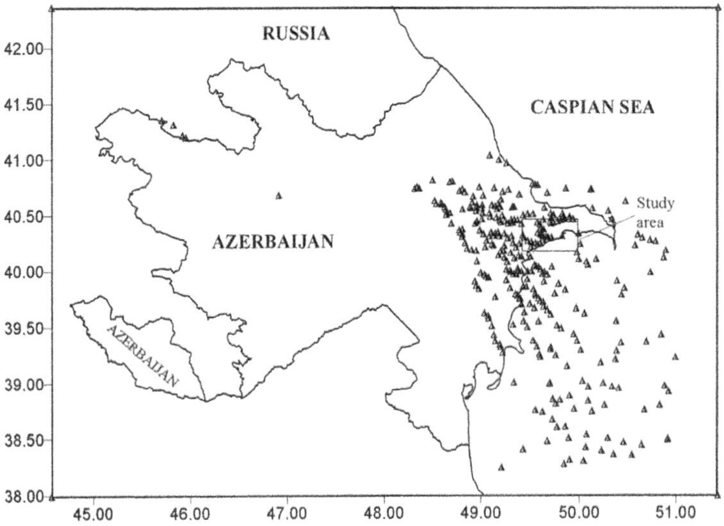

Figure 1. Distribution of mud volcanoes in Azerbaijan and the study area
The Lokbatan mud volcano was chosen because it is very active not only as compared with other volcanoes in Azerbaijan but as compared with all the mud volcanoes worldwide. Lokbatan has been recorded as having erupted 23 times over the last 200 years, assuming no other eruptions have been missed out.

The last eruption took place on October 25, 2001. The present study was started immediately after that eruption.

Together with the gravimetric and geodetic studies, radiometric and geochemical observations were also conducted. Profiles were chosen in different directions to carry out gravimetric and geodetic measurements on the mud volcano areas. One profile, more extended than the others (11 km long), begins near a key gravity-geodetic point at Lokbatan and passes through a line joining the mud volcanoes Lokbatan - Akhtarma - Puta - Gushkhana. The distances between the measurement points are never greater than 250m. To the south of these volcanoes GPS measurements indicate that there is a decreasing platform motion speed, reaching zero velocity in places. Observations of dynamic processes within and upon the volcanoes are currently continuing on a semiregular basis.

2. GEOLOGICAL AND GEOPHYSICAL SETTING

The areas around the Azerbaijan mud volcanoes are reflected by minima of the Bouguer gravity field (from –120 to –40 mGal). The GPS measurements show that the decrease to zero of the horizontal movement velocity vector takes place in the neighborhood of the mud volcanoes (Guliev et al., 2002; Kadirov, 2000).

The majority of the mud volcanoes are located in fracture (fault) zones (Gubkin and Fedorov, 1938; Kadirov, 2000; Peive, 1956; Rakhmanov, 1987). Mud volcanoes and hydrocarbon accumulations have been related to the largest faults considered as deep faults (Lerche and Bagirov, 1999). Such faults are recognized from geological, geomorphological and geophysical data.

3. METHODS AND RESULTS

3.1 Gravity

Measurements of gravity differences between points were undertaken by four gravimeters under a simple closed loop arrangement. The scale interval of the gravimeters was determined at different temperatures using a control instrument. When processing field measurements, corrections were made for relation of scale interval to temperature, for non-linearity of scale of a micrometer screw, and for lunar-solar attraction. The longitudes and latitudes of the measurement locations were determined with the help of GPS, and altitudes of the points with a level produced by the firm Carl Zeiss, Jena (the mean quadratic error on a 1 km traverse measured in both directions was 0.2 mm).

To calculate the Bouguer gravity anomaly, the altitude was read from the lowest level (at the Lokbatan reference point), and the interlayer density taken to be 1.82, 2.0, 2.3 g/cm^3.

Gravity modeling used a minimization condition on a multi-parameter function describing the least-squares difference between modeled and observed gravity fields and involving parameters of the initial structure model. The initial parameters of the model are modified such that the difference between observed and computed fields does not exceed 1mGal. Along the NE–SW profile gravity modeling was performed in order to investigate the depth structure and tectonic evolution of the mud volcanoes.

Along the profile Lokbatan – Akhtarma-Puta – Gushkhana, a chart of Bouguer gravity field anomalies is compiled for various values of the intermediate stratum density (1.8; 2.0; 2.3 g/cm^3). The results obtained show that in zones of mud volcano development (Lokbatan, Akhtarma-Puta, Gushkhana) there are local negative anomalies of -5, -3 and -2 mGal, respectively.

For gravity modeling, the observed gravity values and the geological/ geophysical structure section along the NE-SW profile make up the initial reference information. The initial structure model along the profile was obtained using seismic data, well in-formation, geological information, and the density-depth distribution of major rock units in the study area. The initial geologic-geophysical cross-section of the sedimentary cover along the profile is provided by seven contact boundaries:

a) boundary along the lower Akchagil ($N^2{}_2$);
b) 3 apparent seismic horizon in the Productive Series ($N^1{}_2$);
c) boundary separating the upper-middle Miocene and lower Miocene-Oligocene series ($P_3 - N^1{}_1$);
d) boundary separating Oligocene and Eocene-Paleocene series ($P_1 + P_2$);
e) boundary of the Mesozoic surface (M_z).

The differential density contrasts across the seven contact boundaries (from shallow to deep) are 0.01; 0.04; 0.08; - 0.2; 0.25; 0.15; and 0.3 g/cm^3, respectively. The gravity field calculated on the basis of the described model shows that there is a difference between observed and calculated fields in the mud volcano zones (Figure 2). When computing the gravity model, 10 iterations were first performed on the selection of all the boundary configurations. Then the selection on density was undertaken. The extra mass in zones of mud volcanoes in the initial model is compensated for by the insertion of deconsolidation zones and additional contact boundaries in these parts of the profile. The deconsolidation stretches to roughly 3 km depth. A contact boundary, representing the volcano neck (with a differential density - 0.3 g/cm³), in these areas is raised to depths of just 5 m below the surface.

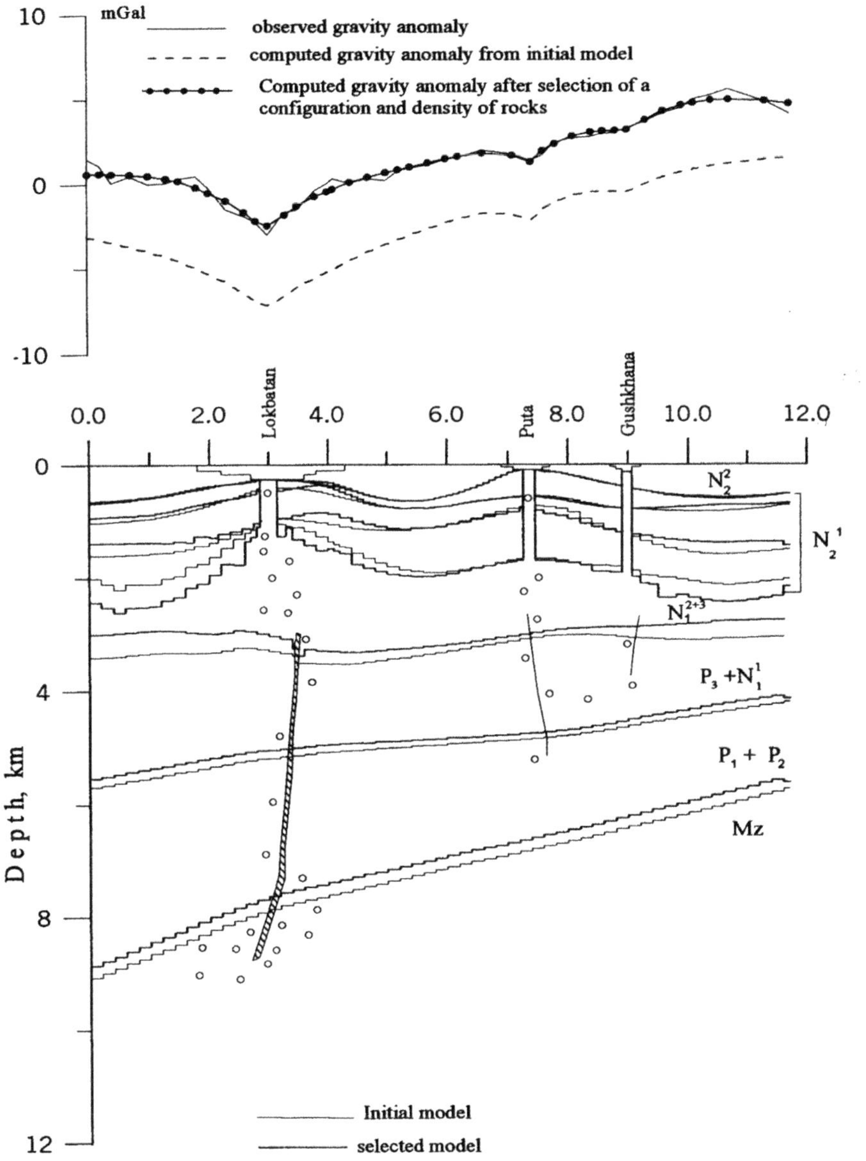

Figure 2. Gravity model of the Lokbatan - Puta-Akhtarma - Gushkhana profile

3.2 Geodetic results

GPS measurements indicate a decrease in the velocity and a significant accumulation of elastic energy in the southern Absheron Peninsula. This phenomenon may be responsible for the activation of seismic events and mud volcanoes in this region, because ten mud volcanoes erupted during the 1998–2002 period in the Absheron and Shamakhy–Gobustan areas. The strong earthquake (approximate magnitude 6-6.3 on the Richter scale) in the Caspian Sea at the end of 2000, and its aftershocks, probably represent a response to the deformational processes and the related stress accumulation, ongoing at the foothills of the Greater Caucasus, the Absheron Peninsula and the middle Caspian regions. However, the tendency to have horizontal motions in the Azerbaijan territory suggests an activation of geologic processes (seismic activity, activation of mud volcanoes and, in adjacent zones, accumulation of elastic stress).

The repeated geodetic leveling on the Lokbatan volcano shows that there are currently active geodynamic processes occurring there. During the period from November 2001 to November 2002, on the NE part of volcano (about 2 km in length), the contact boundary was as shallow as 60 cm (Figure 3).

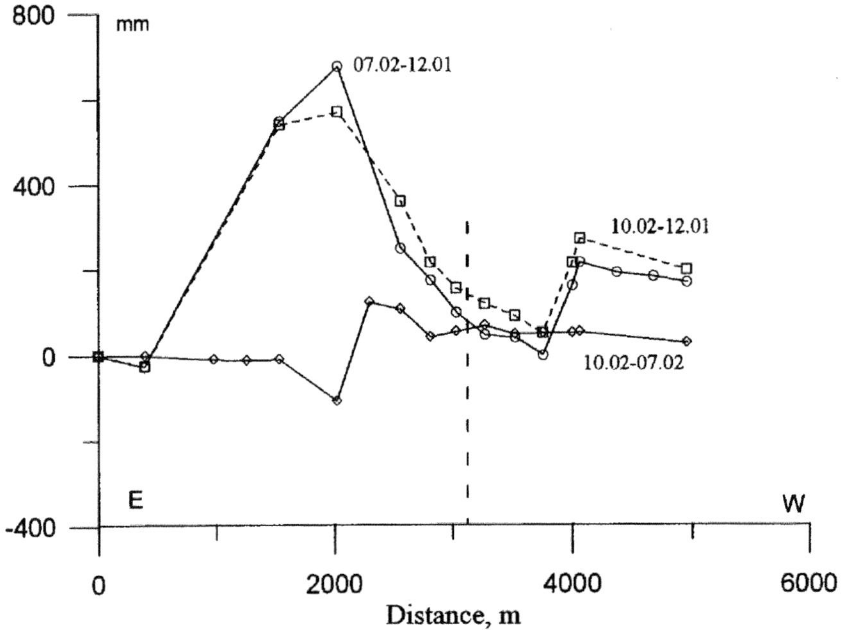

Figure 3. Vertical movement along a profile crossing the Lokbatan mud volcano

3.3 Radiometric results

In addition to the geodetic, gravity and thermal measurements, total γ-ray intensity was also measured with a probe on the surface across the Lokbatan mud volcano on profile about 5 km long. The Lokbatan mud volcano located at the center of the profile (shown by the dashed vertical line at 3 km on figure 4). Two days after the explosion on 25 October 2001, the γ-ray intensity in the crater was as high as 28 μR/h before returning in a few days to around 10-15μR/h. Apart from the isolated high intensity point at the 1.5 km marker on figure 4 (likely to be due to either to radioactive waste being deposited in an old well or an open fault bringing up radioactive material from depth; it is not known which), noteworthy is the large increase in total intensity in the neighborhood of the Lokbatan volcano, associated with the ejected mud, which was transported from depth.

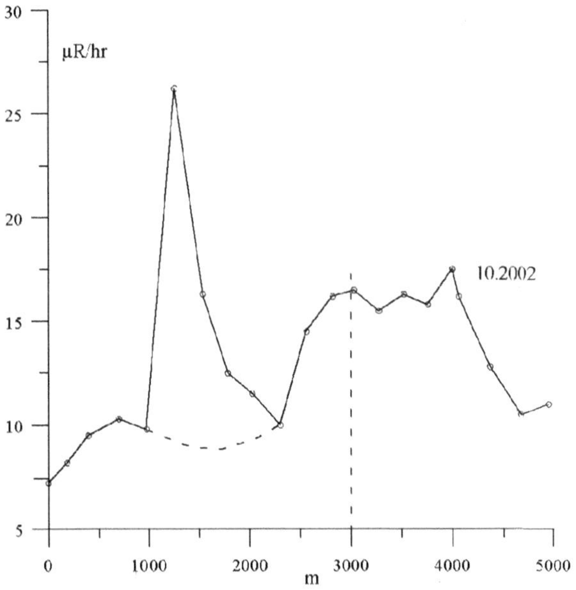

Figure 4. Change in the total γ-ray count (in units of μR/h) along a profile crossing the Lokbatan mud volcano

3.4 Geochemical results

Natural gases play a key role in the energetics of the mud volcanic process. The source of these huge gas volumes is generally considered to be from the sedimentary cover (Dadashev, 1963), but other points of view would have the origin to be far deeper (Valyaev, 1985). There are also different opinions as regards the problem of stratigraphic occurrence of the gas source. One of the first attempts to reach a decision concerning this problem was made by Dadashev (1963), who linked mud volcanoegases to Paleogene deposits on the basis of the methane-to-homologues ratio.

Based on hydrocarbon composition, mud volcano gases belong to the group of “dry” gases (<2% C_2H_6 +). Isotopic composition data of methane carbon shows that it be-longs to the catagenic gases (average isotopic composition of CH4 lying between - 40 to - 50 ‰). Studies of inert gases (He, Ar) and their isotopic ratios show that mud volcano gases are similar to the gases of oil and gas fields, although the stratigraphic sources of the mud volcano gases are older [4]. In the last few decades there have been theoretical investigations that allowing us to determine the dependence between isotopic composition of methane carbon (ICC), its gaseous homologues and vitrinite reflectance (R0). For instance, Stahl notes that (Stahl, 1977):

$$\delta^{13}C\ (CH_4)\ (‰) = 17 \log R_0\ (\%) - 42 \tag{1}$$

and

$$\delta^{13}C\ (C_2H_6)\ (‰) = 22.6 \log R_0\ (\%) - 32.2 \tag{2}$$

Basing ourselves on these theoretical relationships we can estimate the approximate depths from which the hydrocarbon gas components migrate (methane, ethane, propane, and butane). Knowledge of the approximate hypsometric depth of the gas source allows one to estimate the nominal stratigraphic source of the gas (by use of geological profiles).

R_0 values to a depth of 6 km vary in the range 0.33-0.61%; for larger depths R_0 values were defined by extrapolation (Figure 5).

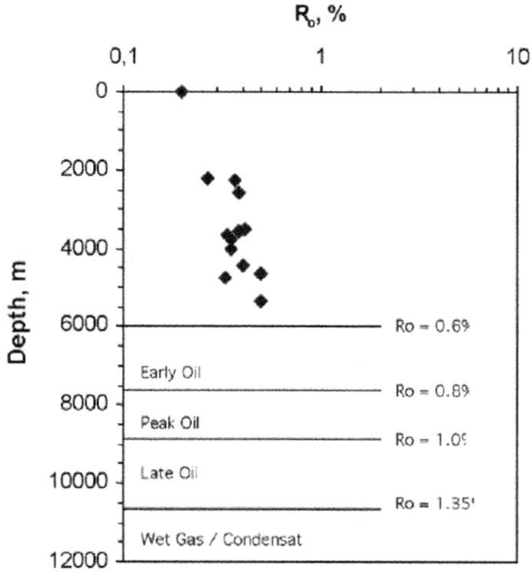

Figure 5. Vitrinite measurements with depth through the Lokbatan volcano and extrapolated trend curve based on the ICC isotopic connection given in text.

Vitrinite reflectance analyses were performed on randomly oriented particles, using a conventional microphotometric method (Stach et al., 1982).

Use of large amounts of available methane ICC data (Dadashev, 1985; Stahl, 1977) for such estimations has not generally proven fruitful because of the mixing of catagenic (relatively heavy ICC) and biochemical (relatively light ICC) methane that distorts the virgin values.

Measurements of gas maturity have been conducted for the mud volcanoes Lokbatan, Puta-Akhtarma and Gushkhana. The measurements have also been conducted for adjacent mud volcanoes with the aim of constructing maps for the changes to this parameter within the area. The stable isotopic composition of carbon in hydrocarbon gases was analyzed using CJS Sigma mass spectrometers.

Ethane isotopic carbon composition (ICC) of these gases varies within -26.2 to -29.4‰. The change in R_0 within the area, calculated according to the above equations connect-ing reflectance and ICC, provides an interesting picture (Figure 6).

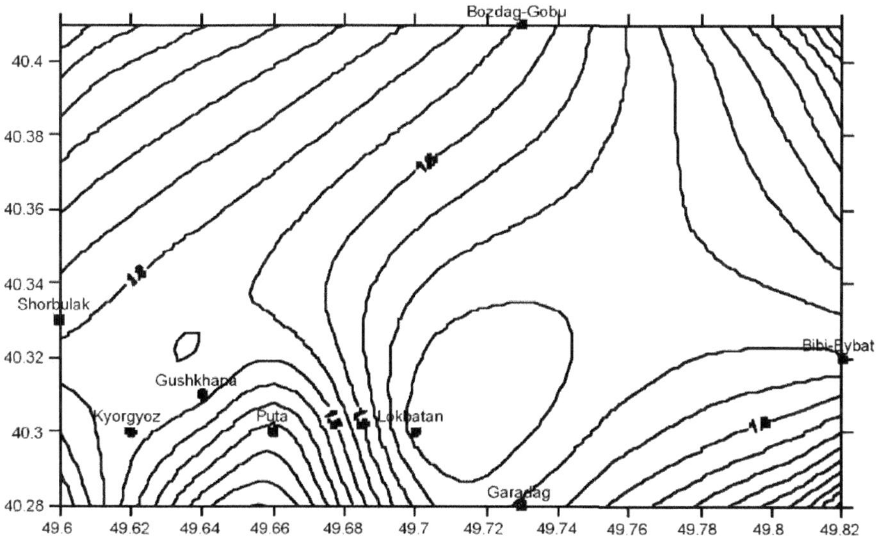

Figure 6. Two-dimensional contours of maturity across the general region encompassing the mud volcanoes

Two neighboring (3-4 km apart) mud volcanoes, Lokbatan and Puta-Akhtarma, are characterized by sharply different values of maturity; maximal on the Lokbatan volcano (R_0 = 1.7%) and minimal on the Puta-Akhtarama volcano (R_0 = 1.35%). Thus, if the deepest gas focus is at volcano Lokbatan (approximately at depth 11.5 km), then the depth of gas focus is minimal on volcano Akhtarama-Puta (approximately 10 km). On the whole, however, there is a tendency (from NE to SW) for a maturity reduction and a corresponding reduction in the depth of the gas focus.

The high activity of the Lokbatan mud volcano can be explained by a larger deep focus for the gases. This explanation presupposes, however, that a rather large interval of sedimentary section participates in the process of gas formation (i.e. the volume of gas generation is maximal and the energy resources of gas here are also relatively large). Support for this explanation comes from the large volumes of gas ejected during eruptions (more than 100 MMm^3), very high frequency of eruptions (approximately every 3-4 years), and the long period (in the order of a year) of gas burning after recent eruptions.

4. DISCUSSION

Upward transportation of hydrocarbons, together with the enclosing clayey plastic mass of the intermediate layer, under convective processes would appear to be a dominant mechanism of migration and accumulation in the upper parts of the series, with a further breakthrough in the overlying permeable series (Guliev and Kadirov, 2000; Kadirov and Kadyrov, 1990). As the clay is raised, the lithostatic pressure lessens, with the result that hydrocarbon phase transitions will occur and hydrocarbons will be in free phase. It is quite likely that the interstitial layer is the main generator of most of the mud volcanoes in Azerbaijan (Guliev and Kadirov, 2000). From December 2001 through June 2002, the following sequence of events took place. The surface rose in and around the Lokbatan volcano. However, the amplitude of rise varied strongly along the structure. To the east, measured from the neck of the volcano, out to a distance of about 1000 m, the surface rise was 677 mm. A secondary maximum of 218 mm was observed on the western part to a distance of about 1000 m from the neck. The area of the crater also participated in the general rise, but lagged appreciably behind the rise of the regional sites. Geodetic observations from December 2001 through October 2002 show that, in the period June 2002 - October 2002, to the east away from the crater there was a relative lowering of the surface, while to the west and also in the crater structure, the surface showed a relative rise. Comparison of the change of amplitude of surface movement with the change of radioactivity along the profile, indicated increased radioactivity in the crater and to the western side, where the differential uplift was largest. An interpretation for this correlation is the enhanced flow of liquid material as a result of microfracture growth in and around the site of the largest uplift gradients.

5. CONCLUSIONS

The results show that there are negative local gravity anomalies (-5, -3 and -2 mGal), respectively, in mud volcano zones. The calculated field based on the model described shows that in mud volcano zones there is a difference between observed and calculated fields. The extra mass in mud volcano zones is compensated for by introduction of de-compaction zones and additional contact boundaries.

Precisely how this continued gas supply is related to geodynamic models of mud flow from depth (6 km?) in the Lokbatan mud volcano, and precisely how the dynamics and recharging of the gas and mud flow operate in a roughly cyclic manner, requires further and more detailed observations than are currently available. Isotope gas composition shows that gas of the Lokbatan mud volcano has very deep roots as compared to adjacent volcanoes; such deep roots evidently account for its higher activity. Focuses of mud and gas flow formations are different, with the mud focus being located at a relatively shallow depth as compared to that for gas.

ACKNOWLEDGEMENTS

The authors are very grateful to the US CRDF for financial support under grant NG2-2285 that enabled this work to come to fruition.

REFERENCES

1. Dadashev, A. A., 1985. *Peculiarities of carbon isotope composition of hydrocarbons on the western flank of the South Caspian Basin.* Abstract of Ph. D. dissertation, VNIIYAG Moscow
2. Dadashev, F. G., 1963. *Hydrocarbon gases of Azerbaijan mud volcanoes.* Azerneshr Baku., (in Russian).
3. Gubkin, I. M. and Fedorov, S. F., 1938. *Gryazevye vulkany Sovetskogo Soyuza i ikh svyaz's genezisom neftyanykh mestorozhdenii Krymsko-Kavkazskoi geologicheskoi provintsii* (Mud Volcanoes of the Soviet Union and Their Implications for the Genesis of Oil Fields in the Crimea-Caucasus Geological Province), AN SSSR , Moscow.
4. Guliev, I., Ismet, A.R., and Djafarova, P.S. , 1996. *Inert components in gases of deposits of Guneshly,* Azeri, and Chirag. Azerbaijan Oil Industry, No 9, 7-11.
5. Guliev, I. S. and Kadirov, F. A., 2000. A mechanism of intrastratal migration of hydrocarbons, *Transactions (Doklady) of the Russian Academy of Sciences/Earth Science Section,* **373**, No. 6, August September, 941-944.
6. Guliev, I. S., Kadirov, F. A., Reilinger, R. E., Gasanov, R. I., and Mamedov, A. R. , 2002. Active Tectonics in Azerbaijan Based on Geodetic, Gravimetric, and Seismic Data, *Transactions (Doklady) of the Russian Academy of Sciences/Earth Science Section.* **383**, No. 2, 174-177.

7. Kadirov, F.A., and Kadyrov, A.H., 1990. *The possibility of the thermal convection in the sedimentary. Layers of Azerbaijan*, Revue Academy of Sciences of Azerbaijan, Earth Sciences. Publishing House ELM, Baku (in Russian).
8. Kadirov, F.A., 2000. *Gravity field and models of deep structure of Azerbaijan.*, Publishers 'Nafta-Press', Baku, (in Russian).
9. Lerche, I., and Bagirov, E., 1999. *Impact of Natural Hazards on Oil and Gas Extraction* , Kluwer Academic/Plenum Publishers, New York.
10. Peive, A.V. , 1956. Sedimentation, Folding, Magmatism, and Mineral Deposits in Relation to Deep faults, *Izv. Akad. Nauk SSSR, Ser. Geol.*, No. 3, 57-71.
11. Rakhmanov, R.R., 1987. *Mud Volcanoes and Their Implications for the Oil and Gas Content of Deep Formations*, Nedra., Moscow.
12. Stach, E., Mackowsky, M.T., Taylor, G.H., Chandra, D., Teichmuller M., and Teichmuller, R., 1982. *Coal Petrology*, 2nd Edition, Elsevier, Berlin.
13. Stahl, W. J., 1977. Carbon and nitrogen isotopes in hydrocarbon research and exploration., *Chem. Geol.* **20**, 121-149.
14. Valyaev, B. M., 1985. *Isotope image of the mud volcano gases.*, Litologiya i Poleznyie Iskopaiemyie, 1, 72-86.

PHYSICAL PROPERTIES OF MUDS EXTRUDED FROM MUD VOLCANOES: IMPLICATIONS FOR EPISODICITY OF ERUPTIONS AND RELATIONSHIP TO SEISMICITY

Achim J. Kopf[1], M. Ben Clennell[2] Kevin M. Brown[3]

1 RCOM, University Bremen, Klagenfurter Strasse, 28359 Bremen, Germany, e-mail: akopf@uni-bremen.de

2 CSIRO Petroleum, Australian Resources Research Centre, 26 Dick Perry Avenue, Kensington, WA 6151, Australia, e-mail: Ben.Clennell@csiro.au

3 SCRIPPS Institution of Oceanography, UCSD 9500 Gilman Drive, La Jolla CA 92093-0244, USA, e-mail: kmbrown@ucsd.edu

Abstract: Scientific drilling into submarine mud volcanoes on the Mediterranean Ridge accretionary complex has documented episodic eruptive activity over the last 1 to >1.5 million years. Mud extrusion is related to plate convergence between Africa and Eurasia that caused backthrust faulting of accreted strata over the seismically active, rigid backstop of Crete (Greece). The domes consist of mud breccia with up to 65% of polymictic clasts embedded in a clayey matrix dominated by kaolinite, smectite and hallyosite. Laboratory measurements of viscosity, permeability and frictional strength of the clay-rich mud from Napoli Dome shed light on the extrusion dynamics and its relationship to seismicity. Viscosities of 10^6 Pa·s lead to predictions of ascent velocities up to 60-300 km/a based on Poiseuille's flow law. Frictional shear strength and permeability were found to have very low values. Friction coefficients (μ) determined during ring shear and direct shear tests are below 0.26. These results point to velocity-strengthening behaviour of both the mud volcano clay and reference mineral standards of smectite, illite, and kaolinite. Permeability of deformed clay-rich matrix measured using a ring shear permeameter, is less than 10^{-19} m^2 at ~1 MPa normal stress. We propose that the low permeability and strength observed during our tests have two important geological implications. First, these properties allow pore pressure build-up at depth, especially within poorly drained fault zones, accretionary prisms, and mud reservoirs. Fault movement is facilitated by the low intrinsic strength and reduced effective stress of material in the fault zones while the elevated porosity, low viscosity and high internal pressure of the mud promotes subsurface mobilization, leading utimatelylink between seismicity and mud volcanism since the mud and clay reference standards tested all underwent stable sliding when sheared under fixed load-point velocity or stress. We believe that seismogenesis occurs at deeper levels than mud mobilization, but still within a kinematically-linked (and perhaps hydraulically-linked) fault system. Increased mud volcano activity may thereby serve as an earthquake precursor, since seismic faulting at depth may cause stress state perturbations along the fault, which in turn may trigger liquefaction, excess pore pressure transients, and ascent/extrusion.

G. Martinelli and B. Panahi (eds.), Mud Volcanoes, Geodynamics and Seismicity, 263–283

Key words: Mud volcano, clay, friction, permeability, shear strength, seismicity, Mediterranean Ridge

1. INTRODUCTION

Mud volcanism and diapirism are well-known phenomena which occur predominantly in collisional tectonic areas (e.g. Barber and Brown, 1986; Brown and Westbrook, 1988; Kopf, 2002). The presence of mud domes and ridges along parts of the northern flank of the Mediterranean Ridge accretionary complex is related to its overall collisional tectonics between Africa and Eurasia (Camerlenghi et al., 1995). Geophysical data suggest that individual mud domes vary greatly in size (Camerlenghi et al., 1995; Kopf et al, 2001). During Ocean Drilling Program (ODP) Leg 160, a transect of five deep holes into Napoli mud dome gave insights into mud volcano formation (Robertson et al., 1996). The mechanism and evolution seem strongly dependent on the consolidation and gas content of finegrained sediments at depth. A model of evolution of the state of consolidation of such sediments following a specific stress path, that eventually leads to the formation of mud volcanoes and sedimentary diatremes following gas expansion in the pore spaces, has been proposed by Brown (1990). In this study, the findings of the ODP drilling campaign (Emeis et al.,1996) together with analyses of facies, clast composition, and physical parameters like viscosity, particle size, strength and permeability on the material recovered (Kopf et al., 1998) were combined with new experimental data from geotechnical tests on mud from Napoli mud volcano. The main objectives included (i) the characterization of permeability and pore pressure evolution during shear of clays and muds, (ii) the frictional stability of such clays, and (iii) the implications of the results for the possible causal relationship between eruptive activity and seismicity in fault-related mud extrusion features.

2. GEOLOGICAL SETTING

The collision zone between the African and Eurasian plates is characterized by NNE-SSW-oriented shortening and consumption of Mediterranean seafloor beneath Eurasia (Fig. 1a). The >200 km wide and 1500 km long wedge consists of Cretaceous (possibly earlier) to Pleistocene sediments (Fusi and Kenyon, 1996), including (regionally restricted) Miocene evaporites (Chaumillon and Mascle, 1997), evidence for which comes from brines and halite clasts (Emeis et al., 1996). There are three provinces of mud extrusion features along the Eastern Mediterranean Ridge (Huguen et al., 2001; Kopf et al., 2001). First, there is a belt of small (≤ 1 km diameter) domes near the toe of the large, arcuate accretionary wedge. Second, there is a series of large (up to >35 km wide), flat mud pies. Third, a series of mid-sized (up to a few km in diameter) mud domes is found some 150 km behind the toe in the area of the backstop to the wedge. The Olimpi mud volcano field (Fig. 1b) is a cluster of mud domes in this backstop area, due south of central Crete. The target feature for our study is Napoli mud dome, which is located in the eastern part of the Olimpi field. Initial site surveys during ODP Leg 160 (Emeis et al., 1996) revealed the flat-topped, slightly asymmetric nature of the volcanoes with depressions (a shallow moat as well as inward-dipping reflectors underlying the flanks; cf. seismic section in Fig. 1c), which have also been documented from mud volcanoes elsewhere (Barber and Brown, 1988). Napoli dome has been shown to be actively extruding mud and pore fluids (Aloisi et al., 2000) and hence is highly suitable for this study, which aims to relate dynamic process to physical properties of the sediments.

3. GEOTECHNICAL TESTING OF MINERALS AND ROCKS AND THEIR IMPLICATIONS FOR MUD VOLCANOES

Only a limited number of geotechnical tests have been reported on mud volcano materials (e.g. Yassir, 1989). However, since the muds are reworked, general principles can be adopted from soil mechanics of unstructured soils (e.g. Lambe and Whitman, 1969; Lupini et al., 1981; Skempton, 1985). Conventional soil mechanics tests are however restricted to low normal stresses since they are designed to investigate soils loaded by civil engineering structures. As a consequence, not much is known about how soil behaves under loads exceeding 2 MPa, equivalent to only an overburden of a few hundred meters of sediment in a natural mud volcano setting. On the other hand, it has been found that some mud volcano clays and shales originate from depths of more than 8-10 km in accretionary complexes or orogenic wedges (see review by Kopf, 2002). As a consequence, rock mechanical tests in the laboratory have to be consulted to understand the mechanical behavior of clay minerals under such high stresses. However, there has been relatively little work published jet.

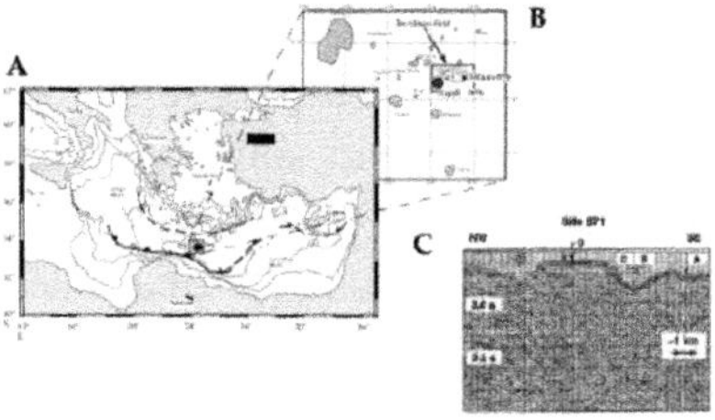

Figure 1. a) Map of the Eastern Mediterranean showing the main features in the area and the large accretionary complex;
b) The Olimpi area at the inner deformation front with its mud volcanoes varying in size and shape. During ODP Leg 160, the Milano mud volcano (Site 970) and Napoli mud volcano (Site 971) were drilled;
c) Seismic reflection profile (migrated time section) across the Napoli mud volcano from which samples have been used in this study. Note the presence of inwarddipping reflectors beneath the flanks of the mud dome as well as a moat encircling the structure.

A growing body of work on constitutive friction laws attests that frictional stability is more important than the absolute shear strength of a material in controlling failure and potential for seismogenesis (see review by Scholz, 1998). Frictional stability is a function of the change in friction coefficient (μ) at a given fault-normal effective stress; a primary control is strain rate (e.g. Saffer et al., 2001). Also, the composition and fabric of a geomaterial is crucial for frictional stability. While high-porosity rocks often show velocity strengthening (i.e. μincreases with increasing shear rate, e.g. during slip), low-porosity rocks typically weaken. In sediments, clays tend to strengthen, whereas framework silicates and lithified rocks weaken (Lockner and Byerlee, 1986; Marone, 1998). In general, earthquakes (events of accelated slip where a significant amount of stored energy is radiated as elastic waves) are believed to nucleate in unstable materials where "run-away" slip can occur. Also, the replacement of clays in "weak" shear zones by precipitation (carbonate, zeolites, quartz) or mineral transformation processes can, while temporarily strengthening the material lead to a decrease frictional stability, as indicated by either unstable stick-slip or conditionally stable behavior (Moore and Saffer, 2001). These processes are suggested to occur in the upper seismogenic zone at temperatures exceeding ~120°C where smectite, opal, and other phases break down (e.g. Tichelaar and Ruff, 1993). Previous studies suggest that the basic level of r and the velocity dependence of the frictional resistance may vary with effective normal stress level (e.g. Lockner and Byerlee, 1986; Saffer et al., 2001). Also, water content has been shown to have a profound weakening effect on the residual frictional of rocks (e.g. Dieterich and Conrad, 1984). Consequently, laboratory studies have attempted to simulate the increasing normal stress (σn) of a seawater-saturated sediment that migrates down dip along the plate boundary fault and is subjected to various slip velocities and confining stresses (e.g. Byerlee, 1978, 1990; Logan and Rauenzahn, 1987; Marone et al., 2001; Brown et al., 2001). These results have hitherto been interpreted largley in terms of rock mechanics phenomenonology (Mohr-Coulomb friction law, small volume and pore pressure changes). In overpressured muds that retain high porosity even at great depth, both volumetric stateand stress history have to be taken into account more explicitly, in keeping with tenets of soil mechanics.

4. SUMMARY OF PREVIOUS RESULTS

4.1. Results from drilling Napoli mud volcano

Drilling Napoli mud volcano (ODP Site 971) demonstrates interfingering between hemipelagites with clast-bearing, matrix-supported mud debris flow deposits (Fig. 1c, d). Overall clast abundance is only 15-25% of the total volume of the mud breccia, and decreases from the base and flank to the crest of the feature. The clayey matrix contains a mixture of Pleistocene, middle Miocene, Oligocene, and Eocene nannofossils. The holes at the crest of the dome (971D, 971E) recovered gaseous (e.g. methane, H2S, higher hydrocarbons) silty clays with scattered clasts of mudstone, siltstone, and angular fragments of coarsely crystalline halite of presumed Messinian age. Mud extrusion here began at about 1.5 Ma (Robertson et al., 1996). Layering in mud breccias was observed and indicates various mud flow events (Flecker and Kopf, 1996).

The massive debris flow found at Hole 971A at Napoli (20-80 mbsf with an age range of > 0.46 Ma to 1.5 Ma; Fig. 3) result in estimated c. 60m/Ma sediment accumulation minimum based on the 1.04 Ma interval.

This value may not be particularly meaningful, however, as the main part of a debris flow is usually deposited very quickly, and may be followed by a period of settling and sedimentation of the finer components. Evidence for rapid deposition comes from oedometer tests on mud breccia samples (Kopf et al., 1998). The compressibility suggests that thin mud breccia intervals should take only 102 to 104 years to consolidate, i.e. consolidation is faster than the geological time available (see Camerlenghi et al., 1995, their Fig. 14).

4.2. Conduit geometries

Although mud volcanism has been studied for about two centuries (see summaries by Higgins and Saunders (1974) or Kopf (2002), scant information exists regarding the geometry of mud feeder conduits. Many researchers have argued that the majority of mud volcanoes is related to faults, or even intersections of faults (Kopf, 2002). However, little is known about the geometry of such faults at depth.

Especially in regions of the former Soviet Union, abundant mud volcanoes disseminate considerable amounts of unlithified rock, oil and gas (Jakubov et al., 1971; Kopf, 2002; Etiope and Milkov, this volume; Judd, this volume). In Azerbaijan, more than 300 mud volcanoes are known and have been described in great detail, showing feeder channels and gryphons between 0.1 to about 20 meters across. Moreover, radial and concentric fractures of 10-150 cm width have been observed in the crestal area before mud extrusion took place. On Sakhalin Island, similar mud domes with feeder channels of 2 m in diameter were found (Gorkun and Siryk, 1968). On Trinidad, Higgings and Saunders (1974) described vents of about 2-5 meters in diameter, situated on the mud volcano crest (i.e. crater area). The Atalante mud volcano at the deformation front of the Barbados Ridge accretionary complex has a wide conduit of c. 200 meters in diameter (Henry et al., 1996). For the Mediterranean Ridge features, feeder geometries of 2-3 m have been mathematically derived (Kopf and Behrmann, 2000). Regardless of these studies, the deep, subsurface portion of the mud volcano conduits is still poorly constrained despite improvements in 3-D seismic imaging of low velocity and gas-charged zones (e.g. Bouriak et al., 2000). The main reason for this is the topography of the domes which represents a severe hinderance to geophysical imaging (see estimates by Ivanov et al., 1996).

4.3. Clay composition and physical properties

4.3.1 Matrix

Calcite and quartz are the main matrix constituents. Kaolinite, sodium-rich smectite, and minor chlorite and illite are the clay minerals present. Detailed information and a full list of X-Ray diffraction results are provided in Robertson and Kopf (1998). A recent study on the Mediterranean MVs confirms that the muds contain largely kaolinite, hallyosite, and smectite (sometimes chloritized), but only traces of illite (Zitter et al., 2001). Wet bulk densities range from 1700-2100 kg/m^3 for pebbly mud breccias, and only 1300-1700 kg/m3 for the virtually clast free, mousse-like silty clays on the crest (Emeis et al., 1996). Average grain densities range around 2600 kg/m3 (Emeis et al., 1996).

4.3.2 Clasts

The different varieties of mud breccia contain clasts of variable composition, size, shape, and abundance, of which a detailed description can be found in Robertson and Kopf (1998). It is important to note that some of the clasts recovered, namely the most abundant, variably lithified calcareous muds and mudstones, are fragmented by non-systematic, dilatant fractures (probably hydrofractures, cf. Behrmann, 1991). Other lithoclasts, dominantly pelagic carbonate and quartzose litharenites of turbiditic origin and variable provenance (Robertson and Kopf, 1998), are generally well lithified and about 5 cm in size. The largest clasts recovered during ODP Leg 160 were ~10 cm across (limited by the drilling devices), but Formation MicroScanner images revealed clasts of up to 50 cm in diameter in the borehole wall (Robertson et al., 1996).

4.4 Viscosity tests

In addition to physical property measurements on mud breccias (e.g. water content, bulk and grain density, porosity, Vane shear strength) as part of the shipboard routines (Emeis et al., 1996), we tested the viscosity of the mud matrix at different water contents. Dynamic viscosity h of the matrix was determined using a Bohlin *CVO* eccentric rotating disk viscosimeter (Fig. 2a; principles outlined in Macosko, 1993). Viscosity tests were run on a set of samples covering the range of water contents from 24% to 65%. The viscosity determined for the sample with the highest water contents was 106 Pa s (Kopf and Behrmann, 2000). This value may range up to one order of magnitude higher for the other three samples (24%, 40%, and 54%) tested from larger depths, and depended on the water content and the number of oscillation tests carried out on the same sample.

On the other hand, degassing of the same immediately after opening thesealed core-liner definately led to underestimating the viscosity. For the above reasons we used 10^6 Pa s as an approximation for calculations regarding the flux rate. Interestingly, this value coincides in broad terms with the viscosities of felsic magmas, which range in the order of 10^5 to 10^{10} Pa s (Petford et al., 1994) and are thus higher than those for fluidized mud. For magmas, the narrowest dykes observed in the field are only several centimeters wide (Reddy et al., 1993), whilst dykes are commonly about several meters to tens of meters wide (e.g. Corry, 1988). Flow rate and conduit geometry also have a strong impact on the surface expression of the extruded mud feature, with narrow conduits favoring domes and wider conduits leading to pie-like features (Lance et al., 1998; Kopf, 2002).

When using Poiseuille's law, the flow of a fluid of viscosity η through a cylindrical conduit with the radius r is

$$v = \frac{\pi \cdot g \cdot \Delta p \cdot r^2}{8\eta} \quad (1)$$

where Δp is the pressure gradient over the pipe, g (= 9.81 m/s^2) is the acceleration due to gravity, and v is the velocity of flux. If this pressure gradient is chosen to be the buoyancy (bulk density difference between fluidized mud [of 1600 kg/m^3] and consolidated claystones [2100 kg/m^3]) alone, ascent velocities in mud volcano conduits can be calculated. For a diameter of 1 m, the rate would be ~ 70 km/a (or 0.022 m/s). For wider geometries, these flux rates drastically (for details see Kopf and Behrmann, 2000). Also, if the sample contains only a few per cent of free gas, ascent rates may rise by orders of magnitude (Brown, 1990).

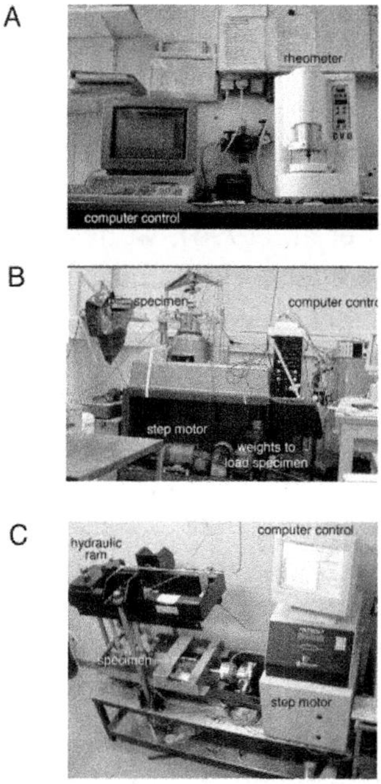

Figure 2. Analytical facilities for geophysical properties measurement: a) Bohlin CVO rheometer, b) Ring shear permeameter, and c) GEOCOMP direct shear apparatus. See text

5. METHODS OF THIS STUDY

5.1 Ring shear permeameter

Ring shear permeameter tests were conducted in a customized *Wykeham-Farrance* large ring-shear apparatus (Fig. 2b). An annular sample of 140 mm outer diameter and 101 mm inner diameter undergoes progressive loading increments up to normal stresses of ~1MPa (for details, see Bishop et al., 1972). The original height of the sample is approximately 25 mm. The sample is constrained to fail along a mid-plane separaring two halves of the annular sample holder. The sample chamber is sealed and can drain to top and bottom through porous stones. At the mid-plane the pore pressure can be monitored with a pressure transducer. Initial shearing was conducted at slow load-point displacement rates of between 0.001 and 0.074 mm/s to maintain conditions as near to undrained as possible. The shear stress is measured continuously to < 1 kPa accuracy during rotation using high-precision linear displacement transducers attached to calibrated proving rings on the paired torque arms. Vertical permeability across the shear zone was measured by pumping fluid at a constant rate from the syringe pump through the sample and monitoring the pressure difference with a differential pressure transducer (0.5 kPa accuracy). While a back pressure of up to 400 kPa can be applied in the modified ring shear, in these tests we first saturated samples fully and then conducted flow tests with an outflow burette open to the atmosphere.

This was necessary to enable the maximum level of effective normal stress (approx. 1 MPa) to be attained.

5.2 Direct shear apparatus

Direct shear tests were conducted in a newly designed direct shear apparatus based on the *GEOCOMP Large Shear Track II* (Fig. 2c), in which the initially 30 to >70 mm-thick samples were carefully loaded to normal stress levels of 10, 20, and 30 MPa for testing. During shear of the then ~15 mm-thick sample, pore pressure across the fault zone is monitored via three porous ports that penetrate the lower half (~10 mm) of the sample and tap into the level of the shear zone. After consolidation, testing was not initiated until the excess pore pressure in the sample was <1% of the total normal stress. During the shear phase, the sample is initially failed under a constant vertical normal force to a near residual state at a load point displacement rate of 0.001 mm/s for 12 mm. The horizontal and vertical loads and displacements as well as pore pressure along the shear zone are continually monitored. Subsequently, velocity stepping (from 0.00001 to 0.1 mm/s) is conducted while the sample is in its residual state.

All tests were carried out over a total displacement of 22-26 mm (i.e. half the test was performed at different shear rates on the residual path).

6. RESULTS

6.1 Ring shear tests

Results from ring shear experiments are summarized by sample composition (simplified percentage of total clay) in Figure 3. It can be seen that at low stresses, a systematic increase in smectite over quartz causes a near-linear decrease in μ_r' with pure smectite being as low as 0.1. The results match the trend from ring shear tests on bentonite by Lupini et al. (1981; see grey line in Fig. 3). Similarly, both pure illite, kaolinite, and chlorite have low coefficients of residual friction (μ_r=0.2, 0.21, and 0.22, respectively) when compared to quartz (μ_r=0.63). Admixture of relatively small amounts of clay already has a profound weakening effect (e.g. μ_r=0.42 for a sample with 40% clay and 60% quartz). Natural mud volcano sediments can simplistically be viewed as a mixture of weak platy minerals (clay minerals, and smectite in particular) and strong granular components (quartz, feldspar, tephra/ash, or rock fragments). In addition, the abundant clast-size particles present in the mud volcanic ejecta would cause a significant increase in μ.

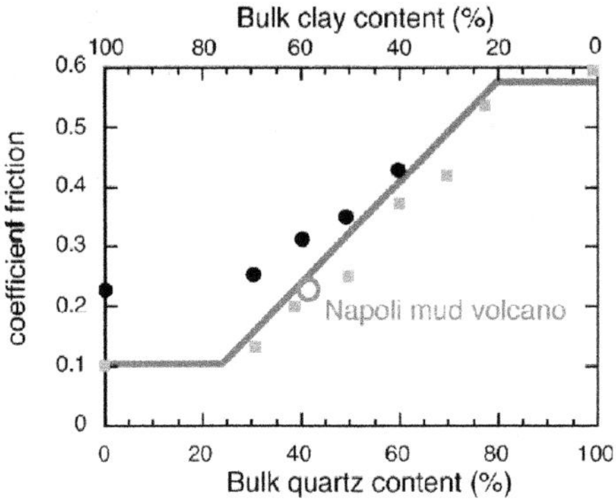

Figure 3. Results from ring shear tests plotted as μr vs. quartz content for seawater-saturated mineral standards and Napoli mud volcano clay.

If we look at results from natural mud volcano deposits from Napoli dome it is surprising to find then weaker than some pure clays such as kaolinite. Friction coefficients (μ_r) around 0.21 reflect the abundant smectite and hallyosite in the extruded materials (Fig. 3), which more than compensates for the non-clay fraction in the mud matrix.

Vertical permeability of the Napoli mud volcano clay is about two orders of magnitude lower than in pure kaolinite, which is largely a result of high smectite contents and low porosity (see upper axes in Fig. 4). When looking at the path of incremental loading (kPa values) and shearing (1 degree of rotary shear is approximately equal to 1 mm of displacement) in Figure 4a, it becomes apparent that large displacements favor an increase in permeability, possibly because a pervasive shear zone can be maintained with time with a quite stable permeability across it. When unloading the sample after shearing at maximum normal load (893 kPa; Fig. 4a), the overconsolidated material dilates slightly and recovers about 60% of its original permeability (compare the two points at 388 kPa; Fig. 4a). This is consistent with results from uniaxial deformation tests by Fitts and Brown (1999) who found irreversible dewatering and changes in smectite mineral structure to occur only at loads exceeding 1.3 MPa. Overall permeabilities in the mud volcano clays sheared at effective overburden stresses of 400-900 kPa are ~10^{-19} m^2. This very low vertical permeability has a profound effect on pore pressure transients during shear, and impacts upon on mud extrusion dynamics and episodicity (see Discussion below).

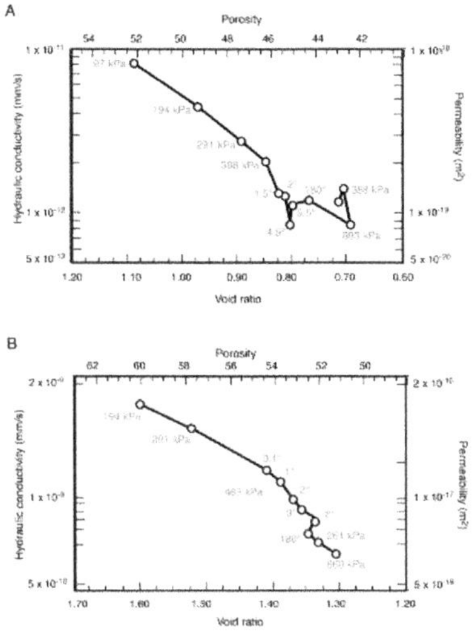

Figure 4. Ring shear permeability results from sample Napoli 971B-3H-5 (a) and pure kaolinite (b) as porosity and void ratio versus permeability. Numbers indicate confining normal stress at each test increment as well as the rotation (in °).

6.2 Direct shear tests

Direct shear tests were usually conducted for about 20-25 mm of displacement over several hours; a sheared sample after the test as well as part of a typical protocol are given in Figure 5. A series of the individual direct shear tests are shown together with ring shear results in Figure 6, where residual friction µr) is plotted against increasing normal load to simulate migration of the sediment down dip of the décollement.

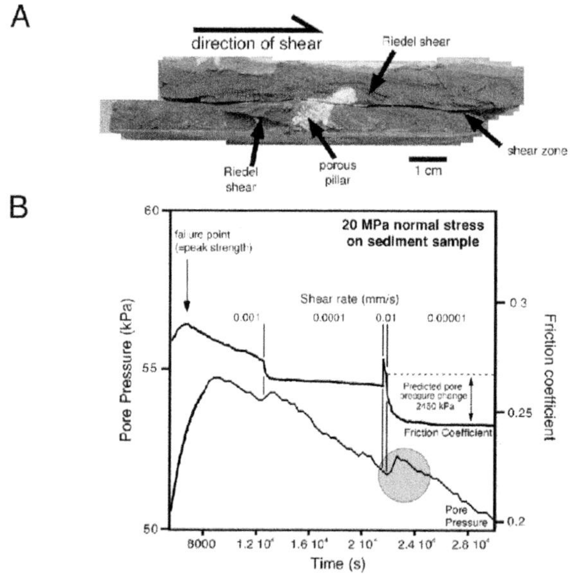

Figure 5. Sheared clay sample (a) and typical protocol (b) from direct shear tests.

During the tests, the clay-dominated samples predominantly show a generally well developed peak and subsequent lower residual shear strength profile (i.e. strain weakening behaviour). The pure clay mineral standards showed a gentle increase within the σ_n range up to 30 MPa.

Smectite remained low (μ_r of 0.11 at 10, 20, and 30 MPa), whereas both illite and chlorite increased to μ_r=~0.26 at 30 MPa (Fig. 6). In contrast, quartz shows strain hardening behavior, but a decrease with increasing normal stress to values of μ_r=0.48 to 0.5. All samples tested showed positive (a-b)-values (=ratio between μr and change in shear rate; see Scholz, 1998) during velocity stepping (Fig. 5b), and hence velocity strengthening behavior (i.e. stable sliding). In general, the pore pressure tended to peak during the period of the initial failure, kept rising for some time immediately after failure, but then steadily dropped during the course of the velocity stepping tests as the sample came to lie in its residual state (Fig. 5b). The quartz-rich and pure quartz samples showed little or no pore pressure response during failure, presumably because their elevated permeabilities allowed them to deform in a fully drained state. Even in the clays and clay-rich samples, pore pressure increases where found to be <3% of the total normal load (maximum value observed was 0.65 MPa at σ_n=25 MPa in pure illite). This relatively small number is most likely a scale effect between laboratory measurements and natural systems. The pore pressure variations observed in the laboratory have to be viewed as perturbations due to shear to be added on top of regional transients, which in mud volcano settings may be quite high prior to an eruption.

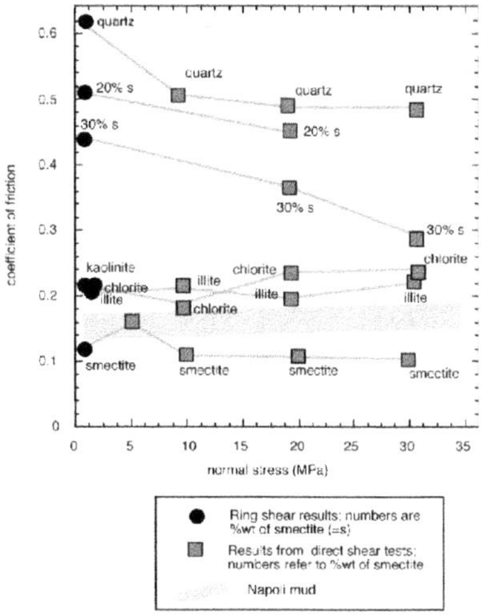

Figure 6. Friction coefficient versus normal stress of water-saturated sediment standards. Grey shaded area indicates that muds from Mediterranean Ridge mud domes are weak and plot between smectite and illite trends. See text for discussion.

7. DISCUSSION

In the discussion, we develop the results of our geotechnical studies into two directions. First, we use earlier data on mud volcanic episodicity and tie them to the permeability and viscosity results. Second, we relate the weakness and frictional behaviour of clays to the stress state of faults in accretionary prisms and discuss the implications for mud extrusion and the onset of seismogenesis.

7. 1 Episodicity and velocity of mud ascent

While historical records of mud volcanic eruptions are very detailed in some onshore mud volcano areas based on direct observation (e.g., Guliyev, 1992), little information exists in the marine realm. For some of the Mediteranean Ridge mud volcanoes, a period of extrusive activity of 10^2 to 10^4 years on the basis on compressibility data from mud breccia samples has been put forward by Camerlenghi et al. (1995). The upper limit of this estimate corresponds to the order of magnitude of estimates by Kopf and Behrmann (2000) concerning the duration of mud extrusive activity based on Stokes' law ($1.2\text{-}5.8*10^4$ years). This indirectly points to small conduit geometries in the meter range, despite these having rather high Poiseuille flux for likely pressure gradients.

On the other hand, the ascent velocities estimated for the mud are similar to those known from magma rise. As shown in various studies, such rhyolitic liquids have viscosities between 10^4 and 108 Pa s and rise through1-5 m wide dykes (e.g. Marsh, 1982; Bacon, 1992).

A comparison of mud and magma has a limited validity due to the differences in scale.

For magmas coming from mantle depth, both viscosities and ascent velocities are initially very low, but can reach values of several tens (and rarely hundreds) of kilometers per hour as a result of pressure release and gas expansion (Baloga et al., 1995).

As mentioned above, gas expansion as a consequence of liberation of methane dissolved in the pore water at depth as well as from gas hydrate dissociation can play an important role in mechanisms driving mud effusion at shallow depth. Therefore, we are confident that for short periods of time in the upper part of the accretionary prism, high ascent rates may well be reached on the Mediterranean Ridge.

The relationship between ascent rate, episodicity, and pore pressure transients has long been identified as a key element in mud volcanic activity (e.g. Yassir, 1989; Kopf, 2002). An important mechanism for generating overpressures is sediment loading (Bethke, 1985; Bredehoeft et al., 1988; Mello and Karner, 1996). In sedimentary basins, sediment is rapidly being added to the basin floor. The new sediment adds additional load to the older sediments beneath and increases fluid pressures so long as drainage is inhibited. Elevated fluid pressures drive fluid flow, and as fluid escapes, the stress between the individual grains of the sediment (effective stress) is increased, bulk volume is reduced, and porosity is lost.

If permeability is sufficiently low, the pore fluids will be unable to escape at a rate comparable to the rate of loading due to sedimentation. When this occurs, the pore fluid pressure will increase above hydrostatic, a condition termed overpressuring, geopressuring, or excess pressuring (Osborne and Swarbrick, 1997; Bjørlykke and Hoeg, 1997; Mello and Karner, 1996).

One of the major controlling parameters in overpressure build-up is the permeability of the geomaterials involved. Accretionary complexes, where rapidly deposited trench sediment is accumulated in an imbricate thrust stack, often show overpressured sediments (Moore and Vrolijk, 1992;Moore et al., 1995; Saffer and Bekins, 1998; Gamage and Screaton, 2001).

The preservation of overpressures in these systems is controlled by low permeability sediments (clay rich wedges such as Barbados tend to be more overpressured), and by a lack of connectivity in fault and fracture conduits.

For the Mediterranean mud volcanoes, we have found the permeability to be rather low even for typical marine muds (10^{-19} m^2; see above and Fig. 4a). Our data are in good agreement with earlier work on similar smectitic material (e.g. Neuzil, 1994). The low permeabilities, especially vertical to shear fabrics and other evidence of preferred particle orientation, may favor localized small-scale compartments where pore pressures approach values in excess of hydrostatic. These areas of weakness may potentially cause hydrofracture, mud ascent, and either diapirs or diatremes which in the case of surface piercement spawn mud volcanoes. Given that many mud volcanoes are fault-related features (e.g. Kopf et al., 2001), and that fluid flow and permeability along such faults is highly discontinuous (Bangs et al., 1999), mud extrusion is both a localized and temporary phenomenon. Among a large number of interacting processes, there are a few aspects which mainly control overpressure and episodic eruption. First, tectonic processes may cause unroofing of mud fluid and mud reservoirs, e.g. by landslides, tectonic movement and associated gas hydrate dissociation, or lateral shortening.

Second, changes in the regional stress field may cause pore-pressure changes, for instance during the seismic cycle in earthquake-prone areas (Hasiotis et al., 1997).

Third, mineral dehydration reactions produce excess water which adds significantly to the pore pressure of the low-porosity rocks, may cause decrease in effective stress, and possibly hydrofracturing (Behrmann, 1991; Moore and Vrolijk, 1992; Moore and Saffer, 2001). Fourth, diagenetic reactions like quartz cementation due to dissolution of biogenic silica tests (diatoms, radiolarians) may seal and/or compartimentalize a fault zone, this way causing profound variability in the pore pressure distribution (Byerlee, 1993; Bangs et al., 1999).

In the second part of the discussion, we try to link the variations in the stress state and frictional stability of faults to mud volcanic processes and their relationship to tectonics and seismicity.

7. 2 Stress state of faults, seismogenesis, and mud extrusion

It has long been known that fault initiation, propagation and slip are depend on fault zone mineralogy and transient pore pressure as well as factors such as temperature, stress regime, and linkage. To separate the effect of intrinsic sediment friction from the influence of pore pressure is a major goal of modern tectonic analyses (Rice, 1992; Brown et al., 2001), and is a key for this study of fault-related mud volcanism. From our data (e.g. Fig. 5b) it appears that pore pressure plays only a minor role in shear zone mechanics in the laboratory, because pore pressure never compensated for more than a few per cent of the total normal load on the sample (e.g. Brown et al., 2001; Clennell, unpubl. data). However, as has been discussed in the previous paragraph, this may be different when looking at a scale of kilometers in a natural system. Several authors have shown that along permeable, sediment-dominated fault zones like subduction zone décollements the pore pressure can climb to near lithostatic values (Moore et al., 1995). Also, the pore pressure distribution down-dip as well as along-strike may be quite variable along these systems (Saffer and Bekins, 1998; Bangs et al., 1999).

If we consider the permeability and pore pressure distribution within a large accretionary complex like the Nankai margin (Saffer and Bekins, 1998, 2002) in the light of our shear results (Fig. 6), the stress state along the plate boundary fault can be calculated. A simplicistic approach of this calculation is shown in Figure 7 from which can be deduced that the shear stress along deep faults in several kilometers depth is still only around 10 MPa. Given that clays dilate only moderately and generally show velocity strengthening, no large stress drops are expected from slip along thrusts with clayey gouge aligned in the fault zone. The relatively large stress drops reported during earthquake rupture require that the fault zones are original strong prior to rupture, and this is likely due to healing of the fault zone at deeper levels. This could be the result either of cementation (calcite or quartz precipitation from deepseated solutions) or mineral transformation and low-grade metamorphism (e.g. Moore and Saffer, 2001). For instance a quartz cemented fault gouge could produce a large stress drop when ruptured because of its dilatancy and high intrinsic friction coefficient (see Fig. 6, or Byerlee, 1978). Such dilatancy promotes further cementation by quartz, perpetuating the cycle.

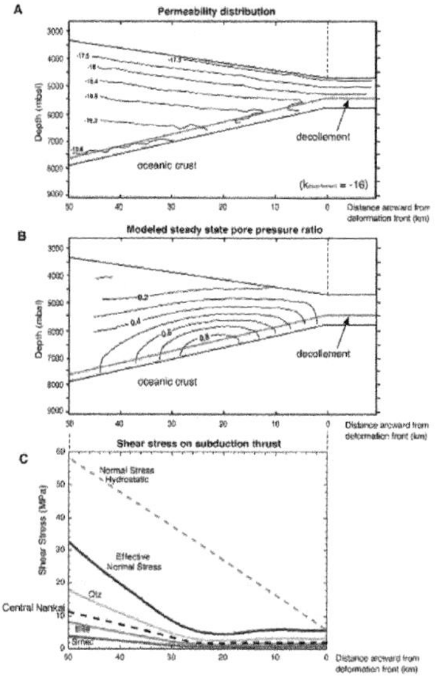

Figure 7. a) Permeability distribution in the Central Nankai forearc wedge (taken from Saffer and Bekins, 1998); data are shown as the log of permeability (m^2);b) Pore pressure ratio ($\lambda^*=[p_{fluid}-p_{hydrostatic}]/[p_{lithostatic}-p_{hydrostatic}]$) as a function of distance from the deformation front modeled for the Nankai accretionary wedge (from Saffer and Bekins, 1998); c) Stress development vs. distance from deformation front based on pore pressure modeling data (b) and shear test results (this study; see Fig. 6). The estimated trend for the Central Nankai décollement zone is illustrated by the black dashed line.

As a consequence of the stress drop and seismic slip, the stress state of the fault zone (or at least parts of the fault zone) may change dramatically. The changes in physical parameters such as pore pressure and in physical properties such permeability may occur both adjacent to the seismogenic zone, and in splays of the fault further up-dip. No matter whether the stress field variation is transferred through kinematic or hydraulic linkage, sensitive systems like mud volcanoes juxtaposing thrusts of the same fault system are likely to respond to the seismic event.

In fact, scientists in onshore as well as offshore mud volcano settings found evidence for pre-, co-, and post-seismic mud volcano activity (Papatheodorou et al., 1993; Hasiotis et al., 1997; Hovland et al., 2002).

Similarly, landslides and slumps triggered by earthquakes have been observed in addition to the mud volcanic extrusions. Anomalously high pore fluid flux of onshore mud volcanoes in Azerbaijan has been reported to preceed regional earthquakes by up to 5-7 days. Gas emission with e.g., CO_2 concentrations twice as high as the background flux was measured up to a month before a major earthquake in the same area (A. Aliyev, pers. comm., 2003). This makes mud volcanoes a potentially powerful precursor for earthquakes if monitored on a permanent basis.

8. SUMMARY AND CONCLUSIONS

The Napoli dome, together with several other similar features in this part of the Mediterranean Ridge accretionary complex, is interpreted as mud volcano with an evolution of episodic eruptions of low viscosity clays with clasts. Shear strength, friction coefficient, and permeability of the extruded materials are low, which is primarily a result of high smectite and hallyosite contents.

A mud volcano conduit of very limited width (<1-2 m) is in agreement with seismic cross sections showing almost continuous reflectors within the domes. A narrow channel is supported by the cone-shaped geometry of the volcanoes. During episodic mud volcanic cycles, the conduit may temporarily be blocked to facilitate pressure build up (periods of inactivity), which ultimately leads to eruptions and extrusive activity, pore pressure drop, plugging of the conduit, etc.

Based on feeder channel geometry of only a few meters, the ascent velocity (i.e. Poiseuille flow) of overpressured mud (at a viscosity of 10^6 Pa s) is estimated to be around 20-300 km/a, which is several orders of magnitude higher than for shale or salt diapirs.

It appears as if no direct relationship between seismogenesis and mud volcanism, because the clays involved deform by stable friction (velocity strengthening behavior). However, given the metastable, sensitive nature of episodically active mud domes, these features present an interesting opportunity to be used as earthquake precursors.

ACKNOWLEDGEMENTS

The authors thank Christian Friedrich and Wolfgang Schemionek for assisting with the viscosity measurements at the Material Research Centre at Freiburg University, Germany, and Jill Weinberger for carrying out ring shear tests at SCRIPPS Institution of Oceanography, USA. Funding for this research was provided by grants through DFG, NERC, and NSF.

REFERENCES

1. Aloisi, G., Asjes, S., Bakker, K., Bakker, M., Charlou, J.L., De Lange,G.J., Donval, J.-P., Fiala-Medoni, A., Foucher, J.-P., Haanstra,R., Haese, R., Heijs, S., Henry, P., Huguen, C., Jelsma, B., De Lint, S.,Van der Maarel, M., Mascle, J., Muzet, S., Nobbe, G., Pancost, R., Pelle, H., Pierre, C., Polman, W., De Senerpont Domis, L., Sibuet, M., van Wijk, T., Woodside J.M., and Zitter, T., 2000. Linking Mediterranean brine pools and mud volcanism. *EOS, Trans. AGU*, 81/51, 625-632.
2. Bacon, C.R., 1992. Partially melted granodiorite and related rocks ejected from Crater Lake caldera, Oregon, *Transactions Royal Soc. Edinburgh, Earth Sciences,* **83**, 27-42
3. Baloga, S., Spudis P.D. and Guest., J.E., 1995. The dynamics of rapidly emplaced terrestrial lava flows and implications for planetary volcanism, *J. Geophys. Res.*, **100**, 24509-24519.
4. Bangs, N.L., Shipley, T.H., Moore J.C., and Moore, G.F., 1999. Fluid accumulation and channeling along the northern Barbados Ridge décollement, *J. Geophys. Res.*, **104**, 20399-20414.
5. Barber A., J., and Brown, K.M., 1988. Mud diapirism: the origin of mélanges in accretionary complexes?, *Geology Today*, **4**, 89–94.
6. Behrmann, J.H., 1991. Conditions for hydrofracture and the fluid permeability of accretionary wedges, *Earth Planet. Sci. Letters*, **107**, 550–558.
7. Bethke, C.M., 1985. A numerical model of compaction-driven groundwater flow and heat transfer and its applications to the paleohydrology of intracratonic sedimentary basins, *J. Geophys. Res.*, **90**, 6817–6828.
8. Bishop, A.W., Green, G.E., Garga, V.K., Andersen A., and Brown, J.D., 1972. A new ring shear apparatus and its application to the measurement of residual strength, *Géotechnique*, **21**, 273-328.
9. Bjørlykke, K., and Hoeg , K. , 1997. Effects of burial diagenesis on stresses, compaction and fluid flow in sedimentary basins, *Mar. Petr. Geol.*, **14**, 267–276.
10. Bouriak, S., Vanneste, M. and Saoutkine, A., 2000. Inferred gas hydrates and clay diapirs near the Storegga Slide on the southern edge of the Vøring Plateau, offshore Norway, *Mar. Geol.*, **163**, 125-148.
11. Bredehoeft, J.D., Djevanshir R.D., and Belitz, K.R., 1988. Lateral fluid flow in acompacting sand-shale sequence: South Caspian Basin, *AAPG Bull.*, **72**, 416–424.
12. Brown, K.M., and Westbrook, G.K., 1988. Mud diapirism and subcretion in the Barbados Ridge accretionary complex: the role of fluids in accretionary processes., *Tectonics*, **7**, 613–640.

13. Brown, K.M., 1990. The nature and hydrogeologic significance of mud diapirs and diatremes for accretionary systems,. *J. Geophys Res.*, **95,** 8969–8982.
14. Brown, K.M., Kopf, A., Underwood, M.B., Weinberger J.L. and Steurer, J., 2001. Frictional Coefficients of Multi-Component Sediments: Implications for the Aseismic to Seismic Transition Zone, W. Nankai, *EOS Trans. AGU (Supplement)*, **82/47**, F1248.
15. Byerlee, J.D., 1978. Friction of Rocks, *Pure Appl. Geophys.*, **116**, 615-625.
16. Byerlee, J.D., 1990. Friction, overpressure and fault normal compression, *Geophys. Res. Lett.*, **17**, 2109-2112.
17. Byerlee, J.D., 1993. Model for episodic flow of high-pressure water in fault zones before earthquakes, *Geology* **21**, 303-306.
18. Camerlenghi, A., Cita, M.B., Della Vedova, B., Fusi, N., Mirabile L., and Pellis, G., 1992. Geophysical evidence of mud diapirism on the Mediterreanean Ridge accretionary complex, *Marine Geophysical Researches*, **17**, 115–141.
19. Chaumillon, E., and Mascle, J., 1997. From foreland to forearc domains: new multichannel seismic reflection survey of the Mediterranean Ridge accretionary complex (Eastern Mediterranean), *Mar. Geol.*, **138**, 237-259.
20. Corry, C.E., 1988. *Laccoliths: Mechanisms of emplacement and growth*, GSA Spec. Pap. 220, Boulder
21. Dieterich, J.H., and Conrad, G., 1984. Effect of humidity on time- and velocity-dependent friction in rocks, *J. Geophys. Res.*, **89**, 4196-4202.
22. Emeis, K.C., and Shipboard Scientific Party, 1966. *Proceedings ODP,* (Ocean Drilling Program) *Initial Reports* 160, College Station, TX
23. Fitts, T.G. and Brown, K.M., 1999. Stress induced smectite dehydration ramifications for patterns of freshening fluid expulsion in the N. Barbados accretionary wedge, *Earth Planet. Sci. Letts.* , **172**, 179-197.
24. Flecker, R. and Kopf, A., 1996. Clast and grain size analysis of sediment recovered from the Napoli and Milano mud volcanoes, ODP Leg 160 (Eastern Mediterranean), in *Proceedings ODP*, (Ocean Drilling Program), Init. Reports 160, Emeis, K.C., Robertson, A.H.F., Richter, C., et al., College Station, TX, 529-532.
25. Fusi N., and Kenyon, N.H., 1996. Distribution of mud diapirism and other geological structures from long-range sidescan sonar (GLORIA) data in the Eastern Mediterranean Sea, *Mar. Geol.*, **132**, 21-38.
26. Gamage, K., and Screaton, E.J., 2001. Permeability measurements and implications for generation of overpressures at the Nankai accretionary prism, ODP Leg 190 Sites 1173 and 1174, *EOS, Trans. AGU*, **82/47**, F1194.
27. Gorkun, V.N. and Siryk, I.M., 1968. Calculating depth of deposition and volume of gas expelled during eruptions of mud volcanoes in southern Sakhalin, *Int. Geology Review*, **10/1**, 4-12.
28. Guliev, I.S., 1992. *A review of mud volcanism*. A report by the Azerbaijan Academy of Sciences, Inst. Geology, Baku.
29. Hasiotis, T., Papatheodorou, G., Kastanos, N., and Ferentinos., G., 1997. A pockmark field in the Patras Gulf (Greece) and its activation during the 14/7/93 seismic event, *Mar. Petr. Geol.*, **13**, 333-344.
30. Henry, P., Le Pichon, X., Lallemant, S., Lance, S., Martin, J.B., Foucher, J.-P., Fiala-Médioni, A., Rostek, F., Guilhaumou, N., Pranal, V. and Castrec, M., 1996. Fluid flow in and around a mud volcano field seaward of the Barbados accretionary wedge: Results from Manon cruise, *J. Geophys.* ***Res.*,** **101,** 20297-20323.
31. Higgins, G.E., and Saunders, J.B., 1974. Mud volcanoes – Their nature and origin, *Verh. Naturf. Ges. Basel*, **84**, 101–152.

32. Hovland, M., Gardner, J.V., and Judd, A.G., 2002. The significance of pockmarks to understanding fluid flow processes and geohazards, *Geofluids*, **2**, 127-136.
33. Huguen, C., Mascle, J., Chaumillon, E., Woodside, J.M., Benkhelil, J., Kopf, A., and Volkonskaia., A., 2001. Deformational styles of the Eastern Mediteranean Ridge and surroundings from combined swath mapping and seismic reflection profiling, *Tectonophysics* , **343,** 21-47.
34. Ivanov, M.K., Limonov , A.F., and van Weering., T.C.E., 1996. Comparative characteristics of the Black Sea and Mediterranean Ridge mud volcanoes, *Mar. Geol.*, **132**, 253-271.
35. Jakubov, A.A., Ali-Zade , A.A., and Zeinalov, M.M., 1971. *Mud volcanoes of the Azerbaijan SSR*, Publishing house of the Academy of Sciences of the Azerbaijan SSR, Baku.
36. Kopf, A., Robertson, A.H.F., Clennell, M.B., and Flecker R., 1998. Mechanism of mud extrusion on the Mediterranean Ridge, *GeoMarine Letters*, **18**, 97-114.
37. Kopf, A., and Behrmann, J.H., 2000. Extrusion dynamics of mud volcanoes on the Mediterranean Ridge accretionary complex, in *From the Arctic to the Mediterranean: Salt, shale, and igneous diapirs in and around Europe*, Vendeville, B., Mart, Y., and Vigneresse, J.L., (eds.), Geol. Soc. London, Spec. Publ. 174, London, 169-204.
38. Kopf, A., Klaeschen, D., and Mascle, J., 2001. Extreme efficiency of mud volcanism in dewatering accretionary prisms, *Earth Planet. Sci. Letters*, **189**, 295-313.
39. Kopf, A.J., 2002. Significance of mud volcanism, *Reviews of Geophysics*, **40/2**, [DOI 10.1029/2000RG000093].
40. Lambe, T.W., and Whitman, R.V., 1969. *Soil mechanics*, Wiley & Sons, New York.
41. Lance, S., Henry, P., Le Pichon, X., Lallemant, S., Chamley,H., Rostek, F., Faugeres, J.-C., Gonthier, E., and Olu, K., 1998. Submersible study of mud volcanoes seaward of the Barbados accretionary wedge: Sedimentology, structure and rheology, *Mar. Geol.*, **145**, 255-292.
42. Lockner, D.A., and Byerlee, J.D., 1986. *Laboratory measurements of velocity-dependent frictional strength*, US Geol. Survey Open-File Report 86-417.
43. Logan, J.M., and Rauenzahn, K.A., 1987. Frictional dependence of gouge mixtures of quartz and montmorillonite on velocity, composition and fabric, *Tectonophysics*, **144**, 87- 108.
44. Lupini, J.F., Skinner, A., and Vaughan, A. E., 1981. The drained residual strength of cohesive soils, *Géotechnique*, **31**, 181-213.
45. Macosko, C.W., 1993. *Rheology - Principles, Measurements, and Applications.* VCH, New York.
46. Marone, C. J., 1998. Laboratory-derived friction laws and their application to seismic faulting, *Annual Rev. Earth Planet. Sci.*, **26**, 643-696.
47. Marone, C.J., Saffer, D.M., Frye, K., and Mazzoni, S., 2001. Laboratory results indicating intrinsically stable frictional behavior of illite clay, *EOS, Trans. AGU*, **82/47**, 1248- 1249.
48. Marsh, B.D., 1982. On the mechanics of igneous diapirism, stoping and zone melting, *American Journal Science.*, **282**, 808-855.
49. Mello, U.T., and Karner, G.D., 1996. Development of sediment overpressure and its effect on thermal maturation: Application to the Gulf of Mexico Basin, *AAPG Bull.*, **80**, 1367–1396.
50. Moore, J.C., and Vrolijk, P., 1992. Fluids in accretionary prisms, *Rev. Geophys.*, **30**:, 113-135.
51. Moore, J.C., and Shipboard Party ODP Leg 156, 1995. Abnormal fluid pressures and fault-zone dilation in the Barbados accretionary prism; evidence from logging while drilling, *Geology*, **23**, 605-608.
52. Moore, J.C., and Saffer, D.M., 2001. Updip limit of the seismogenic zone beneath the accrectionary prism of southwest Japan: an effect of diagenetic to low-grade metamorphic

processes and increasing effective stress, *Geology*, **29**, 183-186.
53. Morrow, C., Radney, B., and Byerlee, J.D., 1992. Frictional strength and the Effective Pressure Law of Montmorillonite and Illite Clays, in *Fault mechanics and transport properties of rocks*, Evans, B. and Wong, T.-F. (eds.), Academic Press, San Diego.
54 Neuzil, C.E., 1994. How permeable are clays and shales?, *Water Resources Res.*, **30**, 145-150.
55. Osborne, M.J., and Swarbrick, R.E., 1997. Mechanisms for generating overpressures in sedimentary basins: a reevaluation, *AAPG Bull.*, **81**, 1023–1041.
56. Papatheodorou, G., Hasiotis, T., and Ferentinos, G., 1993. Gas-charged sediments in the Aegean and Ionian Seas, Greece, *Mar. Geol.*, **112**, 171-184.
57. Petford, N., Liste, J.R., and Ross, R.C., 1994. The ascent of felsic magmas in dykes, *Lithos*, **32**, 161-168.
58. Reddy, S.M., Searle, S.M., and Massey, J.A., 1993. *Structural evolution of the High Himalayan Gneiss sequence, Langtang Valley, Nepal*, Geol. Soc. London, London, Spec. Publ., 74, 375-389.
59. Rice, J.R., 1992. Fault stress states, pore pressure distributions, and the weakness of the San Andreas Fault, in, *Fault mechanics and transport properties of rocks*. Evans, B., and Wong, T.-F. (eds.) London Acad. Press, London, 475-503.
60. Robertson, A.H.F., and Shipboard Scientific Party of ODP Leg 160., 1996. Mud volcanism on the Mediterranean Ridge: Initial results of Ocean Drilling Program, Leg 160, *Geology*, **24**, 239–242.
61. Robertson, A.H.F., and Kopf, A., 1998. Origin of clasts and matrix within Milano and Napoli mud volcanoes, Mediterranean Ridge accretionary complex, in: Robertson, A.H.F., Emeis, K.C., Richter, C., et al., *Proc. ODP* (Ocean Drilling Program) Sci. Results 160, College Station, TX, 575-596.
62. Saffer, D.M., and Bekins, B.A., 1998. Episodic fluid flow in the Nankai accretionary complex: Timescale, geochemistry, flow rates, and fluid budget, *J. Geophys. Research*, **103**, 30351-30370.
63. Saffer, D.M., and Bekins, B.A., 2002. Hydrologic controls on the morphology and mechanics of accretionary wedges, *Geology*, **30**, 271-274.
64. Saffer, D.M., Frye, K.M., Marone, C., and Mair, K., 2001. Laboratory results indicating complex and potentially unstable frictional behavior of smectite clay, *Geophys. Res. Letts.*, **28**, 2297-2300.
65. Scholz, C.H., 1998. Earthquakes and friction laws, *Nature* , 391, 37-42.
66. Skempton, A.W., 1985. Residual strength of clays in landslides, folded strata and the laboratory, *Geotechnique*, **35**, 3–18.
67. Tichelaar, B.W., and Ruff, L.J., 1993. Depth of seismic coupling along subduction zones, *J. Geophys. Res.*, **98**, 2017-2037.
68. Yassir, N.A., 1989. *Mud volcanoes and the behaviour of overpressured clays and silts.* Unpubl. PhD thesis, London.
69. Zitter, T.A.C., Van Der Gaast, S.J., and Woodside, J.M., 2001. New information concerning clay mineral provenance in mud volcanoes. Proc. 36th CIESM congress, Monaco, 23-28 Sept.2001, Rapp. Comm. Inter. Mer Médit. 36, 46-47.

Acknowledgments

The Editors wish to thank the NATO Science Committee and in particular Dr. Alain Jubier for supporting the Advanced Research Workshop on "Mud volcanism, Geodynamics and Seismicity" held in Baku, Azerbaijan, May 19-26, 2003.

Thanks are due to Prof. Adil Aliyev, Prof. Akif Alizadeh, Prof. Ibrahim Guliev, Dr. Vagif Ibrahimov and Dr. Efendiyeva Malahat of the Geological Institute of Azerbaijan. Furthermore thanks are due to ARPA, Environmental Agency of Emilia-Romagna Region, Italy.

Thanks are also due to Prof. Dario Albarello, Dr. Alan Judd, Dr. Leonid Levin and Prof. Francesco Mulargia for encouragement and suggests in carrying out the Advanced Research Workshop and present book.

The Editors acknowledge Henry Monaco and Enrica Manenti for cooperation in editorial handling.

Subject index